THE CUNNING FARMER

"One often hears practitioners speak of 'working with the land' and 'following the Wheel of the Year,' yet modern life often places countless layers of separation between the practitioner and the natural world. Todd Elliott is an exception. As a farmer, he works the land in the most literal sense, and his magical practice is deeply rooted in an intimate relationship with the fecund forces of nature. *The Cunning Farmer* is an outstanding work, distilling more than twenty years of blending esoteric practice with the rhythms of farming. Drawing on a rich tapestry of influences, Todd shares knowledge tried, tested, and refined in the fields themselves. It is a rare and remarkable book—earthy, wise, and utterly authentic."

DARRAGH MASON, AUTHOR OF *SONG OF THE DARK MAN*
AND HOST OF *THE SPIRIT BOX* PODCAST

"Like minerals and animals, plants constitute links in various ontological chains that reach back up to the divine potencies of which they are tokens, recapitulating the gods themselves within the world of matter. To cultivate a plant is therefore akin to harnessing something of the essence of a given deity—and the farmer is a veritable magus, capable of conjuring and containing the vegetative energies of that god. Todd Elliott has ploughed this fertile soil and, true to form, laboriously passes the rich bounty of his discoveries on to us, his ravenous readers."

P. D. NEWMAN, AUTHOR OF *THEURGY—THEORY AND PRACTICE:
THE MYSTERIES OF THE ASCENT TO THE DIVINE*

THE CUNNING FARMER

AGRARIAN MAGICAL PRACTICES, MYTHOLOGY, AND FOLKLORE

TODD ELLIOTT

INNER TRADITIONS
ROCHESTER, VERMONT

Inner Traditions
One Park Street
Rochester, Vermont 05767
www.InnerTraditions.com

Cataloging-in-Publication Data for this title is available from the Library of Congress

ISBN 979-8-88850-131-3 (print)
ISBN 979-8-88850-132-0 (ebook)

Printed and bound in India by Replika Press Pvt. Ltd.

10 9 8 7 6 5 4 3 2 1

Text design by Virginia Scott Bowman and layout by Priscilla Harris Baker
This book was typeset in Garamond, with Frutiger, Gill Sans, Kepler, NixRift, and Sveva used as display typefaces
Creative Commons Agreements: CC BY 2.0: pgs. 16, 192, 230, 301, 433; CC BY-SA 2.0: pgs. 349, 420; CC BY 2.5: pgs. 25, 248; CC BY-SA 2.5: pgs. 257, 339; CC BY 3.0: pg. 179; CC BY-SA 3.0: pgs. 66, 85, 160, 197, 250, 265, 268, 280, 376; CC BY 4.0: pgs. 278, 299, 318, 332, 340, 419; CC BY-SA 4.0: pgs. 108, 198, 202, 256, 303, 308, 310, 427, 431. All other art is in the public domain or by the author unless otherwise noted.

To send correspondence to the author of this book, mail a first-class letter to the author c/o Inner Traditions • Bear & Company, One Park Street, Rochester, VT 05767, and we will forward the communication.

Scan the QR code and save 25% at InnerTraditions.com. Browse over 2,000 titles on spirituality, the occult, ancient mysteries, new science, holistic health, and natural medicine.

CONTENTS

PART THREE

THE DIVINE AND EARTHLY REALMS

....

*Understanding the Cosmological Structure
Underpinning Agrarian Magic*

PART FOUR

PLANT MAGIC AND SACRED AGRICULTURE

FOREWORD

❧

JOHN MICHAEL GREER

Plenty of books claiming to be about earth magic have been published over the last half century or so. Most of them bear the unmistakable hallmark of the city-dwelling author, for whom nature is something to visit on spare weekends and the earth is a distant presence down somewhere past the sewer pipes and subway tunnels. For that matter, though most of these books talk at more than ample length about ancient traditions, few of their authors would recognize a genuine ancient tradition if it leapt up from the soil and bit them on both buttocks. The wholesale manufacture of what some scholars call "fakelore"—alleged folk traditions with conveniently untraceable histories and roots as mobile as those of Birnam Wood—has been a booming industry for decades. Even authors who want something more authentic too often get lost among the thickets of pretentious fakery.

Thus it was a definite shock, as well as a considerable pleasure, to read the manuscript of Todd Elliott's *The Cunning Farmer*. To begin with, Todd is no urbanite; he is a working farmer who makes his living in the time-honored way, with hands plunged into the living soil. Nature and the earth are not distant abstractions to him. His book is informed throughout by the kind of practical experience of nature's cycles and processes that too many other authors in the field of nature spirituality can only dream of having. Its subject is no less focused on realities: Todd has written a practical manual for those who intend to practice farming as a magical art, within the kind of enchanted worldview that the official voices of our culture angrily insist no one can be allowed to experience anymore.

This in itself would make Todd's book worth reading. What makes his work stand out to an even greater degree is the fact that he has done it by drawing intelligently on the last three thousand years of Western occultism and traditional spirituality. From the Pagan traditions of Greece and Rome straight through to the writings of twentieth-century nature mystics Rudolf Steiner and John Michell, with respectful attention to the Native American and African-American traditions that have shaped the spiritual landscape where he lives, farms, and works magic, *The Cunning Farmer* shows an impressive grasp of occult tradition informed by modern scholarship.

Behind Todd's work lies one of the most promising movements in modern occultism, the systematic study and revival of ancient, medieval, and Renaissance magic with as much accuracy as the surviving sources will permit. For many centuries—since the Renaissance revival of classical magic, in fact—occult writers and teachers have routinely rewritten the past to suit the needs of the present. The same habit of thinking that led Renaissance occultists to backdate the *Corpus Hermeticum* to the time of Moses, and convinced English mage Dion Fortune to project her own magical innovations onto forgotten Atlantean priesthoods, has encouraged a great deal of sloppy scholarship in the occult community in our own time.

Why that should have inspired a movement in the other direction in the late twentieth century is a question I don't propose to address here. What matters is that the movement in question happened, sending some modern occultists and Pagans in search of teachings and practices with a verifiable ancient history, and inspiring others to fess up to the modern origins of the material they practiced and taught. That movement had its own excesses, and too often inspired a sort of historical fundamentalism for which "currently accepted scholarship says it, I believe it, that settles it" might as well have been the motto. Despite such vagaries, it accomplished a great deal of good by clearing away the thickets of historical falsification and bringing to light a rich bounty of forgotten magical teachings. These have turned out to be just as relevant now as they were in the distant past.

This work of excavation is still ongoing, but over the last few years it has started to be accompanied by an equally valuable process of synthesis, in which modern practitioners and authors take the legacies of the past as foundation and raw material for new syntheses. It is to this latter process that *The Cunning Farmer* contributes. While it is informed by a wealth of traditional lore, it weaves this together into an original system of working with nature and the earth on inner and outer planes at once. Prophecy is a risky business, but I think it is quite possible that in retrospect, this book's publication will mark a significant turning point in the development of contemporary nature-centered spirituality—and maybe even in the struggle to repair our culture's failed relationship with the Earth itself.

JOHN MICHAEL GREER
PAST GRAND ARCHDRUID
ANCIENT ORDER OF DRUIDS IN AMERICA

JOHN MICHAEL GREER is a highly respected writer, blogger, and independent scholar who has written more than seventy books, including *The King in Orange*, *The Long Descent*, *Circles of Power*, and the award-winning *New Encyclopedia of the Occult*. An initiate in a variety of Hermetic, Masonic, and Druidic lineages, he served for twelve years as Grand Archdruid of the Ancient Order of Druids in America. He lives in Rhode Island.

ACKNOWLEDGMENTS

I would like to take a moment to thank all of those who have supported me in the journey of the writing of this book.

To the Creator and the powers of the cosmos, whether you are called gods, angels, or archetypes, you have provided my inspiration and given me words in my dreams and in the wee hours of the night, and brought me to the right source at the right time. May this work contribute to helping those who have lost their way find their way back to the path and to the Earth.

To Esmee, the love of my life, the true goddess of my heart, my lover, best friend, and life partner, we have been together through many joys and difficulties. She has supported me through every step of this process, listening to the ramblings and musings to which I constantly subjected her and which were the beginnings of this project. She also made time to have me read the entire book out loud, chapter by chapter, and offered kind and practical advice. She has always been my muse throughout our time together, but particularly during this project.

To my boys, Miller and Wagner, who have also been patient, kind, and encouraging, and have endured their share of impromptu lectures on esoteric topics, especially Miller, who is also my right-hand man in farm work. Wagner, a budding wizard in his own right, seems to always be on the same wavelength with whatever I'm studying.

To dear friends Tiffany Stafford and Carden Willis, who have proofread and edited the manuscript, put in many hours on this project, and offered constructive criticism and encouragement. These few words cannot begin to express the depth of my gratitude to them.

To my mother, Lynn, sister, Emily, and mother-in-law, Teresa, as well as Teresa's crew of sisters who have been my devoted readers from the very first blog post I made. I also thank Teresa for the loan of rare books from her personal library to further my research.

I extend my thanks to the subscribers and readers of my blog, *The Cunning Farmer*, for your interest and support, which have encouraged me to pursue the writing of this book.

I also thank friend and author Normandi Ellis for encouragement, for setting me on the path to become a published writer, and for putting me in touch with Inner Traditions International.

Finally, I would like to express my gratitude to all of the people at Inner Traditions International, particularly Jon Graham, Courtney Jenkins, and Lyz Perry, who have been so helpful to me as I navigate the publishing process.

ENTERING THE WORLD OF THE CUNNING FARMER

Writing this book began as a quest to reconcile the tension between two foundational passions of my life: my love of the land, the Earth, and plants, and my deep and abiding need to connect with a transcendent spiritual reality. In the past, I have felt that farming, which is hard, time-consuming, often boring work with long hours committed to rote and tedious tasks, was something that took me away from my intellectual and spiritual pursuits, which were relegated to my scarce free time as a hobby. As I have gotten older and am now entering my fifties, the spirit is calling me more insistently, and I am being urged to pursue this calling. I have also come to realize that my calling as a farmer is absolutely essential to who I am, a calling that is as spiritual as it is physical. This book is about how there is no separation between the two vocations—that it is only through the embodied life, which is symbolized by our relationship with the Earth, that we live into our role as mediator between the heavenly and the earthly.

It has been my life as a farmer that has given me the opportunity to experience intimately the change of the seasons, the flush of new growth in the spring, the scent of rain on freshly plowed soil, the countless rainbows and thunderstorms, the first appearance of ripe seed heads in the midsummer grass that marks the turning and ripening of the year, the burgeoning harvest of July, the dusty reddish haze of the August Sun, and the first clear, crisp day of autumn. All these moods of nature I have come to know as intimately as only a lover knows his beloved.

It has also been my privilege, because as a lover of the night I am a rarity among farmers, to wander among my crops in the bright moonlight of a July

Full Moon, after a thunderstorm when the power of the vegetable world is at its peak, and hear the tree frogs, the screech owls, and the katydids, and see the brilliant flashes of the fireflies, the stars of Virgo, Libra, Scorpio, and Sagittarius whirling overhead, Jupiter and Saturn tracking across the sky, and the immensity of the Milky Way glittering above. The power, life, and spiritual vitality of all living things are palpable and undeniable at moments like these. It has become my practice, there under the stars, under the rising Moon, standing on the Earth, surrounded by that power and beauty, to honor the spiritual forces that are its source with humble and heartfelt prayer in simple rituals of great personal meaning. This is my way, and I make no apologies for it. I commit to no particular sect, school, or religion beyond that of the perennial Hermetic tradition as I see it.

My religious philosophy is informed by the *Corpus Hermeticum*, Neoplatonism, Hellenic and northern European Paganism, Kabbalah, Advaita Vedanta, Buddhism, esoteric Christianity, and animist thought from Indigenous traditions around the world. I believe in one Source for all things, outside of space and time, which pours out its being as life and light, thereby creating matter and the physical cosmos. I believe (although spatial metaphors such as "higher" and "lower" are inadequate and not to be taken literally) that there exists a hierarchy of spiritual beings that "descend" from this one creative Source that we may term "gods" and "angels," as well as lesser spiritual beings we may call "daimones." All of these are limited and finite in comparison with "the One" and are actually aspects of it through which it interacts with the cosmos, administering reality, so to speak.

There are apparently many different kinds of these beings, some creative and healing, which we may call "good," and some destructive and wrathful, which we may call "evil." What their purpose is in the scheme of the cosmos is an interesting subject for speculation but probably above my pay grade. I believe, however, in providence, not as in a naive "everything turns out for the good" providence, but more like what the ancient Greeks called *pronoia*, in which everything has a role to play in the vast cosmic plan that is ultimately indifferent to the finite and ephemeral beings that we are in our physical forms. We are, however, much more than physical creatures; we are gods and sparks of god hidden within dark husks of matter, and the spiritual life consists of truly remembering this and liberating the trapped light.

Matter, the dark husk I spoke of, is made of mind stuff, and has no ultimate reality. It seems solid, but science tells us that each particle contains vast reaches of empty space, punctuated by particles that can be further divided into ever-smaller particles that have no physical existence but are just mathematical probabilities until observed, merely bits of information in the cosmic dreaming mind of the One, the *lila* of Ishvara, the dance of Shiva. This universe, made of mind stuff in the shape of matter, controlled by spiritual beings that administer creation, is somewhat malleable and plastic—under the right conditions—on a spiritual level. And this is where magic comes in.

A warning to the casual reader: Much of this book deals with magic. If this is going to be a problem for you, turn back now. If you think that we are hellbound for exploring such matters, or if you are a skeptic, I ask you to step outside of the comfortable paradigm in which you have been confined and realize the cosmos is bigger and stranger and the Divine more capacious than you think. Keep an open mind.

Magic has been defined as non-normative religious practice, meaning magic and religious practices are very similar, and there are no hard and fast rules to distinguish between the two. Magic uses similar tools to religion, such as ritual and prayer, for purposes or in ways that lie outside the normal religious structure of a culture. Being dissatisfied with the normative religious cultures in which one finds oneself, but still having a strong desire to interact with spiritual realities, many have decided to forge a relationship with the spiritual on their own terms. Much of this is labeled magic and witchcraft, perhaps because it is an attempt to use supernatural channels to cause changes in the material life of practitioners. This said, I think there is plenty of "normative" religious behavior that functions as an attempt to improve the material well-being of worshippers and thus is indistinguishable from magic.

I believe the context of my practice to be in the spirit of the venerable tradition of the rural magical practitioners known as the cunning folk of Great Britain, which was later brought to North America. I borrow magical and spiritual practices from a wide range of sources. I am a dual-faith practitioner; I don't reject the Anglican Christianity of my ancestors, and I use the biblical psalms and verses in my magic. I practice the daily office from the Book of Common Prayer or from the Liturgy of the Hours to stay in touch with both the transcendent and the spirits of my birth tradition. I use prayers and techniques from the medieval and Renaissance Latin grimoire tradition to call upon angelic intelligences. I use Hebrew kabbalistic prayers from the Golden Dawn tradition of ceremonial magic, as well as the practical Kabbalah. I am inspired by the *Orphic Hymns* to the Greek deities, Greco-Egyptian Magical Papyri, as well as the Book of Psalms.

I also find myself at home in the tradition of astral and planetary image magic of the Andalusian grimoire, *Picatrix*, which was brought to the West during the Middle Ages. I don't see any contradiction in borrowing from these diverse systems. Magic, to me, is a continuum; from ancient times the names have changed, but the aims and techniques haven't. The Pagan gods were baptized as angels, saints, or planetary intelligences after the collapse of the ancient Pagan religions in the West. For me, they are all forms of the One and channels of divine energy.

In addition to these practices of what I call "temple work"—or "high magic," as some call it—I also emphasize the importance of working devotionally and magically with the spirits of the land and of the place in which I live. I also honor the deities from various traditions who embody the forces of the natural world, the old gods and goddesses of the Earth, the Sun, the Moon, the planets, the wind, the rain, and the harvest. One of my goals as a practitioner is to reclaim—in spirit anyway, not necessarily by reconstruction—the Indigenous animism of my deep ancestors in northern and western Europe, to honor both the Heathenism of my deep ancestors as well as the Christianity of my more recent ones, which, knowing something of the history of the lands of my ancestors, means about fifteen hundred years of ancestry—not something to be lightly cast aside. It has become important to me to honor the ancestral gods, so that means walking between two faith worlds, making a sometimes uneasy coalition between the ancestral gods. Some of my writing will address that tension, which is not my issue alone but is a core theme.

My eclectic approach has put me outside the fold of the orthodoxy of my birth tradition, for which I have the utmost love and respect, but I believe the divine is infinite love and generosity and is not limited to one culture's version of truth. As such, it is not threatened by individuals following their own inner guidance, wherever it leads. I can't imagine what motivates the fundamentalist sectarianism of all faiths that causes people to demonize those who don't see things as they do. Much of what we discuss here may be threatening to those who think that their own culture's particular scripture or revelation is the only or unique path for all mankind, those people who see what I am advocating here as idolatry. I categorically reject such notions as attempts to impose an imperialism of the spirit by colonizing souls and minds, often for the sake of consolidating political power. God does not belong to any church but rather is the common Source of all.

I am here to share my insights and experiences, both as a passionate student of the esoteric and a tiller of the soil. I am engaged in a learning process to find powerful practices to deepen my understanding of the participatory reality in which we find ourselves. Magic abounds in this beautiful world. We just need to claim our birthright as cosmic co-creators.

In my primary vocation, I am a veteran practitioner of organic farming, which I have practiced for twenty-five years. I have taught various aspects of the trade to many apprentices over the years. My farming style is practical and is something that I learned from some very down-home, practical, salt-of-the-earth tobacco farmers and several far-out hippie farmers/mystics/shamans. It has been honed by experience and the joy of seeing happy, thriving crops as well as the heartbreak of seeing not a few crop failures. You learn to take the good with the bad, to replant, to go on, to wait for next year. You also learn to pray, to petition, to implore, to beg, and to rail against the superhuman powers to which you owe your livelihood.

And this is where the magic comes in, and has always, since the first Levantine farmer planted the first seed of emmer wheat in the dust he scratched up with his hoe and then waited for the rain, asking whatever gods to bless that first crop. He danced, burned fragrant resins, and sang the sacred rain songs—just like farmers all over the world still do today. This is about co-creation, working with the forces of creation to enhance those forces' ability to bring forth life in the specific way that the farm and the human and animal community dependent on it need to thrive and flourish.

In this book, we will explore the idea that the rural people of the premodern world, the farmers and hunters, the midwives and herb doctors, the shamans and witches, the cunning folk of all times, have much to teach us in our modern world of technology and alienation from the natural world. My work will delve into folk magic and high magic, astrology and philosophy, alchemy, mythology, and folklore in an effort to promote a way of seeing, a way of thinking, and a system of practice that helps us reconnect and re-enchant our relationship with our land and our cosmos.

We will also touch upon the very practical aspects of farming, the everyday magic of soil and seed that is at the heart of our life as farmers. We will touch on connecting with land spirits, moon magic, weather magic, the magic of the different cosmological realms, plant magic, and agricultural magic. It is not my primary mission with this book to teach people how to farm; my primary goal is to help those engaged in life on the land to deepen their connection to the elemental and celestial forces of this beautiful planet on which we live in order to bring meaning, wonder, good fortune, success, and the wisdom that comes from being in relationship with these spiritual forces. We are given this life not to acquire material possessions or to distract ourselves with mundane business for its own sake. We are here to love, to gain wisdom, to heal, to renew, and to live in relationship with the community of souls that is our cosmos. Life is precious and short, and we are here to live our best life.

I would like to propose that we undertake a project of rewilding the imagination in order to reinvigorate our relationship with both nature and the world of the spirits, whereby we reclaim our birthright as intermediaries between the gods and nature. We live in the crux and fulcrum of creation, fully enmeshed and embodied, yet capable of communicating our desires through all worlds, the divine, the celestial, and the elemental. We need to reanimate and re-enchant our relationship with the world, which has been rendered sterile by the way of seeing that has captured the imagination of the West since the Enlightenment—the worldview of materialism and scientism, where only the sensible, that which is available to the senses and which can be measured, is real, and anything else is "only subjective."

We will develop this concept of the cunning farmer as a mediator, as the one who consciously actualizes the connection between the spiritual powers of the heavens, the land, the ancestors, and the gods, angels, and earthly spirits with whom we share the land and our lives. There is a greater spiritual ecology that extends through all levels of existence, and it is our task to honor and respect the powers that animate the natural world in which we live. We can do so by engaging in acts of devotion to the land itself, to the divinity inherent in the trees, the springs, the fields, the Sun, Moon, and stars, through which we can awaken from the sleep of materialism that has fallen on our world. The world is alive and imbued with sentience; it is crying out to be loved, honored, and even feared, as an embodiment of the transcendent Source.

There is an archetypal realm, the level of meaning on which the things and events of the physical world participate in some mysterious way that is beyond physical causation but in some way resembles causation. This is the level of the world soul, the divine mind, which organizes events that are linked by meaning and time and space but do not necessarily have any physical connection: how the planets and the Moon influence life on Earth; how words and thoughts can shape reality; how entities which have no body and are pure intellect and spirit can effect meaningful changes in the course of events and consciousness itself. How this works we will probably never know, but how we can use it for our benefit is a different story. If reality is the playing out of the cosmic mind which in some mysterious way creates the physical world, and is not the result of some predictable Newtonian mechanism, how can we use the inherent harmonies and sympathies of that mind to help it work for our benefit? If everything is illusion and the play of *maya*, how can we have a more pleasant illusion? Who and what are the forces and intelligences that invisibly share our world with us, a half-turn away in a different dimension? What do they want and how can we relate to them in a practical and productive manner for our mutual benefit and that of our fellow creatures?

These are some of the questions and concepts that we will explore in the pages of this book. Please join me for this exploration of the world of the cunning farmer.

PART ONE

Introducing Farming Magic

1

CONSULTING THE GENIUS OF THE PLACE

Alexander Pope, the eighteenth-century English poet, giving advice on garden and landscape design, wrote, "Consult the genius of the place in all; That tells the waters or to rise, or fall."[1] These lines have been understood to recommend considering the context of the location when designing as a primary principle in landscape architecture. The poet was invoking a religious idea from ancient Rome, the idea of the *genius loci*, or the spirit or god of the place. In the ancient world, the land was considered to be inhabited by spiritual entities that had to be propitiated with prayer and sacrifice before any kind of work could be attempted on the land. Each grove of trees, each spring, each creek or river, and every hill or mountain had its own spiritual beings, which were variously known as fauns, nymphs, dryads, satyrs, lares, and gods. Proceeding to fell trees, plow ground, or sow without their approval was to court disaster. No respectable Roman would fail to honor the *genii locorum* in any agricultural operation. Cato the Elder, in his farming manual, *De Re Rustica*, stated that it was the sacred duty of a farmer to sacrifice a pig to the god of the grove before cutting down a single tree![2] The care of the local spirit population was as essential to proper agriculture as good plowing techniques and sowing the right seed at the right time.

Such an approach is a far cry from the merely aesthetic sense meant by our poet, Pope. This philosophy entails a deep spiritual relationship with the intelligences of the land and brings a reverence, not without perhaps a bit of fear of incurring the wrath of the land spirits, to the relationship between the farmers and the land they tend. This idea of the natural world as haunted by spiritual entities didn't begin to fade until the Protestant Reformation supplanted the animistic sentiment of the peasantry with the empty theology of a remote God. Later, as the Industrial Revolution largely depopulated the countryside, the widespread adoption of the materialistic worldview after the Enlightenment finally banished the fairy faith in all but the most remote corners of the world.

The Roman poet Virgil, author of the poetic agricultural manual known as the *Georgics*, which is rich in the rural spiritual lore of the ancient Roman farmers, addressed all the gods and goddesses who preside over the farmer's life at the beginning of the *First Georgic*: Ceres, goddess of grain and agriculture; Pan, keeper of the flocks, the fauns, presences of the fields, the dryads, and the spirits of the oak trees; Neptune, who created horses; Minerva, who created the olive tree; Liber/Dionysus, giver of the vine; and Silvanus, god of the forest. He praises,

> *You gods and goddesses who, with such kindness,*
> *Watch over our fields and vineyards and who nurture*
> *The fruits that seed themselves without our labor*
> *And all the crops with rain that falls from heaven.*[3]

So what does "consulting the genius of the place" mean to us? To me, it means first, on a practical level, just as Virgil recommends in the *First Georgic*:

> *And yet, if the field is unknown and new to us,*
> *Before our plow breaks open the soil at all,*
> *It's necessary to study the ways of the winds*
> *And the changing ways of the skies, and also to*
> *Know the history of planting in that ground,*
> *What crops will prosper there and what will not.*
> *In one place grain grows best, in another vines. . . .*[4]

So on the material level, consulting the genius loci could mean to take into careful account the ecological conditions of the land, sun, drainage, soil type, slope, local weather, what crops grow well in your area, and other such considerations, the kind of things that the science of agronomy concerns itself with. This is necessary knowledge, and I myself spent years reading every agronomy book I could get my hands on.

✦ *An ancient Roman depiction of the genius loci, the spirit of place, as a serpent from Pompeii in the first century BCE. From Marisa Ranieri Panetta (ed.),* Pompeji: Geschichte, Kunst und Leben in der versunkenen Stadt, *Stuttgart: Belser Verlag, 2005, p. 111.*

But it's not enough to only heed the material conditions of the land. Consulting the actual spirits who inhabit your land pays on a practical level because they are real, even if you can't see them. I know I'm heading into crackpot territory here, perhaps, but this is my experience. When you take the spirits into account, things just start to go better—crops are healthier, animals thrive, you have fewer problems with pests, weather may be more favorable, and you may even suffer fewer accidents and breakdowns. I'm not saying you're not going to have problems if you neglect basic agronomy—far from it. You must put in the physical work for the magic to work. Careful agricultural technique also honors the spirits of the land. One must work the soil at the right time, when moisture and other conditions are right, and plant at the right season (in the right phase of the Moon in the right sign, but more on this in a later chapter). All of these practices please the land spirits.

A true sage who worked at my side for eight seasons, a gentleman from the mountains of rural Mexico named Salvador, would pause while we were engaged in some manual job like hoeing or weeding and extend his hands in a gesture that took in all the surroundings. He would then say poetically in Spanish, "All this has life—the trees, the sky, the clouds, the earth, the water, everything around us, and we are in contact with this life because we are in contact with Him above who gives it life. And when we need something, all we need to do is ask and He will make it happen if we have faith." That man is the best gardener I have ever seen, and his half-acre patch is a testament to what a prayerful and reverent approach to plants and farming can do. And he took this approach not just to his garden but to his whole life! He would begin every day by commending the day's work to God, and every time we planted, he would bless the first seed or plant in the name of the Trinity and with the sign of the cross. That level of spiritual attention applied to the sacred work of agriculture is what I'm talking about here.

Salvador's devout Roman Catholicism is beautiful and I have immense respect for him and his faith, but it is not my way. However, prayerfully reaching out to the Source in whatever name one finds most appealing whenever we begin any major agricultural operation is an excellent practice. You could use the *Homeric Hymn* or *Orphic Hymn* to Gaia or Demeter if you are of a classical bent and pour out a libation of wine or olive oil before you plow or work ground. Asking the Creator to bless the seeds you are about to plant and honoring the phases of the Moon, the rising of the Sun, the planets, and the stars are all good ways to spiritually connect with the spirits of nature and the spirits of the land. I always make sure to start and end my prayers by acknowledging the Source, the One, from whom all things come, but that's just my way.

A more specific way to connect with the genii locorum is to find a power spot in nature near you—a hollow tree, a spring, a cave, a large oak tree, a hill, something that seems like a gateway. On two farms I have lived on, I have found large oak trees growing next to springs and have gone occasionally to leave offerings and ask favors from the genius there. I meditate briefly and go on my way.

I am wary of adopting overly specific European ways of relating to the land on which I live and also am wary of attempting to imitate the practices of the Indigenous people who once inhabited this land. Gaia and Demeter, in the example I gave above, are universal deities who, while they originated in a particular place, weren't only worshipped in that place. Although the mysteries of Demeter were celebrated in Eleusis, near Athens, she was goddess of agriculture all over the ancient world. To a certain extent your personal imagination colors how you see the inhabitants of the imaginal realm, but I wouldn't expect my fairies to speak with an Irish accent or my lares to wear togas. Robert Kirk, the seventeenth-century Scottish seer, said in his treatise on the fey (also can be written as fae), *The Secret Commonwealth*, that, "their apparel and speech is like unto the people and country unto which they live."[5]

The Cherokee, whose lands are a bit to the south of where I live now, have a well-developed folklore of supernatural beings. I imagine the Shawnee, who used to hunt on what is now my land, and the Adena people, whose arrowheads I find in my fields and whose burial mounds used to dot the creek bottoms in the valley below my farm, did too. One nearby Adena burial mound excavated in the 1950s was found to contain the skeleton of an unusually tall man, who had had two of his front teeth removed so he could insert a "spatula" made from the upper jaw palate and front teeth of a timber wolf![6] Obviously, he incorporated the teeth and jaw of the wolf to take on the powers of a wolf somehow. This was found two miles from here. I call him the Wolf Shaman, and I feel like I have felt his presence before as a protector and guardian of the land. I would like to know what happened to his remains. I heard that they are probably in some drawer in a museum somewhere, poor guy.

In the Celtic countries, the genii of the Romans were known and called by many names in the various districts. The Gaelic people called them the *sídhe*, which is also the word for the Neolithic barrow tombs where they were thought to dwell. Many scholars say there is a continuity between the *sídhe* and the

Gaelic gods, the Tuatha Dé Danann, who were dethroned and driven into the mounds by the advent of Christianity. They are known by the Celts of Wales as the Tylwyth Teg, in Cornwall as piskies, in France as fey, in Germanic countries as elves, perhaps in Muslim lands as *djinn*, and in English-speaking countries as fairies, a word which has been thoroughly Disney-fied, though the process of taming the archetype of the fairy was begun long before Tinkerbell ever graced the silver screen. These beings were much feared by premodern country people, and were variously blamed for illnesses in man or beast, birth defects, insanity, disappearances, and all manner of mishaps and mischief. Stories of misfortune following the violation of a taboo, such as cutting a fairy tree or bush, or plowing a mound or otherwise trespassing on the sacred sites of the fey, abound in folklore and are too numerous to report here. Much of the fairy mischief in the literature is the result of either the innate mischievous tendencies of the good people, as they are called when one wants to avoid offending them, or of failing to properly propitiate them with offerings of milk or butter.

As an aside, several years ago I had a Jersey milk cow just wander off and leave her three-month-old calf (which is unheard of, as these animals are the very archetype of motherhood!) and disappear into the woods, never to be seen again. She had been acting strangely, mooing, pacing, and looking off into the distance. I just thought she was in heat (though in the pasture we had a bull, who was resting with the remainder of the herd under a tree at the time) and would get over it. We had a social engagement to attend and when we returned she had (again, uncharacteristically) charged through our electric fence and disappeared into the hundreds of acres of woods that surround our farm. It was not lost on me at the time that this is exactly what the folk traditions of the Celtic nations were referring to when they spoke of prize cattle (usually milk cows) being "taken" by the fairies or being "pisky led." I should have been leaving out my offerings of butter!

But the fey, as we shall call them, were also helpful bringers of blessings when properly propitiated. The folklore abounds with stories of helping spirits like brownies, who assisted with domestic chores like cleaning and milking cows. I imagine that this help is not to be taken literally. There are times on the farm when a routine job that normally takes a long time goes easily and twice as fast. In moments like that, I always think about how my friend Salvador would pray to St. Isidore, the patron saint of farmers, to send angels to help us with particularly hard tasks, and the job would be accomplished without the problems that so often accompany routine farm tasks. Or maybe they literally meant that upon awakening they discovered that the house had been cleaned, pots scoured, and cows milked by invisible helpers while the farmer and his family slept.

A road being built in County Clare, Ireland, in 1999 was rerouted to avoid having to destroy a sacred fairy tree, the cutting of which which would have had dire consequences according to local folklorists. My question is: Is this custom

✦ *St. Isidore the Farmer. Painting by an anonymous artist of the Cuzco school, c. eighteenth century. Image by Dozenist.*

unique to Ireland? Why don't the people of, say, Kentucky, where I live, honor the land spirits like they do in Ireland? The answer is complicated. Many do honor the spirits of the land. There is a deep culture of love for the land, alive and well here in my home state. Its most famous exponent is agrarian novelist, poet, and essayist Wendell Berry, who lives, writes, and farms only a few miles from me. The Appalachian region also is home to a land-honoring culture that extends beyond the state lines, encompassing the entire mountainous region. Love of land and a spiritual devotion to place is not limited to any one region and is found among good people everywhere.

But many of my neighbors who have earned their living by farming, forestry, or operating land-clearing machinery have had a, shall we say, less-than-reverential attitude to the land. One neighbor told me point-blank that God gave us animals and plants and nature in general for us to do what we want with! I know a young man who has a logging operation that I drove by recently. He had the trunks of several ancient oak trees loaded on the trailer of his log truck. I wonder how mindful of the spirits of the land he is while engaged in turning these majestic beings into lumber. These neighbors are for the most part kind, decent, and hardworking people who are making their living, as I do, by bringing natural products into the modern marketplace to be sold. We all rely upon forest and farm products, as did our more reverent ancestors. The difference is that now these are extracted without piety and without regard for the sacred life of the land. We are increasingly paying the price on every scale for our lack of reverence.

What are the consequences of incurring the wrath of the genii locorum? You only have to examine the lives and communities of those involved in land exploitation to see the consequences of despoiling the natural world. To paraphrase the Buddha, as surely as a cart follows an ox, addiction, crime, abusive relationships, lawsuits, acrimonious divorces, accidents, chronic illnesses, and misfortunes of every kind follow from living a life out of balance with nature. The Greeks believed that hubris (pride and disregard for the divine order of things) is followed by Ate (the personification of folly or delusion), which is followed by Nemesis (the goddess of vengeance who punished evildoers). This is the cycle in which divine punishment is the result of arrogance and pride in humans. And this does not just happen on a personal level. The spiritual life of my region, the nation, and the entire world is out of balance with nature. Many of our social ills can be traced back to this broken relationship between humans and the spirits of the land.

Volumes have been written on the disenchantment of modern life, the

✦ *Nemesis or Fortuna. Engraving by Albrecht Durer, c. 1502.*

meaninglessness and nihilism that is implicit in the materialist worldview and value system of the society in which we live. One way of escaping this pervasive psychic rot is to engage in a life-affirming spirituality that has meaning for you and your life and to drop out as much as possible from the false values of consumer society. I don't want people to think that I am advocating the adoption of antiquated superstitions wholesale; we have modern science and it has given us many gifts. But at what price? I am advocating a rewilding of the imagination by allowing ourselves to see the world as our premodern rural ancestors did, to the extent that it is beneficial and to the extent that it is even possible.

For me, rewilding the imagination looks like adopting a premodern viewpoint as if it described a spiritual reality, if not a physical one. We know that with astrology, for instance, the physical cosmos doesn't literally correspond with the cosmology that underlies traditional astrology. Astrology still describes eternal archetypal correspondences in the world and in the soul. The spiritual world underlies the physical. As Heraclitus said, "Latent structure is the master of obvious structure." The genii locorum—the fey, the gods, and the angels—execute the creative force of the divine here in the world of manifestation, administering the world of nature accordingly. They are forces with whom we can interact and form relationships to our benefit and to that of the world and the human community.

So how to proceed in practice? I've mentioned some practical steps in passing above, which I will repeat here, along with some others that I haven't mentioned, that I consider helpful to those who are interested in entering into a closer relationship with the spirits of the land:

1. Leaving out libations. These are offerings of a drink (often wine, but it could be juice or milk) poured out for specific entities. It was a long-standing custom to leave bread and milk out at night for the fairies in much of the Celtic world.

2. Leaving food offerings outside, on an altar, or at a sacred spring.

3. Performing prayers and blessings any time any significant projects are undertaken on the land, like felling a tree, plowing, or otherwise working ground, planting crops, harvesting, and so on.

4. Marking the turning of the year and seasons, like the solstices, equinoxes, and cross-quarter days, with prayers relevant to the occasion.

5. Finding power spots in nature where you can sense the presence of land beings and enter into communication with them. A pendulum can be helpful for this, by dowsing for yes or no answers, or indicating flows of subtle energy.

6. Practicing geomancy (particularly if you are a bit thick about receiving messages from the nature communication, like I am), which is a method of divination that basically involves making a series of marks in the dirt or sand and deriving answers ranging from either a yes or no to a whole astrological chart from them.

Just getting outside, into nature, night or day, and speaking from your heart in prayer, song, or silence to the power there, with as much or as little ceremony as you are comfortable with, will change your life more than you can imagine. And with repeated practice comes a deeper relationship with the land and its visible and invisible inhabitants.

2
LAND SPIRITS
AND SACRED SPACE

A Pilgrimage to Some of the Earthworks of Southern Ohio

As we will discuss throughout this book, within the shamanic cosmos everything is connected as part of a larger whole. The spirit of a place, therefore, is both highly localized and also part of a larger cosmological structure. Visiting and exploring sacred spaces, such as earthworks and other sacred sites, can help one connect with and understand the spirit of a particular geographical place.

Since I was a child, I have always been fascinated by the idea of ancient ruins. Growing up in suburban Philadelphia, where the oldest buildings in my neighborhood dated to the 1680s, I always longed for contact with true antiquity. I read books about the megaliths of Western Europe, the stone circles of Orkney, the barrow tombs of Ireland, and the pyramids of Egypt and Mesoamerica. I was, and I confess still am, a little bit envious of people who live among the mysterious relics of forgotten civilizations. When I was thirteen, on a family trip to Mexico, I climbed to the top of the Pyramid of the Sun in Teotihuacán, which is the largest pyramid in the New World. It was an awesome experience I will remember for the rest of my life. Other than that, my life has been largely free of opportunities for ruin exploration.

My farming career has been largely spent on two farms in Kentucky. The first was in the Cumberland River Valley on an upper terrace of the bottom, which had been heavily inhabited by prehistoric people. There I found projectile points, stone adze blades, slate gorgets, flint knives, and many more artifacts made by the previous inhabitants of the land. I found something nearly every week.

For the last seventeen years, I have lived on a farm in Owen County, Kentucky, on a ridge overlooking the Eagle Creek Valley. I have also found many projectile points here, mostly from the Adena and Hopewell culture periods. I am often drawn to find out more about the ancient people who lived and worked the same land I do. I feel like I have three sets of ancestors in a sense: blood ancestors, or my actual genetic line; spiritual ancestors, from whom I have received the heritage of spiritual and religious traditions; and land ancestors, or those who lived, died, worked, hunted, and farmed on this land before me.

I think it is important to acknowledge that I am an American of European descent. My culture has had an extremely troubling history of displacement and genocide of the Indigenous inhabitants of what is now the United States, where most of my blood ancestors have lived for over two hundred years. Despite this tragic history, I feel a kinship of some sort and a sense of connection and respect for the people of all kinds that have been sustained by the same ground as I am. They are still here, in the land, and it is our task to make peace with them.

In order to connect with my "land ancestors" and to experience the unique liminality and presence of these ancient sacred ritual sites, I took a trip with my family to south-central Ohio to visit six different Hopewell and Adena culture archaeological sites. We had to travel about two and a half hours from home in order to reach these well-known and better-preserved sites, where people of the same culture who lived on what is now our land also flourished over two millennia ago.

In preparation for and during our trip, my guidebooks to the archaeology of this region were the works of William F. Romain, research associate at the University of Ohio and author of many books and scholarly papers on the Hopewell and Adena cultures and other mound-building cultures. I had come across Romain's work while researching the Wolf Shaman skull, which was unearthed in an Adena burial mound near my home as we discussed in the last chapter. His work fits neatly with the themes of animism, shamanism, sacred space, and the magic of landscape and place explored in this book.

The first stop was Mound City, the center of Hopewell Culture National Historical Park in Chillicothe, Ohio. This place was once a large ceremonial site and burial ground, which is surrounded by a square earthen embankment and contains a number of burial mounds. It is an impressive site, and the visitor center has an astounding collection of artifacts, including hundreds of animal-shaped effigy pipes, bird figures cut from sheets of copper, and the most gorgeous foot-long obsidian spear points. Standing within the necropolis of a vanished culture was a humbling and moving experience. We made sure to make a discreet offering of sacred *rustica* tobacco (which seemed to be a favorite of the original inhabitants, owing to the number of pipes found there) in gratitude to the spirits of the place.

For me, one of the eeriest aspects of the modern Mound City site is that it is located in close proximity to the complexes of the Ross Correctional Institution and the Chillicothe Correctional Institution. It seemed to me to be a really bizarre juxtaposition, an odd choice of location, to surround an ancient sacred site with

✦ *Seip Mound, a Hopewell culture burial mound in Bainbridge, Ohio.*

two prison facilities. In former times, it was customary to place a church, often consecrated to the archangel St. Michael, on top of an older Pagan site, perhaps to keep the Pagan spirits at bay or to draw on the natural power of the site. But why two prisons at this particular site?

Next, we stopped at the Seip Earthworks and mound site located in the beautiful Paint Creek Valley. This site consists of a twenty-seven-acre square enclosure, a forty-acre circular enclosure with a tall rectangular burial mound in the center (pictured on page 23), and an eleven-acre circular enclosure.[1] The mound was excavated and found to contain the remains of about 120 human bodies, as well as a number of artifacts. Five other Hopewell sites in the area contain one square and two circular enclosures of exactly identical acreages![2] The structures were laid out with an extreme degree of precision. Many of the sites were oriented to the sunrises and sunsets at the winter and summer solstices and to the northernmost and southernmost rise and set of the Moon during its 18.6-year cycle.[3]

The Hopewell sites functioned as lunar and solar observatories, burial sites, and ceremonial centers. The degree of astronomical and geometrical sophistication is extremely impressive. Why did these enigmatic people make these fascinating sites? According to Romain, "The driving force behind their monumental architecture appears to have been religion. The essence of their worldview was shamanistic. In this shamanistic worldview, time and space could be traversed by telepathy, it was possible to communicate with all that is, and everything was interrelated and connected in a universal field of magic."[4] This sense of magic, the continued resonance of the rituals of these ancient magicians, was palpable at these sites.

We sat down inside one of the circles at the Seip Earthworks site and imagined what it would be like to be at that place on a summer solstice evening, perhaps when the Moon was full so that it was rising just as the Sun was setting. As the Sun set and the Moon rose higher in the sky, the drumbeats intensified, rattles shook, and the shamans, dressed as bears, wolves, panthers, or deer, transformed into their spirit animals. Pipes of heady *rustica* tobacco were passed around, and gourd bowls of brews containing hallucinogenic compounds of *Psilocybe* or *Amanita muscaria* mushrooms and perhaps herbs such as *Datura* transported the partakers into ecstatic spirit flight. The Hopewell magicians traveled between the worlds, where divine beings taught them the mysteries of the sacred geometry that was essential to their culture. Quoting Romain again, "What distinguished the Hopewell from their predecessors, however, was that through their shamanic

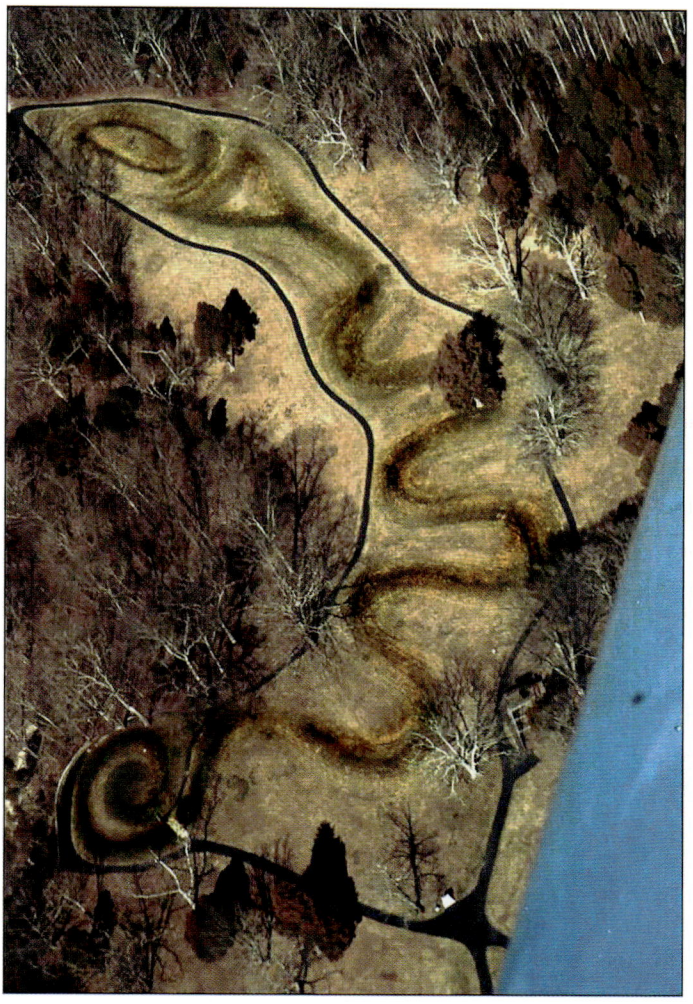

✦ *Aerial view of the Serpent Mound in Peebles, Ohio. Image by Timothy A. Price and Nichole I.*

journeys, the Hopewell came to know something about the fabric of the universe. What they discovered was that this fabric can be described in the language of plane geometry, arithmetic, measurement, and observational astronomy. Using this knowledge, the Hopewell built tremendous symbols of their universe in the shape of their geometric enclosures. Then they tied these symbols into the great lunar and solar cycles of time through a unique alignment system."[5] The idea that the Hopewell and Adena shamans acquired their knowledge of geometry and astronomy through their ecstatic states is extremely fascinating to me, and I will expand on it in later chapters.

The following day, we made an early start and arrived at the world-famous Serpent Mound in Peebles, Ohio. This site features the largest effigy mound in North America. An effigy mound is an earthwork in the shape of a human or animal. The "serpent" is an undulating earthwork in the figure of a snake that appears to be swallowing an egg. The site is old, though estimates by archaeologists vary widely. However, most agree that it is from 2,000 to 2,500 years old and made by either the Adena or the early Hopewell culture.[6] The serpent is laid out with extreme geometrical precision, as with the other earthworks we've discussed, to coincide with celestial events such as the summer and winter solstice sunrise and the varying angles of the moonrise and moonset during the course of the 18.6-year lunar cycle.

Why were the ancient inhabitants of Ohio and Kentucky so fascinated by geometry and astronomy? What motivated them to invest incredible amounts of labor over generations in the production and maintenance of these structures? Why did I keep seeing parallels between the Western esoteric tradition and the shamanism, magic, and sacred geometry of my predecessors in this ancient land? These thoughts were on my mind as we visited the various sites of the region.

Symbology of the Serpent

The lore, symbolism, and mythology of land serpents and celestial serpents, both in the Americas and around the world, is vast and there is not enough space to delve into it here. This particular serpent in Peebles is aligned with the risings of the Sun and the Moon, and it reminds me of the Old-World concept of the eclipse dragon. As scholar Adrian Pirtea states, "The 'eclipse dragon' designates a celestial snake/dragon whose head and tail cover the Sun and Moon and are thus considered to be the cause of eclipses. The identification of the head and tail of this mythical creature with the two lunar nodes, i.e. the two imaginary points of intersection between the orbits of the Sun and the Moon (where eclipses can take place), enabled the mathematization of the eclipse myth and the inclusion of the dragon's head and tail in horoscopes and general predictions." Pirtea also writes, "The idea was unknown to the Greek astrologers of Late Antiquity."[7] According to Pirtea, the origins of the concept are mysterious and the concept became popular with the spread of Arabo-Persian astrology. Could it be that our astute Adena and Hopewell culture astronomers discovered the concept independently in the New World and used similar symbolism to describe it? The serpent is, after all, swallowing an egg. Could the egg symbolize the Sun or the Moon? Archaeologist Romain thinks it's possible: "The Serpent Mound might represent an eclipse. Many

North American Indian legends tell of eclipses being caused by a giant serpent or other animal who swallows the Sun."[8] For more on the symbology of the serpent, see chapter 12.

———

We spent an hour walking around the perimeter of the serpent, taking in the views of the serpent and its surroundings. The whole time we were there, I was impressed by how much the place resembled my own farm landscape, being a narrow, cleared ridge, surrounded by steep wooded slopes and overlooking a fertile creek bottom below. We again made a discreet offering of sacred tobacco to the spirits of that mysterious place and then proceeded to the Fort Ancient Earthworks site.

Fort Ancient is a hilltop site composed of imposing earthen embankments and enigmatic stone-covered mounds aligned with the rising Sun and Moon at the solstices and extreme points in the lunar and solar cycles. The embankments were cut through to allow rays of the rising Moon and Sun to shine on the stone-covered mounds at these propitious moments. The site is huge, and the amount of earth moved by the builders in woven baskets is astounding. There are three and a half miles of earthen walls up to twenty-four feet high![9] The park is home to a mature hardwood forest with massive oak trees. Tired and hungry children limited the length of our stay, but even a brief visit was enough to sense the power of the place.

THE SHAMANIC WORLDVIEW

In addition to having conducted extensive surveys and on-the-ground research, Romain has attempted to place the physical evidence within a cultural context based on what we know about modern shamanic cultures in general, and modern Indigenous North American cultures in particular. In his 2009 work, *Shamans of the Lost World*, he describes what he terms an "archetypal shamanic worldview."[10] He goes on to say that this worldview is biologically based, "an emergent property of our cognitive architecture and interaction with the world during hunter-gatherer times."[11] I believe, going further than Romain into the realm of metaphysics, that there is a psychic metastructure that is beyond the limits of our human cognitive apparatus. Indeed, it seems as though shamans, mystics, and magical practitioners worldwide use altered states of consciousness to enter that psychic metastructure through trance. In these states, we are viewing not only the architecture of our minds but the architecture of the universe.

The fundamental characteristics of the shamanic worldview are the principles that underlie all of the major world religions.[12] Romain says, "The essential aspects of this worldview are essentially the same cross-culturally and deep into history."[13] In his discussion of the worldview of the builders of the Hopewell and Adena sites in Ohio, Romain draws on the work of Mircea Eliade, Michael Harner, and Peter Furst. He describes in detail what he believes to be the essential characteristics of the shamanic worldview in cultures worldwide. His description is fascinating because it details exactly the kind of magical worldview I am promoting in my work.

The shamanic cosmos is one where everything is connected, in which all separate things are part of a larger whole. This is the principle of holism, and it is also found in the Hermetic tradition, where the macrocosm is reflected in the microcosm. The universe of the shamans is also composed of binary opposites, often mediated by a third element which reconciles them. This idea is also found in Western esoteric traditions, at least as far back as Pythagoras. The world picture of shamanic practitioners has a tiered structure, composed of an upper world of celestial beings, a middle world (this everyday world), and a lower world of the ancestors and spirits of the dead. The realms are connected by an *axis mundi*, a cosmic pole (often pictured as a tree) that connects the realms and is used by the shaman to travel between the worlds. The axis mundi is in the center of the world, which is divided into the four cardinal directions. Each direction has its particular gods, spirits, colors, animals, or elements. We see this in nearly every magical tradition as well.

Time in the magical world of the shaman is measured by seasonal events, movements of the Sun and Moon, and the life cycles of the biological world. There are hidden dimensions of reality that can be accessed in special liminal places and times. These portals to the spirit world can be either naturally occurring or constructed features of the landscape. This is the likely reason for the construction of earthworks and stone features the world over.

This worldview is *animistic*, meaning all things (even rocks and tools) are *animated*, possessed of *anima*, or the soul/life force. The cosmos of the shaman is in a constant state of transformation. The beings of the spirit world can enter the human world and take on a variety of shapes, including animal and human form. Shamans, too, are adept at shape-shifting and transformation into animals. They can interact with the spirit world and its inhabitants through the cultivation of altered states of consciousness. These are, it seems to me, the essentials from Romain's list of the characteristics of the shamanic worldview. It is striking that every one of these items is also a characteristic of the magical worldview in general.[14]

Our Hopewell and Adena predecessors were concerned with the making of ritual spaces for the observation of celestial phenomena, probably for the sake of what I would call astrological timing. The work of the shaman has to be done in accordance with the tides of celestial vitality issuing from the Sun, the Moon, and the stars. These ritual spaces were indeed "thin places," portals to the other realms of the shamanic cosmos and access points for the otherworld entities to come into the middle realm. The shamans entered the other planes through the production of altered states of consciousness. These spaces have their exact counterparts in the Western magical tradition.

According to Romain, the sites described above are aligned to the winter solstice sunrise and sunset, in the southeast and southwest respectively, and the summer solstice sunrise and sunset, in the northeast and northwest. This forms an X shape that divides the cosmos into four quadrants, corresponding to the four

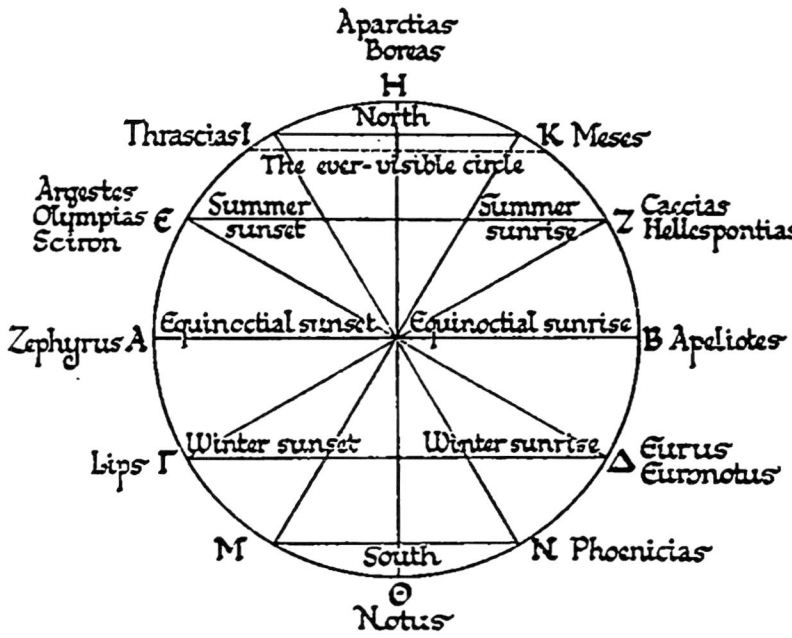

✦ *The winds and the angles of the sun, from Aristotle's treatise* Meteorology *(c. 340 BCE). From* Works of Aristotle, *vol. 3, Oxford University Press, 1931, p. 363.*

directions: north, south, east, and west. This division of the world into quarters, according to the cardinal directions, is common to Native American cultures up to the present day as well.[15]

The concept of a world divided into cardinal quarters is particularly important in hunter-gatherer societies, as knowing the direction the wind comes from helps track rapidly changing weather conditions, allowing for more successful hunting and foraging of wild plant food sources.[16] This next quote of Romain's was particularly resonant with me as a student of Western magical traditions, because you find the identical concepts in Aristotle, and later medieval European magical traditions: "An example is provided by the Sandy Lake Cree of Canada, who believe that spirits live in each of the four directions. The North Spirit brings snow and ice. . . . The South Spirit brings warm weather, which allows for the growth of food plants. The West Spirit brings darkness. The East Spirit brings light. Since the spirits of the Four Directions approach as winds, the Four

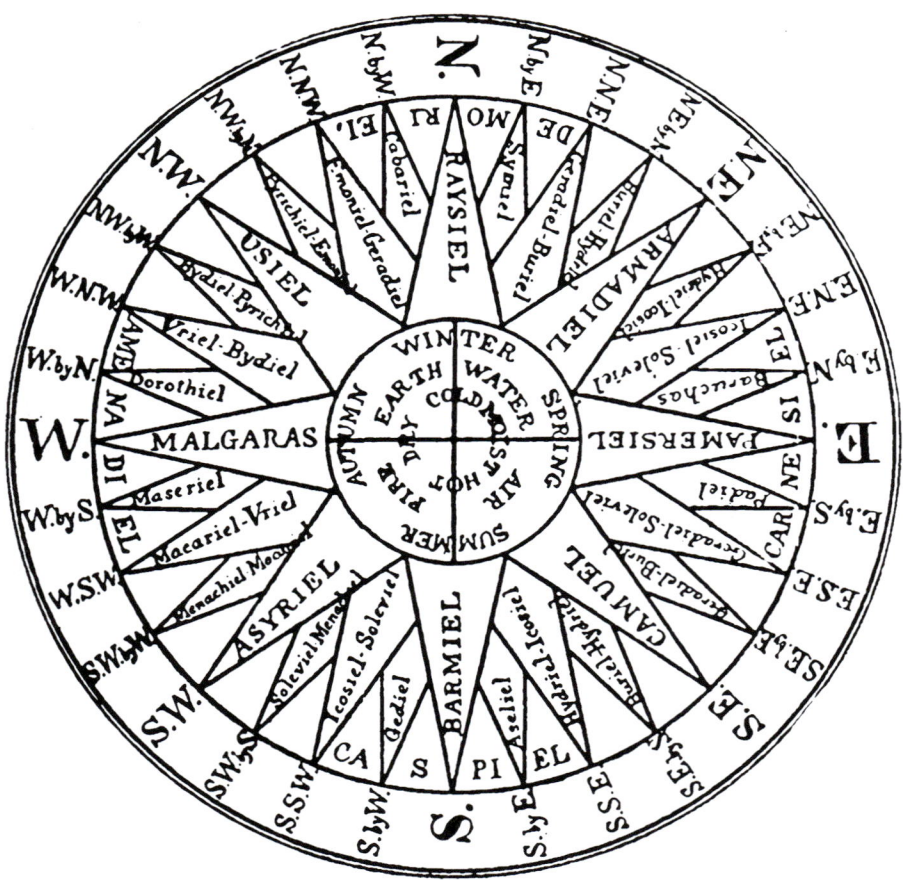

◆ *Diagram of the Spirits of the Air, from the* Ars Theurgia Goetia. *By Aleister Crowley and S. L. Mathers, 1916.*

Directions or Four Quarters are also the Four Winds."[17] The linking of the winds of the directions with spirits, good and bad, is common in other places as well. This is because the winds of the directions have characters or personalities, which are apparent to any perceptive observer. Cold and hardship arrive with the north wind, heat and drought with the south wind, rain and change with the west wind, and storms and snow with the east wind.

The Hopewell sites functioned as magic circles on a grand scale. To quote Romain, "Although the magical symbols used by medieval alchemists and magicians may differ in size from the Hopewell geometric earthworks, the principle is the same. By manipulation of such symbols, the magician or shaman expects to control the forces of nature or the universe. In this sense, the Hopewell geometric earthworks can be thought of as magical symbols. They are just far bigger than the magical symbols we are used to seeing."[18] Indeed, many of the sites feature a square enclosure with two circular ones.

And what is a magic circle? A magic circle is a kind of sacred space, a specific place in the material world, where spiritual forces can interact with the human world. It is not unlike a church or a temple in that way. A magic circle is symbolically set apart using geometry and symbolism. According to William Kiesel in his book, *Magic Circles in the Grimoire Tradition*, a magic circle sets the "boundaries of the sacralized place of working,"[19] and is "a space in the material world . . . set aside so that congress with the spirit world can be attained." Kiesel says, "The circle ranks among one of the most frequently used symbols in the Western esoteric tradition. Geometry, astrology, and other arts employ the circle as a map, guide, or tool for the various ways in which the tradition manifests itself. It is a symbol for the totality of existence, of the divine, of the infinite, the ouroboros-like cycles of the seasons, the wanderings of the stars and the divine order of the celestial realm. Magic circles exhibit qualities related to navigation and orientation. Ceremonial magic has employed the circle as the nexus point between the earthly realm and divine sphere."[20]

As we have seen, the Hopewell magicians/astronomers divided their cosmos into quadrants. Likewise, in the European tradition, according to Kiesel, "The circle in and of itself implies eternity, without beginning or end, whereas the square delineates the divisions or quadrants into the well-known directions, East, West, North, and South. The cardinal directions were important places where the operator had to face the quadrant that fell under the spirit's jurisdiction. This places the center of the circle in the center of the universe, as well as giving the square the meaning of manifest world or kingdom. In this scheme, the macrocosmic and the microcosmic meet in the magic circle placing the operator in a position, which connects the two worlds."[21]

The Hopewell people of Ohio and Kentucky were a nation of astronomers, geometers, magicians, and shamans. They forged a unique culture that was completely integrated into the cosmos, an enchanted world, where the magician sat at the center of the cosmos, holding the worlds in balance with his ability to travel between worlds and converse with the spirits of the winds, the elements, the Sun, the Moon, the stars, the animals, and the unity of all life. That way of life is lost to us. All that remains are enigmatic and haunting earthworks and artifacts. But the way of the shaman, of the earth mystic, of the astrologer and magician is the common archetypal heritage of all humanity, arising from our very human nature and calling us to take our place, in the words of the ancient *Orphic Hymn*, as the children of earth and starry heaven.

3
FIRST PRINCIPLES OF
A MAGICAL WORLDVIEW

In this chapter, I would like to take a moment to give some basic philosophical background in the worldview of traditional astrology and magic. This is a brief summary of my personal understanding—a mini manifesto, if you will. I'm trying to lay out my framework for conceptualizing a magical, ensouled cosmos. There are other ways, but this is the model I'm working with. It is a map, and it must be remembered that "the map is not the territory."

Understanding astrology and this magical map is key to practical organization and vibrancy on the land. The successful practice of magic of any kind requires an understanding that we live in an interconnected cosmos in which influences can flow across all levels—from the most abstract and conceptual to the most concrete and material, between our mind and the world. This practice requires a symbolic framework that expresses the way events in the material world arise and propagate in accordance with thematic archetypal patterns that originate in the Divine Mind and flow out into manifestation on the Earth. This framework provides a model that we can use to direct these archetypal forces by working with the right symbols, the right mindset, and at the right time. Such a system is found in the traditional geocentric model of astrology, in which we see the world as it appears to an earthbound being—a human, a plant, or an animal—with a Sun and a Moon that rise and travel across the sky, and the planets and stars moving around us in a vast circle from our vantage point here on Earth. As terrestrial beings, this is our starting point, our foundation. The accepted scientific cosmology is irrelevant because we are interested in adopting this point of view as a necessarily provisional and incomplete description of the *spiritual* cosmos,

✦ *Alchemical illustration in the "*Basilica Philosophica*" section of Johann Daniel Mylius's 1618* Opus Medico-Chymicum, *showing the procession of occult virtues from the divine into the elemental realm of nature.*

which can't be precisely known; we are interested in the astrological model of the cosmos because it *works* magically. For guiding our magical and agricultural operations, we need *astrology*.

UNTANGLING THE COSMOS

Most of us have been raised in a culture that devalues the spiritual at the expense of the physical, to the extent that the spiritual is deemed to be nonexistent or is given a grudging nod as something completely personal. It is as if it has no place in helping form our understanding of the world. Culturally, for the last few hundred years of intellectual history, this has been a necessary corrective to the overvaluing of the spiritual during the Middle Ages and early modern period.

However, times have changed and humanity is now facing a crisis. We have to change our way of thinking and feeling about the cosmos in order to reclaim our position as caretakers of creation, before we destroy the very Earth from whom we draw our nourishment and every part of our material needs. I intend to briefly lay the intellectual foundations necessary for a magical understanding of the cosmos, or, in other words, an enchanted world. This will not be an exhaustive treatment of these topics but rather an introduction of new ideas for those who might be unfamiliar and a suggestion of resources for those wishing to study in greater depth.

I have always been fascinated by the idea of a hidden reality, a world of spirits behind the apparent solid physical world we inhabit. Could there be more to our world than meets the eye? Many traditional or Indigenous cultures around the world have never lost their sense of enchantment, and now many modern "civilized" people all over the world are reclaiming the enchanted world of spirits and an animated, ensouled world. And not a moment too soon, for our world is teetering on the brink of ecological collapse and societal breakdown. It is imperative that we as humans reclaim our ancient role as caretakers and mediators between the gods and the earth.

The mysteries of creation are lost in the distant past, in mythic time, or in the massive numerical abstractions of science. Science has as its goal the description of the physical cosmos and describes the Earth as four billion years old and the universe as fourteen billion years old, more or less. But what do these unimaginably large numbers really mean? We, as spiritual beings, don't live in a cosmos that is only physical. Equally untenable because of its literalism is the biblical age of the cosmos, which is supposed to be about six thousand years old. We live in the present in our own experience in a living cosmos. I propose that we look on creation as having happened in a time out of time, the *dreamtime*, if you will, the time of the gods and ancestors, the creators who shaped our cosmos. Mythographer Mircea Eliade loved the phrase "*in illo tempore*" from the Vulgate Bible, meaning "in those days," to refer to this time. This mythic creation is eternally present, creating and recreating the world every minute, renewing its being, as if by the very breath of the creator, inhaling and exhaling being into our cosmos.

I further suggest that we posit the idea of a *spiritual* realm, an archetypal realm beyond time and space, which contains the myths and sacred stories, the gods and angels, the pure mathematical forms. From this realm proceeds the logos, the living information that shapes and underlies our experience of the material world, which lives and breathes and occasionally erupts into our own world. This is the realm of the world soul of the ancients, the great dreaming mind of God from which all emerges. We humble humans draw our meaning and our inspiration from this place. It shapes our myths, religions, dreams, art, and all of our creative endeavors, for better or worse.

This concept is close to what Henry Corbin, the gifted exponent of Sufi philosophy, referred to as the *mundus imaginalis* (the imaginal realm), "a world that is [as] ontologically real as the world of the senses and that of the intellect. This

♦ The Untangling of the Cosmos, *Zeus as demiurge in the act of creating, by Hendrick Goltzius (Holland, Mülbracht, 1558–1617).*

world requires its own faculty of perception, namely, imaginative power, a faculty with a cognitive function, a *noetic* value that is as real as sense perception or intellectual intuition. We must be careful not to confuse it with the imagination identified by so-called modern man with 'fantasy,' and which, according to him, is nothing but an outpour[ing] of 'imaginings.'"[1] According to Corbin, this is an intermediary realm between the realms of the senses and the archetypes, and it mediates between the two. "Ontologically, it ranks higher than the senses and lower than the purely intelligible world; it is more immaterial than the former and less immaterial than the latter. This approach to imagination [provides] a basis for demonstrating the validity of dreams and of visionary reports describing and relating 'events in heaven' as well as the validity of symbolic rites," he writes.[2]

I propose that *spiritual forces* are responsible for much of the phenomena that occur in our world on many levels. This bold statement will require, perhaps, a bit of unpacking. First, what do I mean by *spirit*? The ancient Greeks used the word *pneuma*, and the Latin equivalent is *spiritus*, both of which have the additional meaning of breath or wind and became our English word *spirit*. Ancient people often used the word *spirit* to refer to a principle that linked the soul, wholly nonmaterial, to the material body. It was considered to be the fifth element, a subtle form of matter similar to air but more rarified.[3] In a sense, spirit is like Corbin's imaginal realm, a mediator between the physical and the soul that ancients used to explain how the nonphysical acted on the body. In accordance with the Hermetic maxim, "As above, so below," what is true of the individual is true on a cosmic scale; the cosmic spirit joins the *anima mundi*, or the world soul, to physical matter, enlivening and animating the whole cosmos.

HERMETIC AND PLATONIC COSMOLOGIES

And what do I mean by *spiritual forces*?

This question is a bit more difficult. By spiritual forces, I mean independent active intelligences that underlie and shape material phenomena, from hurricanes to political movements, from the collective life of a nation to the collective life of a farm or a forest. All these things, I believe, have nonmaterial organizing principles that are conscious on some level and can be communicated with. Furthermore, living and nonliving things alike are animate, having souls (and spirits). These souls are "made" of mind or consciousness (as is matter, for that matter), which is the primal substance of all that is. These forces are often called archetypes, and they appear to organize and structure reality on the level of meaning. I am seeking to promote, in the words of Richard Tarnas (a gifted exponent of this way of seeing), "a conception of the universe as a fundamentally and irreducibly interconnected whole, informed by creative intelligence and pervaded by patterns of meaning and order that extend through every level, and that are expressed through a constant correspondence between astronomical events and human events."[4]

When I say that there is a level of meaning to reality, I am saying that there is an aspect of reality that is undetectable on a physical level but that nonetheless pervades and organizes it according to intelligible principles. This is what Swiss psychologist Carl Jung was talking about when he introduced the concept of synchronicity, which is the phenomenon in which two events that occur simultaneously in time are not causally linked but united only through meaning, usually involving some sort of archetypal symbolism.[5] Examples are endless, and when one begins to look into this kind of thing, one starts to see them everywhere. This is just the way the universe actually is: Everything is connected, and magic is real.

This insight requires a rehabilitation of a more ancient way of seeing the cosmos, one that has its origins in the mists of prehistory and forms the backbone of the esoteric traditions of the West but shares commonalities with premodern spiritual traditions around the world. I am referring to the Hermetic and Platonic thought systems that flourished in late antiquity and burst into bloom during the Renaissance before fading away during the early modern period and the so-called Enlightenment.

The Hermetic system is the philosophical and mystical system that originated in Egypt during the Hellenistic era, which was the last phase of the Egyptian religion, although it contained much material borrowed from the Greek philosophical schools of Platonism and the Stoics. It is called Hermeticism because it derives from the body of teachings of a sect founded by the legendary teacher Hermes Trismegistus, or Thrice-Greatest Hermes, who was considered to be the embodiment of the Egyptian god Thoth, whom the Greeks equated with their god Hermes. Many of the texts ascribed to Hermes were lost in ancient times, but some that survived have come to be known as the *Corpus Hermeticum*, or more broadly, *Hermetica*. The *Hermetica* can be further subdivided into the practical, which includes works on metallurgy, alchemy, and astrology, and the philosophical, which includes works on cosmology, spirituality, and theology.

The Platonic system is the philosophical school deriving from the teachings of the Greek philosopher Plato (428–348 BCE). Plato's writings consisted

◆ *Hermes Trismegistus, legendary Egyptian sage.*
From an engraving by Pierre Mussard, Historia Deorum Fatidicorum,
Venice, 1675.

of a series of dramatic dialogues in which he would try to prove philosophical concepts by logical argument. His characters, including his teacher Socrates (469–399 BCE), would debate various students on topics such as morality, the immortality of the soul, love and sexuality, the good life, and the ideal form of government. Though his teachings were massively influential in the ancient world and were hotly debated for centuries, Plato never really laid out a systemic philosophy of Platonism. In fact, his writings show signs of a continuing evolution of understanding throughout his life.

Perhaps the most influential of his doctrines is the theory of the forms, which, briefly, is the idea that outside of space and time, there is a realm of perfect archetypes of everything that exists imperfectly in our material world, and that all created things approximate the perfect forms or ideas there. Think about geometry or numbers, for instance. There is, on Earth, no such thing as a perfect circle or square. No matter how perfectly drawn, there will be some departure from the ideal proportions of the ideal circle. However, perfect circles exist in the spiritual/archetypal realm. Or, take the number *one*: This is a pure idea that exists only as an intellectual reality, or in other words, as an abstract quantity.

The most sublime *one* of them all is the idea of the primal unity of all things, which gives its being and light to the whole cosmos. This doctrine has been very influential in the history of theology and philosophy.

Plato said in his creation story, the dialogue *Timaeus*, that the Creator, whom he called the demiurge, made the cosmos from perfect forms and perfect proportions but from imperfect material, which explains the imperfections we see in the world in which we live. We indeed see the forms of perfect geometry everywhere in nature, but everything is born or created, decays, dies, and loses its perfect form. Generation and corruption is the way of the material cosmos, while only the ideal forms endure.

Later philosophers went on to organize Plato's thought into a more coherent system and work out some of the contradictions. These thinkers conceptualized a cosmology of emanationism, whereby a transcendent source of pure being and light overflowed its being into the void, creating the cosmos. As the transcendent light gets farther from the source, it gets weaker, dimmer, and less perfect, more prone to generation and corruption, more concrete. In some ways, if taken the wrong way and to extremes (and let us remember one of the principles of Greek thought was the Delphic maxim, "Nothing in excess," the golden mean), this line of thinking in the Platonic tradition has led to the devaluation of the material and the physical, to dualism. But to me, the ancient tradition is telling us that the whole cosmos is suffused with the light of the Source that breathes life and being into this material realm, which strives ceaselessly toward the good although fallen. Procession and return. All good, all beauty, all harmony is the transcendent goodness of the One, the Good, the Beautiful.

According to the philosopher Plotinus (204–270 CE), as he wrote in his fifth *Ennead*, the One, as the perfect source, "in a way overflows and its superabundance has made something else,"[6] which he called the *nous*, translated as

✦ *The process of emanation as pictured by the kabbalistic school of Isaac Luria.*

♦ *Anima mundi, or world soul, as pictured by Renaissance English*
occult philosopher Robert Fludd, 1617.

mind or *intellect*, and then "pouring forth abundant power,"[7] produces *psyche*, or *soul*. "Soul made all living beings by breathing life into them, those that are nourished by the earth and the sea and the divine stars in heaven. Soul itself made the Sun and this great heaven and ordered it, and makes it circulate in a regular way."[8] Soul "supplies life to the whole universe [by] flowing or pouring everywhere into immoble heaven from 'outside,' inhabiting and completely illuminating it."[9] "Heaven," continues Plotinus, quoting Plato, "moved with an everlasting motion by the 'wise guidance' of soul became a 'happy living being.'"[10] Remember, this soul is also known in the tradition as the anima mundi, or world soul. Plotinus says further that "all things are alive by the whole of it, and all soul, being the same as the father who begat it, is present in each thing and every thing. And though heaven is multiple and diverse, it is one by the power of soul, and this cosmos is a god because of this. The Sun is also a god—because it is ensouled—and for this reason are we, if indeed anything is (a god)."[11] So, one soul pervades the whole universe and makes each and every part divine, from the Sun and stars to human beings.

In this tradition, the divine power of the world soul overflows into the cosmos from without, flowing inward through a series of spheres representing the parts of the cosmos, as pictured in the diagram from 1524 on page 43. It shows a series of ten concentric spheres with the highest and outermost being the Empyrean Heaven, home of God and Angels (in the Christian version here, but this was the sphere of the One in Plotinus's day.) Next is the sphere of the Primum Mobile, the first mover, home of the spiritual powers of the decans (which are ten-degree segments of the zodiac associated with fixed stars and thought to be gods in their own right), and the pure forms of the zodiac. Then comes the sphere of the constellations, the firmament of the actual stars. Let us note that the ancients knew that the signs and the stars didn't exactly line up and that the signs of the zodiac are the pure Platonic forms, eternal and unchanging, whereas the stars are gradually drifting, resulting in the phenomenon known as the precession of the equinoxes.

Having passed through the outermost spheres, the divine influence of soul passes through the spheres of the seven classical visible planets in their respective heavens. These planets are likewise divine beings endowed with souls, and they also contribute their own archetypal energies to the Earth below. These spheres are in the following order from outermost to innermost, according to the ancient system of the Chaldeans, the astronomers of ancient Babylon: Saturn, Jupiter, Mars, the Sun, Venus, Mercury, and the Moon.

Schema huius præmissæ diuisionis Sphærarum.

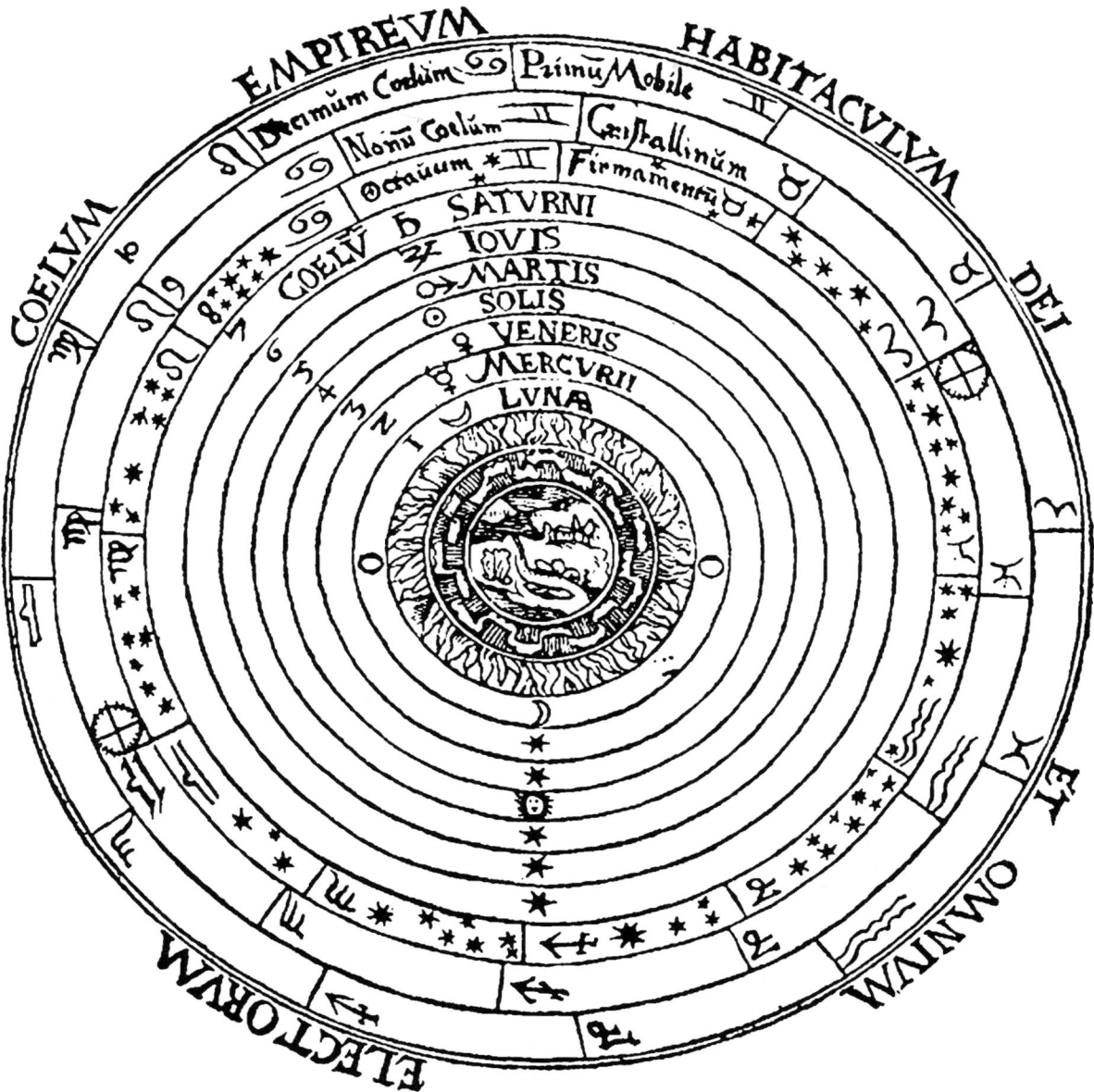

✦ *The cosmological diagram of Peter Apian (1524), showing the concentric spheres of the planets, the zodiac, the fixed stars, and the Empyrean Heaven. From Edward Grant, "Celestial Orbs in the Latin Middle Ages," Isis 78, no. 2 (June 1987): 152–73.*

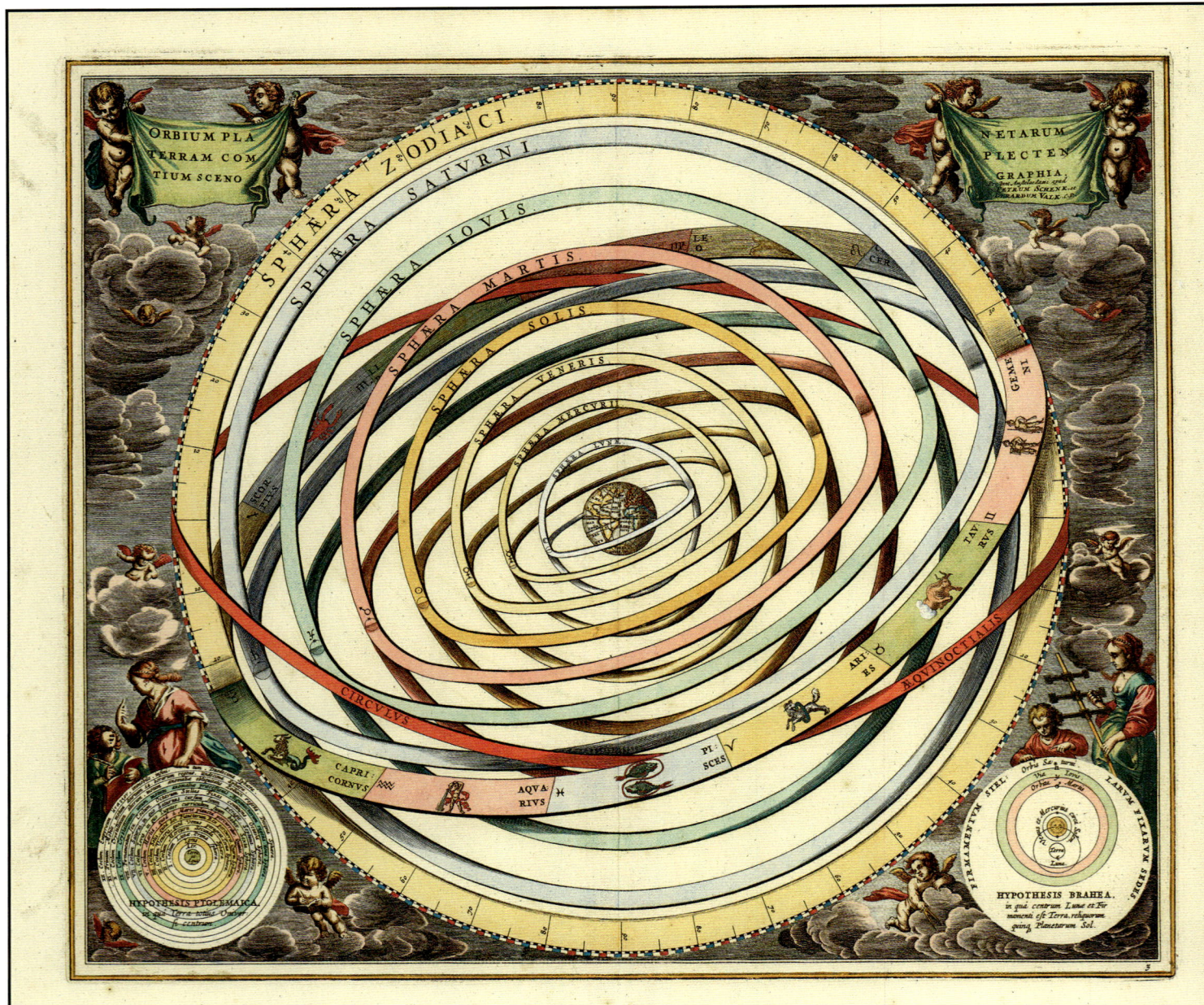

◆ *The Ptolemaic cosmos as pictured by Andreas Cellarius
in his 1660 work,* Harmonia Macrocosmica.

Below the Moon, in the sublunar realm, are the elemental realms of the four classic elements, again from outer to inner, heaviest to lightest: fire, air, water, and, at the center, earth.[12] As astrologers Helena Avelar and Luis Ribeiro put it, "Everything terrestrial is subject to change and will eventually encounter deterioration and decay. These changes do not occur randomly, but according to a rhythm dictated by the movements of the planets. As they move, the planets act upon the four terrestrial elements, changing them in different ways, and consequently causing changes on Earth and on everything it contains, including human beings. This is the fundamental principle of astrology: There exists a correlation between celestial movements and events on Earth that permits us to interpret and foresee these changes."[13]

⌐

You may wonder, "Why should we, in the twenty-first century, care about ancient science, and what does all this have to do with magical farming, the whole point of this book?"

In closing this chapter, let's remember: This is the cosmology upon which the astrological and magical worldview is based. It is the spiritual archetype of the physical cosmos in which we live. It corresponds to the observable cosmos human beings have always seen from our vantage point on Mother Earth. Geocentric cosmology is the way a plant or an animal in the wild sees the world, and is the basis of the seasons. The Sun, planets, and stars all rise in the east and set in the west. The Sun moves south into Capricorn in the winter and into Cancer in the summer (at least in the Northern Hemisphere). The lengthening days of spring bring the Sun into fiery Aries, and the shorter days of fall bring him into cooler Libra. This is the tropical geocentric zodiac based on the seasons and the eternal patterns of life. Astrology isn't an obsolete and discredited version of astronomy; it is a branch of philosophy and a science of the spirit. It describes the playing out of archetypes in time and the lives of human beings.

Astrology is important to farming and a life on the land because the plants and animals we tend are in relationship with the forces of the heavens, not just in a basic material sense as with the seasons and tidal forces of the Moon, but also in a subtle spiritual sense. They are subject to astral (as in, *of the stars*) and planetary forces. People of the ancient cultures who lived closer to the land and who could sense its moods and feel its tides understood this. The movements of the Moon and Sun, the stars and planets, were of such paramount importance to our ancient ancestors, as we have seen, that they devoted an astounding amount of their collective resources into understanding and predicting them. Ancient people were not fools exhausting hard-won resources needlessly in fruitless efforts. They were pragmatists and did only what helped them survive.

In the coming chapters, we will learn the basics of using the principles of astrology to time operations on the land, as well as the art of channeling the lunar and planetary forces to bring about meaningful change in our lives.

PART TWO

FARMING MAGIC PRACTICES

4
THE MOON, MISTRESS OF MAGIC
Understanding Lunar Energies

This chapter introduces the basics of lunar magic in the Western magical tradition, touching on Moon signs and phases for spells and other magical work as well as lunar petitions, lunar mansions, and more.

But first, a brief digression on magic, including a definition and a discussion about similar ideas that are perhaps more familiar to the general reader.

There are so many definitions of magic in Western esotericism that it is difficult to pick just one. One of the most famous is the definition offered by the notorious English magician, Aleister Crowley, which states, "Magick is the Science and Art of causing Change to occur in Conformity with Will."[1] For Crowley, therefore, "Every intentional act is a Magickal Act."[2] This definition, in my opinion, is too broad and lacks the idea that magic is understood to act by *occult*, or hidden, means not currently understood by science. Much of what was considered magic in premodern times has been subsumed under the category of science, like magnetism and chemistry (although it is my understanding that exactly how magnetism works and what matter fundamentally is remain open questions). What continues to be considered magic, then, are the truly occult phenomena that are usually grouped in the category of the *paranormal* or the *psychic*. So a better definition might be: Magic is the science and art of causing change in reality, according to the will, by occult means.

✦ *Opposite: The Full Moon illuminates the skeletal form of an oak tree in winter.*

48

To expand on this, take the proposed idea of a cosmos, which is fundamentally conscious, perhaps "made" of consciousness itself—a mind, in other words. In this cosmos, all parts are bound by hidden chains of sympathy, and through these links, distant parts not physically connected are in communication across all scales: "As above, so below."

What makes Hermetic magic as I am presenting it here different is that we don't just try to change our minds with a simple phrase or mantra, although that is an element of our practice. We construct dramatic rituals with harmonious symbolic elements to link our minds to the archetypal forces of the cosmos, and we perform them at particular times and seasons when we believe the Earth, ourselves, the celestial world, and the divine realm are in a particularly harmonious arrangement, thereby linking our imagination with our cosmology and the cosmos itself and giving force to our intentions. It is not a question of practicing one or the other of these techniques of manifestation. Most magicians practice some form of prayer and use intention or affirmation as well.

In constructing our rituals, it is not enough to just have the right intention. We need more. We use the right incense, the correct herbs, the right-colored candles. We make particular symbols to connect with the proper spiritual forces and offer prayers to the right angels or deities, using the right Divine names at the right times, and—germane to the topic of this chapter—under the corresponding phase and sign of the Moon. Agrippa says, "Man's mind, when it is most intent upon any work, through its passions and effects, is joined with the mind of the stars, and intelligences, and being so joined is the cause that some wonderful virtue be infused into our works."[3] When the mind of the magician "is exposed to the celestial influences, as by the affections of his mind, so by the due applications of natural things . . . binds and draws the inferior into admiration, and obedience."[4] Connecting our mind, through symbols and propitious times, to the celestial intelligences, binds them and brings them "down," as it were, into our sphere to aid us in our work of manifestation.

Agrippa tells us that faith is a necessary component in effective magic: "We must therefore in every work, and application of things, imagine, hope and believe strongly, for that will be a great help. And it is verified among physicians that a strong belief, and an undoubted hope, and love toward the physician, conduceth much to health."[5]

Magical Ethics

Let's discuss, for a moment, magical ethics, because the next section will touch on, shall we say, the *dark arts*, and I don't want anyone to mistake the folklore and historical sources I am passing along for my opinions and practice. I don't believe magic needs any special ethics, because wrong acts are wrong acts, whether committed magically or physically. Magic has often throughout history been the last resort of the powerless in the face of a more powerful opponent or an oppressor—for example, curses directed at abusive ex-spouses or workplace or neighborhood bullies. There are legitimate cases of self-defense, where the law can't help and physical means of defense are illegal, impractical, or impossible. I can also add that the powerful elites throughout history, including recent history up to the present, have used magic or paranormal techniques to gain an edge over their opponents. The older grimoires, such as the *Picatrix*, are full of curses to be employed by kings, generals, and courtiers. But magic is potentially also a way for the downtrodden to level the playing field, and has long been the last recourse of the poor and the powerless. And the Moon as celestial mediatrix and Deity of the Night has been called upon to aid and empower those spells and petitions, throughout history and worldwide.

Moon magic is ancient, with roots stretching into prehistory. There are clay tablets from Mesopotamia containing spells to lend strength to the Moon during lunar eclipses. Classical sources describe practices such as "drawing down the Moon" in order to capture the Moon's power for erotic magic (love and attraction spells) and for the production of lunar foam, or *virus lunare*.[6] One source, Lucius Apuleius, the Roman magus and Platonist philosopher, drops this tantalizing hint, "By sorcery and enchantment the swift rivers might be forced to run against their courses; the sea to be bound immovable; the winds to lose their force and die; the sun to be restrained from his natural journey; the moon to drop her foam upon the earth; stars to be pulled down from heaven; the day to be darkened; and the night be made to continue for ever."[7]

Dew was believed to fall to Earth from the Moon, which, as was mentioned previously, governs moisture and all waters. Dew and foam were thought to have life-giving properties and were symbolically linked with semen and menstrual blood.[8] As scholar Karen ní Mheallaigh writes, "Ancient magical practice suggests that the Moon may actually have been conceptualized as a celestial uterus, with

◆ *The practice of "drawing down the Moon" as practiced in ancient Greece, from
William Hamilton's 1791* Collection of Engravings from Ancient Vases.

its monthly shrinking and swelling, its life-giving moisture, and occasional red-
dening to the color of blood."[9] In ancient India, the Moon was a vessel of *soma*,
the holy drink of the gods, that empties as the Moon wanes and fills as she waxes.
In the *Kauśītaki Upanishad*, the souls of the dead rise to the Moon, which "is
the door to the heaven-world," and fall to the Earth as rain, being reborn as the
animals and plants nourished by that rain.[10]

MAKING MOON WATER

Many practitioners of modern witchcraft make something called "Moon water"
by leaving water out under a Full Moon to charge it with lunar power—in particular, the power of the Moon in the zodiacal sign it occurs in. It is likely that
this recent practice derives from a desire to revive the methods of the witches
of classical antiquity. Moon water also relies on the idea, common in esoteric
circles, that water as a medium can absorb and hold the intentions sent into it

by the practitioner. Holy or blessed water has a similar rationale. Your intention, and the divine influence summoned by your prayer, is absorbed by the water and saved for later use. Chemically, holy water is just salt water, but magically, it has the power to dispel evil spirits and aid healing.

To make Moon water, at moonrise on the night of the fullest Moon—you will need to consult an almanac or astrological calendar to be sure of when this is, as the Full Moon, her opposition to the Sun, is an astronomical instant, which often occurs during the daylight hours in a given place—fill a basin with clean rainwater (for most people this will require some advance planning), and place it where it will be exposed to the light of the Moon all night. As the Moon rises, burn some mugwort or copal, which are lunar incenses, and recite the *Orphic Hymn* to Selene. Here's Thomas Taylor's classic 1792 translation:

> Hear, Goddess queen, diffusing silver light,
> Bull-horned, and wand'ring thro' the gloom of night.
> With stars surrounded, and with circuit wide
> Night's torch extending, through the heav'ns you ride:
> Female and male, with silvery rays you shine,
> And now full-orb'd, now tending to decline.
> Mother of ages, fruit-producing Moon,
> Whose amber orb makes Night's reflected noon:
> Lover of Horses, splendid queen of the night,
> All-seeing pow'r, bedecked with starry light,
> Lover of vigilance, the foe of strife,
> In peace rejoicing, and a prudent life: Fair lamp of Night, its ornament and friend,
> Who giv'st to Nature's works their destined end.
> Queen of the stars, all-wise Diana, hail!
> Decked with a graceful robe and ample veil.
> Come, blessed Goddess, prudent, starry, bright,
> Come, moony-lamp, with chaste and splendid light,
> Shine on these sacred rites with prosp'rous rays,
> And please accept thy suppliants' mystic praise.[11]

As the Moon and the smoke of the herbs rise into the sky, holding your hands over the water, voice your intentions for the water, such as for healing, personal growth, healthy plants, or whatever you need. Meditate on the desired effect, and leave the water to absorb the lunar dew overnight. Pour into jars, and place in a dark place to use for whatever magical purpose may need a little Full Moon energy. Remember that the energy of the Moon is very influenced by what sign she is passing through. Therefore, be sure that the sign the Moon is in is conducive to your magical goal. For example, Moon water made while the Moon is in Aries isn't going to be conducive to promoting peace and tranquility; Libra may be a better Moon for that.

WORKING WITH MOON PHASES

In magical work, as previously discussed with agricultural operations, a Waxing Moon has the power of increasing, augmenting, and enhancing energies that you want to bring into manifestation—for example, it may be used in money work, spells for getting a job, love magic, fertility work (both land fertility and human), and other works for building and growing. In such cases, you find a spell you want to do and perform it during the period beginning the day after the New Moon until the day before the Full Moon. Simple, right? Just wait a few paragraphs, because we're going to make it more complicated by introducing more factors to enhance the lunar virtue.

A Waning Moon has the effect of withering, drying, and suppressing unwanted forces in the cosmos, so the Waning Moon is conducive to operations like warding, in which the operator wishes to protect a space from dangerous spiritual or physical forces, banishings, and yes, maleficia, or curses. Any operation in which you want to decrease, limit, constrict, or reduce a force that is currently being manifested, or could be coming into being, you would want to perform during the Waning Moon, in order to bring your work into sympathy with the energy of the Moon and lend it tidal power.

We can add additional factors to enhance our workings, but these additional factors complicate finding times when we can actually perform our practices. We can't all be Gandalf in our tower waiting for the exact moment when the stars and planets align. We may have to be at work all day or take the kids to soccer practice. So, in the modern world, we do the best we can.

Additional factors include using a sign conducive to the magical work we are doing. *The Greater Key of Solomon* has this advice:

> For those matters which appertain to the Moon such as the invocation of Spirit, the Works of Necromancy, and the recovery of stolen property, it is necessary that the Moon shall be in a Terrestrial Sign, viz.: Taurus, Virgo, or Capricorn.

For love, grace, and invisibility, the Moon should be in a Fiery Sign, viz.:
Aries, Leo, or Sagittarius.

For hatred, discord and destruction, the Moon should be in a Watery Sign,
viz.: Cancer, Scorpio, or Pisces.

For experiments of a peculiar nature, which cannot be classified under any
certain head, the Moon should be in an Airy Sign, viz.: Gemini, Libra, or
Aquarius.

But if these things seem unto thee difficult to accomplish, it will suffice thee merely to notice the Moon after her combustion or conjunction with the Sun, especially just when she quits his beams and appeareth visible. For it is then good to make all experiments for the construction and operation of any matter. That is why the time from the New Moon until the full is proper for performing any of the experiments we have spoken above. But in her decrease or wane it is good for War, Disturbance and Discord. Likewise the period when she is most deprived of light, is proper for experiments of invisibility and Death.

But observe inviolably that thou commence nothing while the Moon is in conjunction with the Sun, seeing that this is extremely unfortunate, and that thou wilt then be able to effect nothing; but the Moon quitting his beams and increasing in Light, thou canst perform all thou desirest. . . .[12]

Some of this is a bit counterintuitive. Why would you want to do curses when the Moon is in Pisces or Cancer, when those are also good signs for growth and fertility? Because she is powerful in those signs, and during the Waning phase, her powers, as the ancients would say, "conduce to corruption."

You might also want to pick a time when the Moon is in a barren sign, a Mars- or Saturn-ruled sign, or even Scorpio, where, although fertile, she has a certain dark dynamism. I like to time the operation for Mars or Saturn hour* if I'm trying to enhance the malefic, withering, or drying power of a Lunar operation.

---

*Planetary hours are an ancient system of dividing the hours between sunrise and sunset and the hours between sunset and sunrise into twelve equal portions, or hours, and assigning them to the planets in the standard, Chaldean order. The first hour, beginning at sunrise, is assigned to the planet that rules the day. These hours, unlike the hours of the clock, vary in length according to the season. The idea is that the planet of the hour rules that hour and its spiritual force is more potent, magically.

Let's break open our Agrippa text again for an example of what we mean. Agrippa discusses the magic square of the Moon, a numerical diagram believed to exemplify and capture the sacred geometry and mathematical principles of the intelligences of the Moon. All you need to know is that the ancients placed great faith in magic squares for capturing and channeling spiritual forces for particular goals. They are known from ancient China and are contained in the I Ching and were popular in ancient Rome. Every planet had its square, and this is the square of the Moon from *Three Books of Occult Philosophy*, as well as the sigils of the spirits and intelligences of the Moon.

Agrippa says that during a fortunate Moon—a Waxing Moon in Taurus or Cancer, when she is rising over the Eastern horizon, on the Ascendant, or directly overhead, on the Midheaven (he doesn't say this here; you are just supposed to know basic astrology)—if you engrave these symbols on silver, the resultant talisman

renders the bearer thereof grateful, amiable, pleasant, cheerful, honored, removing all malice, and ill will. It causeth security on a journey, increase of riches, and health of body, drives away enemies and other evil things from what place thou pleaseth; and if it be an unfortunate Moon, [in Scorpio or Capricorn, fall or detriment, and waning, rising or culminating, the proverbial Bad Moon Rising from the song] engraven in plate of lead, wherever it shall be buried, it makes that place unfortunate, and the inhabitants thereabouts, as also ships, rivers fountains, mills and makes every man unfortunate, against whom it be directly done, making him fly from his country, and that place of his abode where it shall be buried.[13]

The magi of the Renaissance believed that you could capture the spirits and energy of a particular Moon by carving these diagrams into a metal plate at a particularly potent time. This is not too different from the Moon water, except that the lunar force is contained in metal instead of water.

A lead tablet, inscribed exactly as described above, was found at the Lincoln Inn in London during some drainage excavations that were being carried out in 1900. The plate was inscribed with the nine-by-nine lunar square and the sigils and spirit names of the lunar spirits, as well as the following inscription, "That Nothinge maye prosper Nor goe forwarde that Raufe Scrope take the in hande." The tablet was constructed by the cunning man John Wellington Wells against the unfortunate Mr. Scrope, who was owner of the inn from 1543.[14]

The Magic Tables, Seals and Characters of the Planets, their Intelligence and Spirits.

Table of the Moon in her Compass.

37	78	29	70	21	62	13	54	5
6	38	79	30	71	22	63	14	46
47	7	39	80	31	72	23	55	15
16	48	8	40	81	32	64	24	56
57	17	49	9	41	73	33	65	25
26	58	18	50	1	42	74	34	66
67	27	59	10	51	2	43	75	35
36	68	19	60	11	52	3	44	76
77	28	69	20	61	12	53	4	45

The same in Hebrew.

לז	עח	כט	ע	כא	סב	יג	נד	ה
ו	לח	עט	ל	עא	כב	סג	יד	מו
מז	ז	לט	פ	לא	עב	כג	נה	טו
טז	מח	ח	מ	פא	לב	סד	כד	נו
נז	יז	מט	ט	מא	עג	לג	סה	כה
כו	נח	יח	נ	א	מב	עד	לד	סו
סז	כז	נט	י	נא	ב	מג	עה	לה
לו	סח	יט	ס	יא	נב	ג	מד	עו
עז	כח	סט	כ	סא	יב	נג	ד	מה

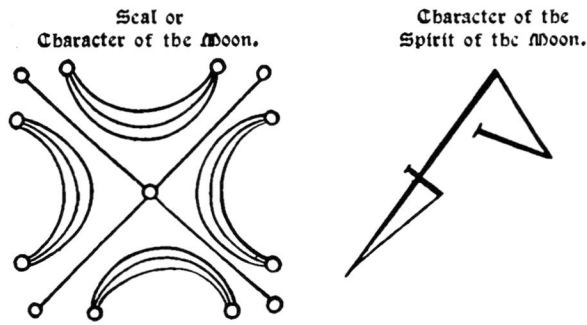

Seal or Character of the Moon.

Character of the Spirit of the Moon.

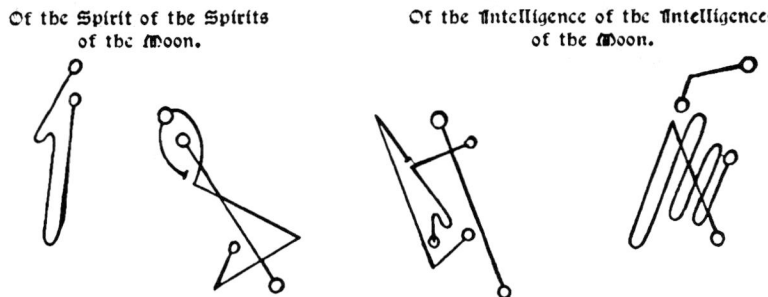

Of the Spirit of the Spirits of the Moon.

Of the Intelligence of the Intelligences of the Moon.

✦ *The table of the Moon showing the sigils and magic square to be inscribed on lunar talismans. From Francis Barrett's pioneering textbook of Western occultism,* The Magus *(1801).*

Magic to increase fertility, psychic abilities, or dreams, to call rain, or to help the growth of plants should also be done during what Agrippa calls a fortunate Moon, when she is waxing in Cancer, Taurus, Pisces, or any of the fertile signs with the exception of Scorpio. There are many ways to use the power of the

Moon in magic. For the cunning farmer, since the Moon regulates the growth of the crops and animals with which we make our livelihood, lunar petitions are a good place to start. A basic lunar petition might involve lighting a purple or white candle, because those are colors assigned to the Moon, and burning some lunar incense, opening the sacred space by your usual method, and writing your intention on a new piece of nice stationery. It may also involve drawing the lunar square and the sigils pictured above, possibly with the name or sigils of the archangel Gabriel or the goddess Selene, or both. Recite your intention in the form of a prayer, and recite the *Orphic Hymn* to Selene. Perform the operation under a fortunate Moon and on a Monday, if possible.

You can use lunar phases to time any magical operation, but some, which fall under the influence of one of the other planetary powers, are best also performed with both the Moon phase and the appropriate day and hour for the particular planet whose power you want to enhance. We will discuss planetary hours in greater detail in chaper 10.

Using the Moon and the Planets

Here's a brief example of how to use Moon phases with planetary powers. Let's say I want to perform some wealth magic, which falls under the influence of Jupiter. I would then use Agrippa's square and sigil of Jupiter and its spirits, but I would perform the operation on Jupiter's day, Thursday, and, if I could manage it, during the hour of Jupiter as well. I would also want Jupiter in good condition astrologically, which means that he is in one of the signs he rules, either Sagittarius or Pisces, or where he is exalted, like Cancer, and also not afflicted by being retrograde, too close to the Sun, or in a hard aspect with Mars or Saturn. Waiting for Jupiter to be in good astrological condition is not always possible because he moves rather slowly, spending about a year in each sign on average. Sometimes important magic can't wait and a Waxing Moon, on a Thursday, at sunrise, is the best you can do. The more factors you add, the harder it is to find an auspicious time to do your magic, but you gain the cooperation of the celestial forces of the cosmos by doing so.

THE LUNAR MANSIONS

No discussion of lunar magic would be complete without a discussion of the lunar mansions, which are essentially an alternate lunar zodiac made up of 28 asterisms, or groups of stars, which lie along the ecliptic, through which the Moon passes during her synodical cycle. Each tropical mansion represents 12 degrees and 51 minutes of the 360-degree tropical zodiac and represents the distance the Moon travels on average during a 24-hour period. We will be discussing the tropical mansions here, as opposed to the constellational mansions used by some other systems, notably the Nakshatras used by Hindu astrology. The theory of the tropical mansions is that, as with the zodiac, each section of the Moon's path around the heavens has a different quality and is home to different celestial spirits (or angels, also known as *rūḥāniyyāt*, the Arabic word used for celestial spirits in the *Picatrix*), and can be channeled through images made and rituals performed while the Moon is passing through that mansion.

This system was originally developed in ancient China and spread to India and the Arab world before entering into Western magical practice in the Middle Ages.[15] This practice was considered suspect, as was much of medieval magic, by the ecclesiastical authorities because it calls upon spirits not admitted as angelic by the Catholic Church. There were debates as to whether image magic or talismanic magic of this kind was "natural" magic—which merely involves the occult forces of nature—or demonic magic, also known as necromancy—which involves the agency of unknown spiritual forces. Basically, if you were burning incense and praying to something, it was considered idolatry and therefore forbidden. Just making images at auspicious times was permitted, however.[16]

In the tradition I practice, we consider the spiritual forces of the mansions to be celestial spirits or angels. They are the energies of stars or groups of stars. Some of them are used for constructive purposes and some can be used for destructive purposes, but they are not evil. They represent the qualities of energies issuing from the Moon into the sublunar world and are there as a tool for the wise to use.[17] As the Bible says:

> *To every thing there is a season, and a time for every purpose*
> *under heaven:*
> *A time to be born, and a time to die; a time to plant and a time*
> *to pluck up that which is planted;*

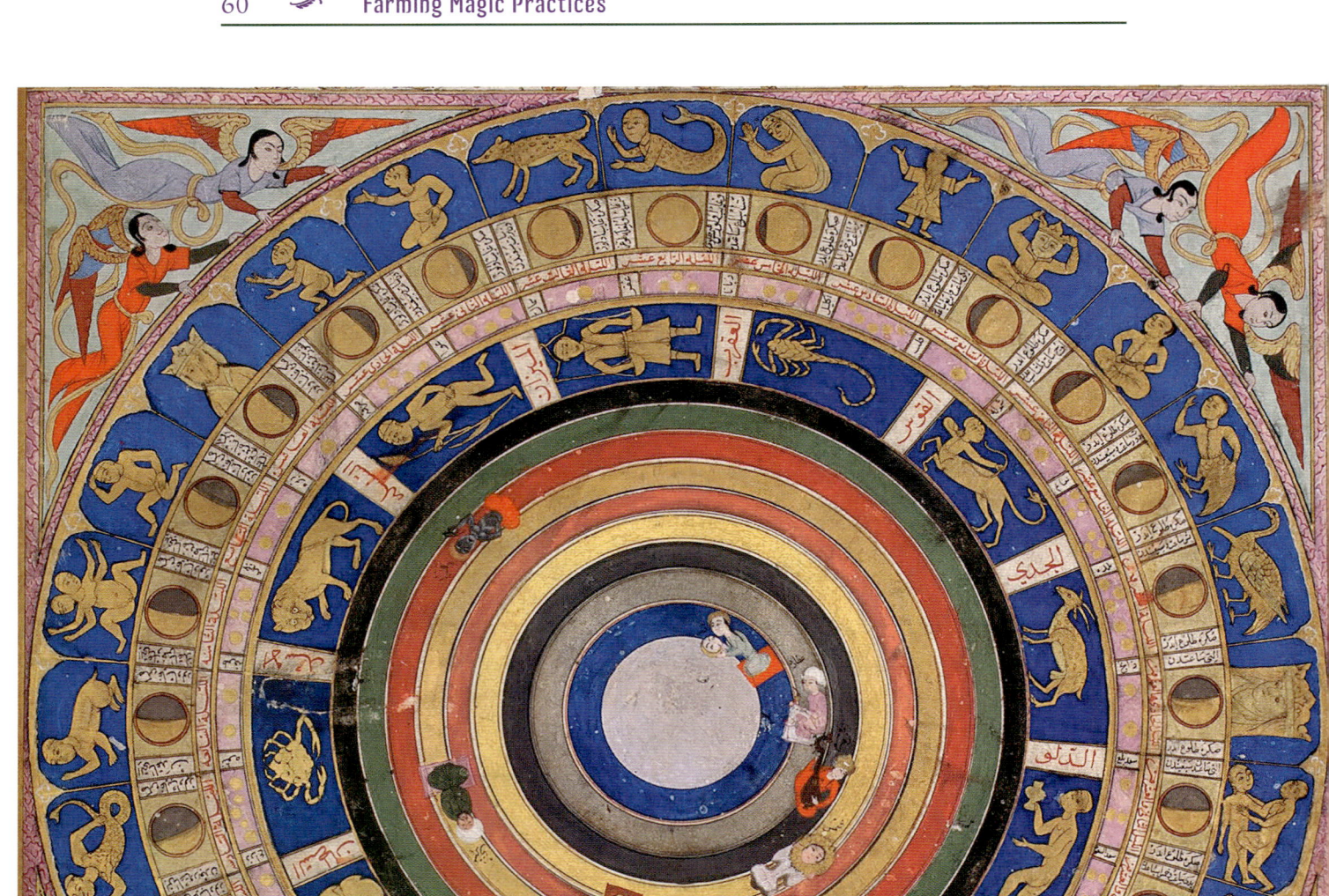

✦ *Celestial map showing the lunar mansions and signs of the zodiac. From the 1583 Ottoman manuscript* Zubdat-al Tawarikh, *written by Seyyid Lokman.*

A time to kill and a time to heal; a time to break down and a
time to build up.[18]

It seems to me that it is no coincidence that there are twenty-eight of these "times" in the biblical Book of Ecclesiastes and that its author, reputed to be none other than the master magician King Solomon, was aware of the tradition of the lunar mansions and their changing qualities and powers. Indeed, as it is written in the apocryphal Wisdom of Solomon, Chokmah or Sophia, the Wisdom of God, had given King Solomon:

Unerring knowledge of what exists
To know the structure and activity of the world
And the activity of the elements;
The beginning and end and middle of times,
The alterations of the solstices and the changes of the seasons,
The cycles of the year and the constellations of the stars,
The natures of the animals and the
Tempers of wild beasts,
The powers of the spirits and the
Reasonings of men,
The varieties of plants and the
Virtues of roots;
I learned what is secret and
What is manifest,
Wisdom, the fashioner of all things
Taught me.[19]

Both Agrippa and the *Picatrix* list the mansions and their powers, and I encourage the curious reader to get copies of these traditional sources or check out the Renaissance Astrology website if interested. A few examples from my own practice will suffice to explain what use may be made of the lunar mansions by the cunning farmer.

The eighth mansion is called Annathra and is for gaining victory. When the Moon is passing through this mansion, fashion from tin the image of an eagle with the head of a man and on its breast inscribe the name of the lord of this mansion. Suffumigate it with sulfur and say: "You, Annediex, do such and such for [this is where you are to insert your own petition] and grant my petition." When this image has been completed in this fashion, take it with you into battle and you shall be victorious and shall prevail. Know that Annediex is the name of the lord of this Mansion.[20]

I made an eighth lunar mansion talisman for a friend who was entering a major competition, and, coincidentally, he won.

The sixteenth mansion is Azebene, and it is for profiting in business. When the Moon is wandering through this mansion, make the figure of a man sitting on a throne and holding scales in his hands on a silver lamella. Suffumigate this with odorous things, and show it to the stars every night until the seventh night, saying: "You, O Azeruch, make such and such happen for me, and accomplish my request." Ask for things pertaining to the buying and selling of merchandise. Know that Azeruch is the name of the lord of that mansion.[21]

After making the sixteenth mansion talisman, the following market season had excellent sales, oddly enough. As I have said, favorable coincidences are your best indicator of success.

The seventeenth mansion is Alichil and it is for preventing a thief from breaking into a house in order to steal anything therein. When the Moon is sitting in that mansion, make the image of a monkey on an iron sigil with his hand above his shoulder. Suffumigate it with the hairs of a monkey and the hairs of a female mouse and wrap it in the pelt of a monkey [I used the fur from a toy monkey and didn't do the last two!]. Afterward, bury it in your house, and say: "You, Adrieb, guard all my things and everything that exists within this house, nor let it be entered by thieves." When the foregoing has been done thieves will flee from your house. Know that Adrieb is the name of the lord of that mansion.[22]

The seventeenth mansion talisman can be classified as warding or protective magic, and as such, it is a bit difficult to discern success, but so far, so good!

We have made several other lunar mansion talismans, notably the twenty-fifth mansion for protecting harvests, and the seventh mansion, Al-Dhira, a helping spirit who brings good luck and "all good things," as the *Picatrix* puts it. I have had good harvests and good things, thank God(s)!

Making lunar mansion talismans is a bit more complicated than other lunar operations because you have to use the principles of electional astrology to cast a chart for a time when the Moon is in the desired mansion, rising or culminating, and not making any hard aspects to Mars or Saturn. This takes a little practice and some basic astrological knowledge. It helps to use astrological software to cast and evaluate the chart for the operation. Christopher Warnock at Renaissance Astrology has a helpful book on the lunar mansions, listed in the bibliography, and also has a course on how to use them in magical operations.

5
MOON WORK

Using the Power of the Moon for Magic and Planting

This chapter will explore the complex topic of the Moon in the life of the cunning farmer. I'll be discussing some of the classical sources of the tradition of astrological magic, because they help teach us a different way of seeing the world that looks beneath the surface of things in order to grasp the energetics underlying the physical world. I would like to give you some idea of how to work with lunar energies, both agriculturally and magically, to help bring about meaningful changes in life circumstances. Accordingly, this chapter will mostly focus on how lunar energies are understood in astrology and how the Moon relates to the material world, specifically agriculture and planting crops.

THE MOON AS A DIVINE BEING

Let's get started with a quote from Renaissance mage Henry Cornelius Agrippa (preserving the seventeenth-century spelling) so we can get a feel for the contours and qualities of the lunar archetype. It is a beautiful piece of poetry in its own right and suitable as an invocation and therefore appropriate to begin our discussion:

> The Moon is called Phoebe, Diana, Lucina, Proserpina, Hecate,* menstruous,
> of a half form, giving light in the night, wandering, silent, having two horns,

---

*Phoebe, Diana, and the other names are epithets of the great Moon goddess of classical antiquity.

✦ *The Full Moon rises over a barn on the author's farm.*

a preserver, a night walker, horn bearer, the queen of heaven, the chiefest of the deities, first of the heavenly gods and goddesses, the queen of spirits, the mistress of all the elements, whom the stars answer, seasons return, elements serve; at whose nod lightenings breathe forth, seeds bud, plants increase, the initial parent of fruit, the sister of Phoebus,* light and shining, carrying light from one planet to another, enlightening all powers by its light, restraining the various passings of the stars, dispensing the various lights by the circuits of the Sun, the lady of great beauty, the mistress of rain and waters, the giver of riches, the nurse of mankind, the governor of all states, kind, merciful,

---

*Phoebus is an epithet of Apollo, god of the Sun.

Classical relief of the goddess Hecate, from Kinský Palace, Prague. Image by Zde.

protecting men by sea and land, mitigating the tempests of fortune, dispensing with fate, nourishing all things growing on the Earth, wandering into divers woods, restraining the rage of goblins, shutting the openings of the Earth, dispensing the light of heaven, the wholesome rivers of the sea, and the deplored silence of the infernals, by its nods: ruling the world, trending hell under her feet; of whose majesty the birds hastening in the air are afraid, the wild beasts straggling in the mountains, serpents lying hid in the ground, fishes swimming in the sea.[1]

To modern science, the Moon, our Earth's natural satellite, is a grayish, spherical ball of rock and dust orbiting the Earth at a distance of 238,900 miles. It has a radius of 1079.6 miles. It has an orbital period of 27 days. It has a mean orbital velocity of 2,287 miles per hour.[2] These are merely physical qualities of the Moon. More important to our tradition is the soul of the Moon—the intelligence, as it was called by the medieval philosophers.

The Moon, at least in many cultures, has been considered a divine being, often associated with the divine feminine (with notable exceptions worldwide). She is the luminary of the night, the sometimes-light and sometimes-dark goddess who governs the flux of all that lives on the Earth.* Anyone who has ever had the experience of watching the May Full Moon—the Scorpio Moon, the most fertile and fecund of all the Full Moons—rise on a warm evening, after a rain, in the countryside, with the frogs, toads, and all amphibians singing a droning chorus, punctuated by the calls of owls, howls of coyotes, and songs of night birds, while watching the illuminations of fireflies and the flash of distant lightning, will begin to understand the fertile power of Lady Selene, the personification of the Moon in Greek mythology.

BASIC THEORY OF LUNAR ASTROLOGY

You've likely heard of planting by the Moon, or by the signs, and heard urban legends about the effect of the Full Moon on crime rates, children's behavior, mental health, fertility cycles, and more. Where do these legends come from? What is the mysterious power the Moon exerts over the Earth and all her creatures? The influence of the Moon on the tides in the oceans is obvious to

---

*I will be using the pronouns she and her to refer to the Moon throughout, following traditional astrological practice.

anyone who has ever seen a spring or king tide, where the gravitational pull of the Moon in her opposition and her conjunction with the Sun (Full and New Moons) causes extreme tidal fluctuations. The tides are an obvious, unquestioned power of the Moon over terrestrial life, but what are some of her hidden or "occult" influences?

The Moon is the very archetype of changeability. Constantly moving, it appears as the fastest celestial object and is in an eternal state of flux. To the ancients, her changes and transformations symbolized the instability of the physical world. According to astrologers I. Alejandro Virgilio and H. Andres Villavicencio in their work, *The Power of Asteria*, "The Moon dictates the basic energetic cycles the earthly plane is being subjected to, cycles that repeat themselves over a relatively short time span and are felt with greater intensity. These cycles are based on the amount of sunlight, and therefore on the celestial energy the Moon is reflecting at any given time based on its location in space relative to the Earth and Sun."[3]

For Claudius Ptolemy, the Hellenistic Egyptian astronomer, "The Moon distributes her effluence to us, this effluence being greatest in the direction of the Earth, since the Moon is nearest the Earth, though most inanimate and animate things are sympathetic to her and change along with her, the rivers increasing and decreasing their flow with her light, while the seas curb their own onrush with risings and settings, and plants and animals either in whole or in part wax or wane along with her."[4] Ptolemy states, further, that the Moon has "power in moistening" and in "the outright softening of bodies and causing them to putrefy." For the ancients, the Moon ruled the moisture that is the largest part of the makeup of all living things. Since we are 70 percent water, it is through that water that the Moon exerts her influence on us.

For Agrippa, the Moon is:

the receptacle of all the heavenly influences, by the swiftness of her course is joined to the Sun, and all the other planets and stars, every month, and being made as it were the wife of all the stars, is the most fruitful of the stars, and receiving the beams and influences of all the other planets and stars as a conception, bringing them forth to the inferior world as being next to itself; for all the stars have influence on it, being the last receiver, which afterwards, communicateth the influences of all the superiors to these inferior, and pours them forth on the Earth.

◆ Luna, *from the astronomical treatise* Sphaerae Coelestis et Planetarum *(c. 1470).*

The moon acts as a medium between superiors, as Agrippa calls the planets and stars of the higher spheres, and the inferiors here below the Moon. He continues, "Her motion is to be observed before all the others, as the parent of all conception, which it diversely issues forth in these inferiors, according to the diverse complexion, motion, situation and different aspects to the planets and other stars; and although it receiveth powers from all the stars, yet especially from the Sun; as oft as it is in conjunction with the same, it is replenished with vivifying virtue."[5]

The Moon communicates the stellar and planetary influences present in the heavens, the aspects and dignities of the planets, to the Earth, but especially she communicates the power she receives from her aspects with the Sun, which we call the phases. Each of the phases has different elemental qualities. Perhaps we should call them energies. From the New Moon to the First Quarter, her energy, or "complexion," as Agrippa's seventeenth-century translator has it, is hot and moist. She is hot and dry from First Quarter to Full, cold and dry from Full to Last Quarter, and cold and moist from Last Quarter to the New Moon.

The fascinating medieval grimoire of astrological magic, the *Picatrix*, written in tenth-century Andalusia, most likely by the Islamic scholar Maslama al-Qurtubi, has a highly developed theory on the changeable energies of the Moon during the different phases of her cycle. It states that the different phases of the Moon in its aspects with the Sun have different elemental qualities

from the time she separates from the Sun until her first square with the Sun [the First Quarter]. During that time her power increases moisture and warmth, but she affects moisture more than warmth. During that time her effects appear in the growth of trees and plants, and her power of increase is more apparent in the herbs that grow in the ground than in the trees that rise above the ground. The second of her qualities is from the end of the first quarter until her opposition with the Sun. During this time her influence is more apparent in increasing heat and moisture equally. During this time her influence is well shown in the increase of moisture and heat in plants and minerals. When she recedes from opposition to her second square [last quarter] with the Sun, at that time her power increases moisture and heat, but heat more than moisture. Her influence appears more in increasing the bodies of animals, vegetables and those minerals that grow, in all their parts; this is why at this time she works more by heat than by moisture. From her second square to her combustion by the Sun [the moment of the new Moon], the effects, motions, and results of her heat are much abated . . . being moderately drying and strongly cooling."[6]

These two systems are slightly contradictory in the specifics, but both are trying to tell us that the Waxing Moon increases growth and the Waning Moon inhibits growth. So, as we have mentioned, in agriculture, as in magic, if you want to increase something like crops, money, or love, and make it grow, perform your operation in the Waxing Moon. If you want to inhibit something like weeds or pests or your gossiping coworkers or psychopathic boss, perform your operation during the Waning Moon. As the *Picatrix* states, "Pay very close attention in all your works to the waning of the Moon, since when she wanes, she reveals and shows harm, detriment, delay and impediment, upon all the affairs of this world."[7]

✦ *Diana/Luna as pictured by Sebald Beham, 1539.*

The Moon in ancient times was thought to be the intermediary for astral and planetary forces to enter the "sublunary world" here below. In other words, in order for cosmic influences to enter the earthly realm, they had to pass through the sphere of the Moon, where they picked up lunar influences according to the phase of the Moon and what sign she was in. The *Picatrix* text says, "Heed the Moon in all your works since she is more important than all the other planets. The Moon has more noticeable effects and authority over everything in the world. Hers is the power of generation and corruption, and she is the mediatrix in those processes. She receives the influences and impressions from the stars and planets, and she pours them into the lower things of this world."[8] Ancient philosophy had a great love of intermediaries, which linked opposites or extremes. The Moon was seen as one of these linking principles, carrying influences from heaven to what they called the sublunar realm, that part of the cosmos below the Moon.

The Moon, like the Sun and the planets, travels through the heavens along the path of the ecliptic. She passes through all the signs of the zodiac during the course of her 27-day, 7-hour sidereal month, all the while changing phases during her revolution around the Earth. The period between the New Moons is about 29 days and 12.7 hours, and this is known as the synodic month.[9] As noted, from a geocentric point of view, the Moon is the fastest heavenly body, moving on average 13.2 degrees per day and taking 2.5 days to pass through each sign of the zodiac.[10] According to astrological tradition, she transmits the powers and energies (or perhaps the angels and *daimones*) of that sign to the sublunar sphere here below during her time in each sign.[11] In her course through the zodiac every month, the Moon is happier and more constructive in some signs and more destructive in others, as she contributes to generation in some and corruption in others. As we have seen, the energies of the Moon change as she goes through her phase cycle. Hence, she transmits a different energetic quality each time she passes through each sign, about thirteen times in a calendar year, because she is in a different phase each time. Astrologers Virgilio and Villavicencio state, "Although the transit of the Moon is one of the oldest observations in astrology, there is no general consensus regarding the attributions specific to each transit. To a certain extent, this is due to the fact that the energies of the Moon change not only according to the sign it transits, but also according to the phase."[12]

THE ZODIAC AND THE MOON

Now that we have some theoretical background about the Moon and how she fits into the cosmology of traditional astrology, how can we as plant workers and magic practitioners put all this Moon lore into practice to empower all our work, both magical and physical? Let's discuss the twelve signs of the zodiac as they relate to the Moon.

As we have said, as the Moon passes through the twelve signs of the zodiac, she transmits something of their character and quality to the Earth below. The signs each have their own unique energy, deriving from the four elements of classical physics: earth, air, fire, and water, which themselves are combinations of the primary qualities, according to Aristotle. These qualities are hot, cold, dry, and moist.[13]

Assigned to the elements, then, we have fire, which is hot and dry; air, which is hot and moist; water, which is cold and moist; and earth, which is cold and dry. These elements and qualities were considered to be the basic building blocks of matter in the realm of the senses. They don't refer, however, to the physical objects of the same name. You can't put elemental water in a glass or cook your food on a campfire of elemental fire. Rather, the physical substances partake of the qualities of the elements without being identical to them.

The twelve signs of the zodiac are divided into triplicities, or groups of three, that are governed by one of the four elements. The elemental triplicities are as follows:

Fire: Aries, Leo, Sagittarius
Earth: Taurus, Virgo, Capricorn
Air: Gemini, Libra, Aquarius
Water: Cancer, Scorpio, and Pisces[14]

The zodiac is actually an ancient calendar system, and the signs are the months of the zodiacal year. In keeping with this calendrical system, the signs are further subdivided into quadruplicities, or groups of four, according to their order in the season. Those that begin the season are cardinal signs. Those that form the stable pattern of the middle of the season are fixed signs. Those that end the season are known as mutable signs, because they mark the changeable weather of the end of one season and the beginning of the new. Virgilio and Villavicencio again: "During [her] transit, the Moon is expressed according to the nature [of] each sign it transits. There is a common energetic imprint in the manifestation of lunar energies when the Moon is transiting through cardinal signs. Other common imprints are based on the particularities of fixed and mutable signs. Each of the four elements has different energies, resulting in twelve possible energetic combinations."[15] These signs are as follows:

Cardinal (or movable) signs, which begin on one of the turning points of the solar year, or solstices and equinoxes: Libra, Capricorn, Aries, and Cancer

Fixed (or solid) signs: Taurus, Leo, Scorpio, and Aquarius

Mutable (or common) signs: Gemini, Virgo, Sagittarius, and Pisces[16]

In astrology, the planets, which in this case includes the Sun and Moon, have signs that they are considered to rule or in which their power can express itself most fully, known as *domiciles* and other signs where they are very potent, or *exalted*. In other signs, their power is thwarted or limited due to the contrary energetic nature of the sign in which they find themselves, known as their *detriment* or *exile* and *fall*. In this case, the Moon is exalted in Cancer, exalted in Taurus, in detriment in Capricorn, and in fall in Scorpio (which is confusing because, as we will see, Scorpio is one of the most fertile signs for planting, and Capricorn isn't too bad either).[17]

LUNAR ASTROLOGY AND FARMING

For agricultural purposes, another important division of the signs is whether they are fruitful or barren. Originally, in natal astrology, it was observed whether the signs placed in the natal chart in the fifth house, which signifies the native's children, or the first house, which signifies the native themselves, were fruitful or barren. A fruitful sign rising or on the fifth house cusp would signify the potential of abundant offspring for the native. Conversely, a barren sign in those same places in a natal chart would be a testimony of few children, or none. In the lunar astrology practiced in agriculture, the Moon passing through a fertile sign is considered an excellent time for planting or breeding animals, because the Moon transmits the fertile nature of that sign to the Earth.[18]

Fertile signs: Cancer, Pisces, Scorpio

Barren signs: Gemini, Leo, Virgo

Moderately fertile: Taurus, Libra, Sagittarius
Moderately barren: Aries, Capricorn, Aquarius[19]

Since antiquity, farmers have timed their operations by the phases of the Moon according to various rationales. For the ancient Roman farmers, following the most common lunar logic, any operation in which decrease, limiting, or restriction was desired was carried out under a Waning Moon, and operations where growth or gain was desired were carried out under a Waxing Moon. This seems to be the perennial understanding of lunar energies: The Waxing Moon causes or influences growth or augmentation, and the Waning Moon restricts or retards growth and causes withering and drying. Furthermore, the Waxing Moon tends to pull the energies up from the Earth, and the Waning Moon tends to pull down. So, when a farmer wants large root crops, often it is desirable to plant during a Waning Moon, but when the part of the plant harvested grows above the ground, generally speaking, it is desirable to plant during a Waxing Moon. In all cases, planting in a fertile sign is preferred, and moderately fertile signs are the second choice.

In the sublunar realm, out on the land, however, it is important to not let the perfect be the enemy of the good. When one is dealing with the weather, it often happens that weather or soil conditions will be too wet to allow one to plant when the Moon is in a fertile sign. Remember, the Moon is the mistress of moisture, and the fertile signs are all water signs.

The reverse is customarily true for weed and pest control: If you want to kill undesirable plants, it should be done during a Waning Moon in a fire sign or a barren sign. Many people like to do their ground preparation during an earth sign such as Taurus, Virgo, or Capricorn. Tillage is more sensitive to soil moisture than many operations on the farm and, in my opinion, needs to be done at the right time regardless of the Moon's condition. It is possible, however, that soil prepared in less-than-ideal moisture conditions (for instance, when it's too wet) may make a better seedbed if the soil is worked while the Moon is waning in an earth sign, because at those times it tends to be more dry.

According to Scottish folklorist F. Marian McNeill, "In Scotland, as elsewhere, the phase of the Moon was always observed when certain kinds of work had to be done. The crescent Moon was believed to encourage growth in substances, and it was the time for sowing and planting—an exception was made for plants such as onion or kail, which tended to run to seed if sown in the increase."[20] Quoting an old man in Eigg, Scotland, McNeill continues:

The Anatomical Zodiac Man, *an example of astrological melothesia, which is the assignment of body parts to the twelve signs of the zodiac, as pictured in the fifteenth-century French manuscript* Très Riches Heures *by the Limbourg brothers.*

The men of old would not kill a sheep nor a goat nor an axe-cow at the wane of the Moon," said an old man in Eigg. "The flesh of the animal is then without taste, without sap, without plumpness, without fat. Neither would they cut withes of hazel or willow for creels or baskets, nor would they cut trees of pine to make a boat, in the black wane of the Moon. The sap of the wood goes down into the root, and the wood becomes brittle and crumbly, without pith, without good. The old people did all of these things at the waxing or the full of the Moon. The men of old were observant of the facts of nature, as the young folk of today are not."[21]

McNeill states that in Scotland, "The waning Moon was propitious for ploughing, reaping, and the cutting of peat, for only then would the natural juices depart and drying be expedited. Eggs laid in the wane were used for hatching, rather than those laid in the increase, and birds hatched in the increase were considered difficult to rear. Animals were gelded in the wane."[22]

These same beliefs and practices regarding agricultural operations and the Moon can be found across the Atlantic in the Appalachian region of the United States, where they were brought by transplanted settlers from the Old World. Eliot Wigginton's 1972 classic book of American Appalachian folklore, *The Foxfire Book*, discusses the practice of planting by the sign in some detail. I will quote a section of the chapter here, because it exemplifies the tradition I have heard from some of the old-timer farmers and gardeners in my twenty-five years of farming in Kentucky. One comment I'd like to add is that the practice of planting by the signs in America is tied to the ancient idea of astrological melothesia, whereby different parts of the body are assigned to the different signs, which is the basis of medical astrology. This tradition dates back to the Babylonians, where the basic structure was recorded in tablets dating as far back as 400 BCE. The parts of the body assigned to the various signs have changed little in the intervening centuries.[23] These are some of the basic prescriptions from *The Foxfire Book*, of which there are many more.

PLANTING—Planting is best done in the fruitful signs of Scorpio, Pisces, Taurus or Cancer (when the signs are in the loins, feet, or breast).

Plow, till or cultivate in Aries.

Never plant anything in one of the barren signs. They are good only for trimming, deadening, and destroying.

Always set plants out in a water or earth sign.

Graft just before the sap begins to flow, while the moon is in the first or second quarter, and while it is passing through a fruitful, water sign, or Capricorn [or better, Taurus]. Never graft or plant on Sunday as this is a barren, hot day (the sun's day).

Plant flowers in Libra, which is an airy sign that also represents beauty. Plant them while the moon is in the first quarter, unless you need the seeds, in which case use the period between the moon's second quarter and full.

Corn planted in Leo will have a hard, round stalk and small ears.

Crops planted in Taurus and Cancer will stand drought.

Plant beans when the signs are in the arms [when the moon is in Gemini].

Root flower cuttings, limbs, vines, and set out flower bushes and trees in December and January when the signs are in the knees and the feet [when the moon is in Capricorn or Pisces].

Never transplant in the head or the heart as both these signs are the "Death Signs."*

If you want a large vine and stalk and little fruit, plant in Virgo—"bloom days."

Don't plant potatoes in the feet [when the moon is in Pisces]. If you do, they will develop little nubs like toes all over the main potato. The best time is a dark night in March.

Plant all things which yield above ground during the increase or growing of the moon, and all things that yield below the ground (root crops) when the moon is decreasing or darkening.

Never plant on the first day of the new moon, or on the day the Moon changes quarters.

In the fourth quarter turn sod, pull weeds, and destroy.[24]

Through these chapters regarding the role of the Moon in both magical and agricultural practice, we have surveyed both occult literature and folkoric sources in order to provide a foundation for seeing the Earth's mysterious satellite and nocturnal luminary as a potent transmitter of celestial forces. As terrestrial beings, the Moon and her tidal rhythms regulate the periodicity of our organic and energetic lives. Knowledge of these rhythms and the occult forces they carry is essential for all forms of practice, whether magical or mundane, that seek to harness the Moon's ancient powers of increase and decrease. This

---

*My note: Aries and Leo are both considered hot, dry, and barren signs.

knowledge can be used for wisdom, healing, personal gain, and the withering and weakening of harmful physical and spiritual forces. As both farmers and magical practitioners, a lunar calendar, an ephemeris, as well as a keen eye for observation or an almanac are essential tools for the timing of all kinds of operations on the land.

6

WEATHER MAGIC

Using Magic to Influence Storms, Wind, and Rain

The voice of the LORD breaketh the cedar trees;
Yeah the lord breaketh the cedars of Lebanon.[1]

PSALMS 29:5

And you will stop the power of the tireless winds which sweep
  all over the earth
And destroy the crops with their blowings,
And again, if you wish, you will bring on compensating
  breezes.
And after a black rain you will produce a seasonable drought
For men, and after the summer drought you will produce
Tree-nourishing streams that live in the ether.[2]

EMPEDOCLES OF AKRAGAS

As we can see from this quote by the fifth-century BCE pre-Socratic philosopher/shaman Empedocles of Akragas, control of the weather has been a perennial concern of the magician. And perhaps no aspect of the life of the farmer is more vexing than coping with the vagaries of the weather. I have seen many of the ugly moods of Mother Nature in my career tending crops and animals. It seems like the weather gets more extreme with each passing season, as humans continue to desecrate and damage the ecosystem. Our divine Mother Gaia is not going to sit back and allow all of this sacrilege to go on unchallenged. Our weather is increasingly showing her in her aspect of Nemesis, the punisher of the wicked and irreverent.

As a farmer and magic practitioner, weather magic is one of my major interests, both folkloric and practical. During my research, anytime I come across a spell or practice in global magical literature or folklore, I put a bookmark on it for future reference. In this chapter, I intend to outline some of the historical practices people have resorted to to save the crops from drought or excessive rain or to abate tempests or floods. I also will relate some of my personal experiments, with the caveat that I will make no extravagant claims to be able to control the weather. I have had a number of favorable coincidences, as well as a few duds (or experiments that took longer to come to fruition).

It was tornado season here in Kentucky when I wrote these words and my region was under a severe thunderstorm and tornado watch. There were also dozens of tornadoes last night in the southern United States. What better place to begin than with magical practices for averting storms?

✦ *A magician mastering the elements. From the 1582 work*
Historia Mundi Naturalis.

Storm magic is a type of magic that is needed immediately and can't wait for the planets to align with the proper Moon sign. At the time I was writing this chapter, a Waxing Moon in Cancer, a very powerful Moon, was probably lending her force to the storm whirling its way across the United States. As we will see, these lunar conditions are precisely the ones we look for when we want to make rain. In other words, we have to work with the energy we have right here in the elemental sphere, the power of our will, and the names and symbols we use to call in the power of the universe.

Weather magic is a type of sympathetic magic in which it is believed, according to Daoist (his spelling) magician Jerry Alan Johnson, that

> the creation of a representation or symbol of the desired outcome, along with the sorcerer's intent, will manifest that intent in reality. The underlying premise is that the universe is one, or a unified field, and thus the microcosm can be used to manifest an event within the macrocosm. This is the basic principle of all sorcery. What differentiates a child's game of make-believe from a sorcerer's magical manifestation is that vital occult factor of the subconscious mind's intention, focused on the application of the work. Unless that underlying stratum of "deep mind" is penetrated, cleared and available to focus the intention, the "magic" remains on the surface of the personality and lacks true power.[3]

WEATHER MAGIC IN THE SCRIPTURES

In various traditions of folk magic (Mexican Ensalmeria, American Hoodoo, Jewish Kabbalah, and Pennsylvania German Braucherei, to name a few) there is a consistent theme of using the Bible as a book of magic (a grimoire) and using psalms in the Book of Psalms as spells. Psalms are liturgical poems, originally composed in Hebrew, that were sung or recited in communal worship. They are ascribed to the biblical King David but were actually composed throughout various periods of Jewish history, some dating back to Canaanite, Egyptian, or Mesopotamian influences. Psalm 104, for instance, has much in common with Pharaoh Akhenaten's hymn to the Sun.[4] Psalm 91 is reminiscent of spells from Babylonian and Sumerian tablets for repelling demons and evil spirits, and even names several of them. They are divided into psalms of praise and thanksgiving, of lamentation, of imprecation, and of cursing. They have a long history of use in both the magical and religious traditions of many nations.

There is also a tradition in which all of the 150 psalms have particular magical uses, stemming from a medieval Jewish text, the *Sefer Shimmush Tehillim.* This text was extremely influential in folk magic because of its inclusion as an appendix in the *Sixth and Seventh Books of Moses,* a German grimoire that drew on diverse strands of medieval magical traditions. These included the Practical Kabbalah and the Faustian tradition of central Europe, as well as elements from Solomonic magic.[5]

Salvador, whom I introduced in the first chapter, the elderly gentleman from rural Mexico, claimed he once dispelled a tornado that was headed toward the trailer park in which he lived by chanting Psalm 91 with intention and making the sign of the cross. He said it simply dissipated! The ninety-first psalm, numbered as ninety in the Catholic Bible, is considered to be one of the most powerful for averting evils of all kinds. I'm not suggesting that anyone can or should leave their shelter and go out and stand in the face of an EF5 and chant a psalm to make it dissolve. My friend Salvador had been practicing these things for years at that point and is probably constantly surrounded by a whole flock of angels!

I have used Psalm 91 for this purpose and actually have committed it to

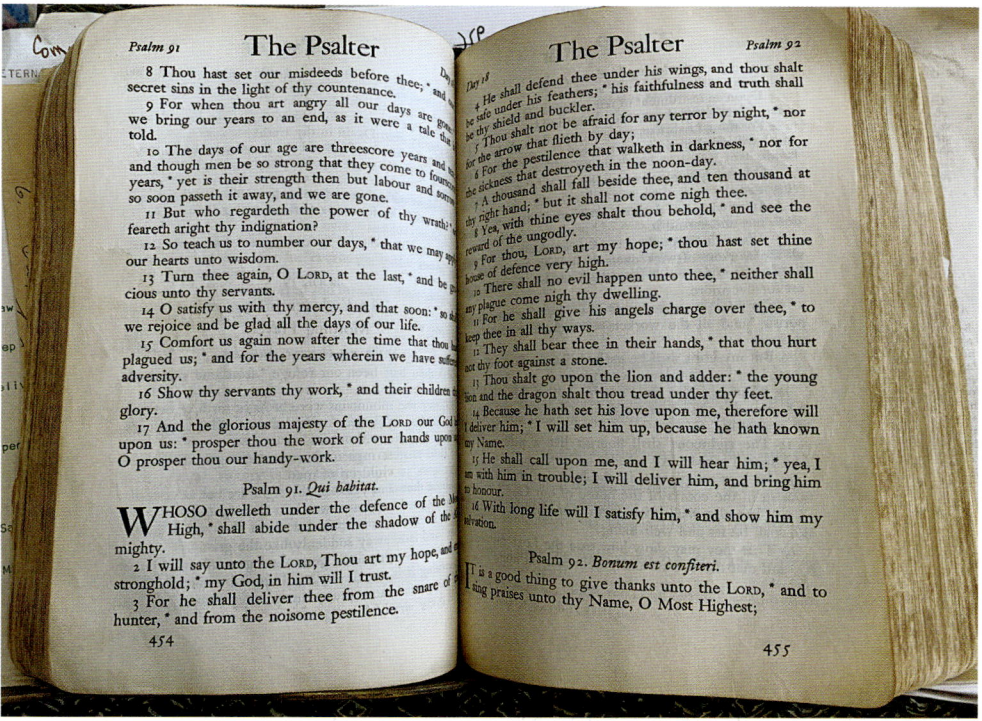

◆ *The author's grandfather's "grimoire," his 1928* Book of Common Prayer, *open to Psalm 91. It is stuffed full of original prayers he wrote for all occasions.*

memory, following Salvador's suggestion—he said that you won't always have a Bible or prayer book with you when you need protection, so it's best to memorize really important prayers. I also find it helpful to put important prayers and spells in the Notes app of my cell phone for quick access wherever I may be if I haven't memorized them yet. I find that, for me, memorization comes with frequent recitation.

Psalm 29 is reputed to be specifically indicated for calming storms at sea or on land.[6] There is a practice of reciting Psalm 29 as well as the Divine Name, Adiriron, ten times each. This is derived from the Jewish grimoire, the *Shorshei Ha-Shemot*. Kabbalist Jacobus Swart's *Book of Sacred Names* gives the text in King James Version English as well as phonetic Hebrew and the Hebrew alphabet. This psalm is one of the oldest texts in the Hebrew Bible. Scholars believe it was originally composed when Yahweh, whose name is repeated eighteen times in the psalm, was conflated with Baal, the Canaanite god of storms, and had not yet become the universal Christian God, worshipped by billions worldwide.[7]

I have opted for reciting in Hebrew because I believe, like the Neoplatonic philosopher Iamblichus, that "the gods have shown that the entire dialect of sacred peoples such as the Assyrians and the Egyptians is appropriate for religious ceremonies and for this reason . . . our communication with the gods should be in an appropriate tongue."[8] Ancient languages like Hebrew are believed to have a natural *occult virtue*, or inherent power. After all, the Kabbalists believe that Hebrew was the original language that God used to "speak" creation into existence.[9] Swart says:

> Every letter of the Hebrew alphabet is considered to be the embodiment of a Spirit Intelligence, each being an "Angel," so to speak. Divine Names are understood to be more than just personal identifications. It is believed that they actually are the perspective qualities or aspects of divinity hidden within their inner meanings, thus the incantation of "Divine Names" releases the spiritual powers they embody. These Names are considered to be "Words of Power," since they are both expressions of the "nature" of Divinity, as well as "Angels" or Spirit Intelligences, who direct those aspects of the Divine Nature encapsulated within them into actual existence.[10]

In theory, reciting these prayers can actually change how reality unfolds! The literature is full of warnings about the dangers of messing about unprepared with these powers.

The Tetragrammaton

There is a taboo in the Jewish faith about the pronunciation of this particular Name, YHVH, which is known to the Western magical tradition as the Tetragrammaton, or the name of four letters. Those who look to traditional Jewish sources online for Psalm 29 in Hebrew will find that in the transliteration, this name, also known as Ha Shem, has been changed to Adonai (Lord), per ancient custom. The true pronunciation of the Name has been lost, and in the Second Temple period, it was only pronounced once a year in the Temple in Jerusalem on the Day of Atonement, or Yom Kippur.[11] Following the destruction of the Second Temple, even this legitimate use of the Name faded. Ironically enough, some scholars believe that one of the main reasons for the secrecy and taboo around the name was to prevent it from being used in magic. It has been pointed out that the only thing more powerful than a name of God for magical purposes is a secret name of God! After the Rabbinic period, the only people pronouncing the Name were magicians. The Jewish variety of magicians were known as Baalei Shemot, or Masters of the Names.[12]

The Name, formed of the four Hebrew letters, Yod, Heh, Vav, and Heh, יהוה, is cognate to the name revealed to Moses in the epiphany he experienced, which is reported in Exodus 3:14, where a voice from the burning bush declared

✦ *Stained glass window with the figure of the Tetragrammaton in the Winchester Cathedral. Image by Oddworldly.*

"EHYEH" (or אהיה), meaning "I am that I am." Yahweh, therefore, means something like "He Who Is" or "The Existent One."[13] Swart says that the Name is generally pronounced "Yahweh" in academic circles. He goes on to say:

> It is however interesting that the letters comprising this Divine Name are the very ones in the Hebrew alphabet used to indicate vowels (the only ones, for that matter, as Alef and Ayin are "silent" letters and not vowels). The letter "Yod" is used to indicate the vowel "i" (pronounced "ee"); the letter "Heh," for example at the end of a word to indicate the feminine gender, indicates the vowel "a" (pronounced "ah"); the letter "Vav" indicates either "o" or "u" (pronounced "oo"). Thus the first three letters of the Divine Name could be read "IAO," a divine name familiar to the Gnostics of both Judaism and Christianity.[14]

The name, IAO, is also common to the Hermetic Magic of the Greco-Egyptian Magical Papyri. Swart adds: "The Tetragrammaton is an expression of vowels, i.e., IAOE (EE-AH-OH-EH), the frequencies of which were considered to act on the four levels of existence (Four Worlds), but also upon the fourfold expression of human existence."[15]

This name is thus a word of power that if uttered in a ritual context, with respect and reverence, is believed to set up a resonance across the four worlds and the human energy body, sending forth a flow of energy between the microcosm and the macrocosm. This is the reason for its vain utterance being forbidden by the Third Commandment of Moses. To speak this word out of its sacred context is to profane a holy mystery. I in no way use this name lightly or in vain but to call upon divine power in dire circumstances.

WEATHER MAGIC ON THE FARM

I have had some extremely uncanny synchronicities around recitations of scripture and divine names in Hebrew, which I take as evidence that some mysterious forces indeed are unleashed when such sacred words are pronounced. The day of the windstorm was no exception. I had decided to pause my writing to perform the recitation described here. I finished and happened to check my notifications, and a new episode of the excellent YouTube channel Esoterica had been released just moments before, while I was performing the ritual. The title: "Who is Yahweh—How a Warrior-Storm God became the God of the Israelites and World Monotheism."[16] The episode mentioned Psalm 29 specifically as being originally

✦ *A cedar tree snapped by the gale-force winds on the author's farm, shortly after the ritual described in the chapter.*

a Canaanite hymn to Yahweh—Yahweh, whose name I had just chanted 180 times! It was a classic synchronicity in which two events that are not linked by causality occur simultaneously and are connected by symbolic meaning!

Incredibly enough, the day's ritual reading of Psalm 29 was accompanied by yet another synchronicity. Verse five from Jacobus Swart's version reads, "The voice of YHVH, breaketh the cedars; yea, YHVH breaketh in pieces the cedars of Lebanon." As I watched, a particularly forceful wind gust cracked a cedar tree right outside of my house! Ten times I chanted, *"Kol YHWH shoveir arazim vaishaber YHWH et arzei ha-l'vanon,"*[17] and the universe (or Yahweh, or both?) said, "Watch this!"

Emergency weather magic sometimes doesn't give you much time for purification. One evening during a tornado warning I recited this psalm standing on my porch under the unnaturally green sky, which is common in severe thunderstorms. Usually when a storm approaches, the wind is pushed out in front of it by the downdraft of cold air falling out of the storm. In this case, however, the wind was being pulled into the storm, as if by suction. This seemed to me to be a bad sign. I recited the psalm and did the accompanying visualization described below, and the storm passed without any real harm, while tornadoes passed a bit too close for comfort. I don't recommend standing out in the path of an advancing severe thunderstorm; it is dangerous, and people should definitely take cover. But I am one of those people who gets a charge of energy, storm *qi* perhaps, from a powerful weather system.

My practice with this psalm varies depending on how much time for preparation I have. If I have more time, I might light candles and incense, open a sacred space, invoke the angels of the elements—the full ceremonial—and put the recitation of the psalm and the Name in the middle of the ceremony as it were. I also visualize a massive being of light with a sword and a shield, the angel of the psalm, perhaps, standing over the land and protecting our family, land, and community from the storm. We have had some big storms pass close by, and by the grace of God(dess) have escaped unharmed. Other times, like today, when I do not have much time, I recite the psalm ten times followed by the Name ten times, visualizing the being of light while repeating the Name. It is important to allow the repetitive recitation of the psalm and the Name to bring the mind into a trance state. Regular practice of these techniques adds to their effectiveness.

In this case, the windstorm eventually passed by, leaving us soggy and wind tossed but mostly unscathed. The only casualty was my children's trampoline. My five-year-old son, a budding wizard himself, had a bit of a crisis of faith, not being much consoled by the fact that our house still had a roof!

I have done some version of this little ritual in the face of some really nasty weather, including Hurricane Ian, which was headed directly toward the South Carolina beach where I was on vacation with my family. It suddenly weakened and took a turn to the north, keeping the worst of it away from us.

Which leads me to an ethical issue: By doing weather magic, are we just making bad weather someone else's problem? If we can turn aside a potent storm, are we just directing the problem at other innocent people elsewhere? If these practices work, then it is our responsibility to apply them with wisdom and to consider the consequences. I can envision a network of community-minded magic practitioners, applying what my friend Martin called "essential technologies of spirituality," to protect their communities from these ever-increasing adverse weather events. The visualizations, in this case, should aim at weakening, as well as deflecting the storm, to reduce its potential for harm downwind.

Weather Magic in Chinese Culture

The Taoist magicians of China also have a rich tradition of weather magic. According to author Jerry Alan Johnson, the magicians of ancient China

> believed that the incredible powers of a raging storm could be calmed and sedated by performing certain magical rituals. These ancient sorcerers believed that the physical world was simply a playground for consciousness, and that each individual could magically act as an energetic conduit (existing between heaven and Earth) and affect any and all forms of matter. The sorcerer's imagination creates his or her inner-verse (through thoughts and intentions) and contains the esoteric realms of magical powers. The sorcerer's inner-verse then creates actions within his or her outer-verse (i.e., spirit and intention creates energy which manifests as matter). With this spiritual understanding, the sorcerer first imagines the raging storm as a large child (a baby demon) that is throwing a temper tantrum. Within his or her mind, the sorcerer gives the "child storm" a name and personality. The sorcerer then begins to gently calm the storm's spirit using soothing words and gestures.[18]

Summoning Rain

There is also a wealth of material in the magical traditions of many cultures relating to practices for summoning rain or storms in times of drought, or simply to create havoc for an enemy. Droughts, as any farmer can tell you, are frightening, slow-moving disasters in which you slowly watch your livelihood wither, dry up, and blow away in the dust. Therefore, rainmaking has been a preoccupation of nearly every society that practices agriculture.

The land on which I farm is a high ridge which has no ponds, lake, or streams from which to draw water. I raise vegetables, many of which are pretty thirsty as crops go. I practice minimal irrigation, because I have to truck in water. Good organic farming practices go a long way toward building resilience in soils and crops, to be sure. But when the rains stop, the prayers start, here on my farm as in most of the world.

As Sir James George Frazer sums up the matter in his monumental study of global mythology, *The Golden Bough*:

Of the things which the public magician sets himself to do for the good of the tribe, one of the chief is to control the weather and especially to ensure an adequate fall of rain. Water is an essential of life and in most countries the supply of it depends on showers. Without rain vegetation withers, animals and men languish and die. Hence in savage* communities the rain-maker is a very important personage; often a special class of magicians exists for the purpose of regulating the heavenly water supply. The methods by which they attempt to discharge the duties of their office are commonly, though not always, based on the principles of homeopathic or imitative magic. If they wish to make rain they simulate it by sprinkling water, mimicking clouds.[19]

This practice of imitative magic is seen in the following two examples from two different magical traditions. Notorious English magician Aleister Crowley, whose vivid career straddled the nineteenth and twentieth centuries, addressed the topic of summoning a thunderstorm as a theoretical example in his influential work, *Magick in Theory and Practice*. Crowley's thoughts are very much in harmony with the Taoist theory of magical action presented by Professor Johnson and are worth quoting at length, not least to give readers unfamiliar with him a sense of Crowley's colorful and entertaining prose, and lucid thought:

> Suppose I wish to produce a thunderstorm. This event is beyond my control or that of any other man; it is useless to work on their minds as my own. Nature is independent of, and indifferent to, man's affairs. A storm is caused by atmospheric conditions on a scale so enormous that the united effort of all of us Earth-vermin could scarcely disperse one cloud, even if we could get at it. How then can any Magician, he who is above all things knower of Nature, be so absurd as to attempt to throw the Hammer of Thor? Unless he be simply insane, he must be initiated in a Truth which transcends the apparent facts. He must be aware that all Nature is a continuum, so his mind and body are consubstantial with the storm, and are equally expressions of One Existence, all alike of the self-same order of artifices whereby the Absolute appreciates itself. He must have assimilated the fact that Quality is as just much a form as Quantity; that as all are modes of the One Substance, so their measures are modes of their

---

*Note: Frazer was a product of his age and often used terms that are no longer considered appropriate in his work.

relation. Not only are gold and lead mere letters, meaningless in themselves, but appointed to spell the One Name; but that the difference between the bulk of the mountain and that of the mouse is no more than one method of differentiating them, just as the letter "m" is not bigger than the letter "i" in any real sense of the word.[20]

Crowley's reasoning might require a bit of unpacking here: I understand him as saying, in so many words, that on the level of Mind, whereby all things are connected, mass and size are simply concepts, and small things can move quite large things by contacting the level of Mind.

He continues:

Our Magician, with this in his mind, will most probably leave thunderstorms to stew in their own juice; but should he decide (after all) to enliven the afternoon, he will work in the manner following:

First, what are the elements necessary for his storms? He must have certain stores of electrical force, and the right kind of clouds to contain it.

He must see that the force does not leak away to the earth quietly and slyly.

He must arrange a stress so severe as to become at last so intolerable that it will disrupt explosively.

Now he, as a man, cannot pray to God to cause them, for the **Gods are but names for the forces of nature themselves.**

But, *as a Mystic*, he knows that all things are phantoms of One Thing, and that they may be withdrawn therein to reissue in other attire. He knows that all things are in himself, and that he is All-One with the All. There is therefore no theoretical difficulty about converting the illusion of a clear sky into that of a tempest. On the other hand, he is aware, *as a Magician*, that illusions are governed by the laws of their nature. He knows that twice two is four, although both "two" and "four" are merely properties pertaining to The One. He can only use the mystical identity of all things in a strictly scientific sense. It is true that his experience of clear skies and storms proves that his nature contains elements cognate with both; if not they could not affect him. He is the Microcosm of his own Macrocosm, whether or no either one of them extend beyond his knowledge of them. He must therefore arouse in himself those ideas which are clansmen of the Thunderstorm; collect all available objects of the same nature as talismans, and proceed to excite all these to the utmost by a Magical ceremony; that is by insisting on their godhead, so that they flame within and without him, his ideas vitalising the talismans. There is just a vivid vibration of high potential in a certain group of sympathetic substances and forces; and this spreads as do the waves from a stone which is thrown into a lake, widening and weakening until the disturbance is compensated. . . . He transmits his particular vibration as a radio operator does with his ray. . . .

In practice, the Magician must "evoke the spirits of the storm" by identifying himself with the ideas of which atmospheric phenomena are the expressions as his humanity is of him; this achieved he must impose his Will upon them by virtue of his superiority of intelligence and the integration of his purpose to their undirected impulses and uncomprehending interplay.[21]

Crowley describes tuning the consciousness of the magician to the idea of the storm by constructing a ritual composed of sympathetic symbolic elements, which unites the microcosmic idea of a thunderstorm with the macrocosmic idea of a thunderstorm. Sorcerers all over the world construct their rain rituals in a myriad of ways, but common to all of the various practices is the idea of what Crowley calls the Magical Link, which unites the mind of the operator through the operation to the desired effect in the so-called "real" world. Let's take a moment to examine some specific examples from a few different traditions, holding these principles in mind.

Rain Spells from Across the World

I actually haven't come across many rain spells from notoriously rainy Great Britain and especially Scotland. One exception is this one, which is relatively recent (published in 2008), but perhaps from an older tradition, from Gemma Gary's excellent book *Traditional Witchcraft: A Cornish Book of Ways*:

> When rain is needed, an iron vessel and the knife are taken to a high place, be it a rocky carn, hill, or towering cliff top. Within the vessel, a small fire is made and upon this is set a good bundle of ferns and some henbane to burn with much smoke. . . . The aid of spirits is invoked with dances made against the sun around the iron vessel of rising smoke. The blade is held aloft as conjurations for rain are muttered into the smoke as it rises into the sky. Further circumambulations may be made around the smoking vessel while sprinkling water upon the earth from a dipped branch of heather.[22]

✦ *Witches performing a rain spell. Woodcut, c. 1489, artist unknown.*

I have not tried this one, but it does sound delightfully witchy. It draws on the archetypal image of the witch as a raiser of storms, in this case for securing needed rain. I imagine it being performed in an evocative setting, such as atop a cliff in Cornwall by the sea. Please don't breathe the smoke from burning henbane; it is toxic. This spell might be good to combine with elements of the rain spell from *Ars Rabidmadar*, a seventeenth-century Italian grimoire, which gives more specific directions:

> To make it rain:
> Take sea water, or *marinata*, which I will teach you how to make artificially afterwards. Sprinkle this on a circle which you have drawn on the ground, in the middle of this circle place a bloodstone (heliotrope). Then with the wand, namely the magic one . . . draw on the right hand side the character of Beschard and on the left Elelogaphatel and standing over the wand recite these words: "[+Aerlea] Helelogaphaton" and you will see the clouds fill with sky, be released as water.

An illustration of the circle and the seals of the spirits is given in the text. The recipe for the *marinata*, or artificial seawater is as follows:

Take river water and salt and a little clay and boil for a quarter hour on a fire into which you have thrown ground pumice.[23]

Gary also gives this lovely spell for raising a wind:

To make a "Wind Stone" thread a good length of cord through a hag stone and tie eight knots along its length. Take this out to some exposed place and begin to whirl the stone in the air above your head, whilst invoking the spirits of the air. The speed at which the stone is whirled must be adjusted in accordance with the speed of the wind to be conjured. To slow the wind and conjure calm weather one must begin by whirling the stone at great speed and gradually slow it down. To conjure great winds and gales, start to

◆ *Witches causing a storm at sea. Woodcut, c. 1896, artist unknown.*

whirl the stone slowly and gradually increase the speed to conjure the level of wind desired.[24]

The next bit of storm-raising magic I include for purely folkloric and historical interest is from Portugal, from *The Great and True Book of St. Cyprian or The Sorcerer's Treasure*, translated from the Portuguese by José Leitão and found in his massive tome *Opuscula Cypriani*:

Great Black Magic Conjuration, so as to revolt the Weather, Darken the Stars, See Lightning, Hear great thunder and storms, great phantoms and columns of fire coming from the earth, great chasms open, which will seem to want to swallow the conjurer! It is a terrible spectacle, equal to the last day of the earth!

FIRST CONJURATION

Serpent, that in paradise tempted Eve and thou wert the perdition of humankind, that with thy perverse cunning condemned men to the captivity of perdition, by which cause the God of the Universe condemned thee to be trampled and obedient to man.

As such, in the name of the Divine Spirit I conjure thee and request that without appeal, thou rises from the abyss and makes rain fall upon the earth, and make the waters of the sea rise and the stars in Heaven move, and the firmament be wounded with lightning and thunder. Cover the whole earth with thick darkness, make amazing screams to be heard, given out by the legions of demons, who were for fifteen hundred years imprisoned by the order of the Custodian angel.[25]

This is only a brief excerpt of the ritual, which is to be done at midnight, on top of a tall mountain, with the conjurer holding in their hand a human bone. This ceremony too draws upon the trope of a sorcerer who summons infernal powers to raise a violent thunderstorm. It goes without saying that this is not a good way to break a drought and bring some much-needed rain. Personally, the idea that diabolical powers can be appealed to for the manipulation of forces of nature does not mesh with my understanding of metaphysics and cosmology, because I don't see the spiritual forces of the natural world as inherently evil. Powerful, and perhaps indifferent to human suffering, but not opposed to divine purposes. Indeed, it seems to me like the awesome power of the forces of nature reflects both the providence of divine intelligence and the inexorable play of fate in the material world. We must say with Marcus Aurelius, "The Universe loves to make whatever is about to be. I say to the Universe, 'I love as thou lovest.'"[26]

CONJURING AND SOLOMONIC MAGIC

When Christianity replaced the older worldview with its own, the entire metaphysical schema had to fit into its dualistic framework in which, essentially, there are only two sides, the side of God and that of the Devil. The Devil was the god of this world (for example, in John 14:30, or 2 Corinthians 4:4). Any attempt to negotiate or command the elemental powers of nature, outside of the intercessory prayer permitted by the Church, necessarily involved pacts with demonic entities, which was heresy.[27] Conjurations of this type were an attempt to get around this conundrum.

The ceremony above is a true conjuration in the older sense of the word, in which evil spirits are called upon and constrained by the power of God, Jesus, and the Archangel Michael. In ancient and medieval times the office of the exorcist, who freed people from the power of evil spirits, often blended into the work of the conjurer, who put those same spirits to work to magical ends, using the same techniques—those of religion—but for different aims.[28] In fact, the practice is basic to the branch of magic known as Solomonic magic. The earliest text of the tradition of Solomonic magic, the pseudepigraphic Christian work known as the *Testament of Solomon*, which may date from as early as the first century CE, depicts King Solomon being given by God a magic ring, with a pentagram on it, and with it summoning demons who tell him their powers and by which angels they are constrained.[29]

Conjuring, literally meaning "binding by oath" (something like a *subpoena*), is the practice of binding one lesser spirit, like an elemental spirit or demon, by the power of another more powerful spirit, like an angel or god. It is an ancient practice, featured especially in exorcisms. Exorcism is derived from a Greek word meaning the same thing as conjure, to bind by an oath. The practice dates from ancient Egypt and is mentioned in the *Greek Magical Papyri* from Greco-Roman Egypt. The *Stele of Jeu the Hieroglyphist*, PGM V. 96–172, is a famous example, and one which had an important influence on the history of modern Western magic. The *Stele of Jeu* is an example of an exorcism in which a daimon is driven out of a person by the power of a more powerful daimon, in this case Akephalos, the Headless One, also known as the Hornless One, who is invoked by the exorcist. It is exactly the practice that Jesus is accused of practicing by the Pharisees in Matthew 12:22–29. Here is an excerpt from the dramatic climax of the ritual, which follows the recitation of a string of Divine names and words of power in Aleister Crowley and S. L. Mather's evocative (if slightly inaccurate) translation:

> I am He! the Bornless Spirit! having sight in the feet:
> Strong and the immortal fire!
> I am He! The Truth!
> Who hate that evil should be wrought in the world!
> I am He, lightningeth and thundereth.
> I am He, from Whom is the Shower of the Life of Earth:
> I am He whose mouth ever flameth:
> I am He, the Begetter and Manifester unto the Light:
> I am He; the Grace of the World:
> "The Heart Girt with a Serpent" is My Name!
> Come Thou forth, and follow me:
> And make all Spirits subject unto Me,
> So that every Spirit of the Firmament, and of the Ether:
> Upon the Earth and under the Earth:
> On dry Land or in the Water:
> Of whirling Air or of rushing Fire:
> And every Spell and Scourge of God,
> May be obedient unto me!
> IAO: SABAO:
> Such are the Words![30]

No discussion of Western magical practices of rain magic would be complete without mention of the Sixth Pentacle of the Moon from the *Greater Key of Solomon*. The *Greater Key* is a grimoire dating from Latin versions from the fourteenth or fifteenth century, and it involves complicated conjurations for the production of "pentacles" or magical diagrams related to the powers of the seven traditional planets, similar in a way to the astrological talismans found in texts like the *Picatrix*. Unlike the *Picatrix*, however, the astrological elections are much more basic, involving mostly considerations of the Moon phase and sign, as well

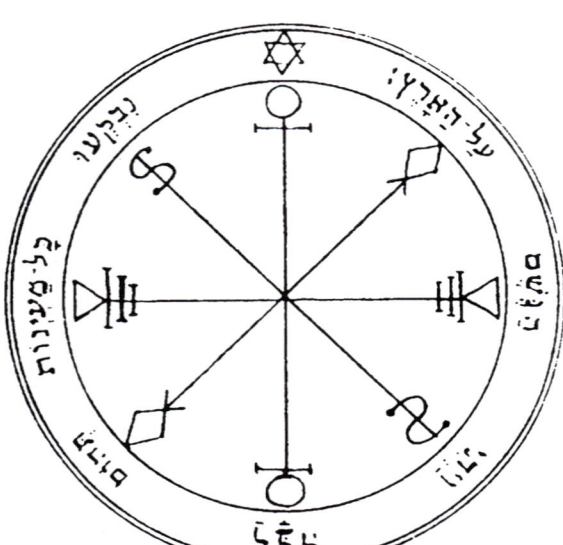

✦ *The Sixth Pentacle of the Moon. From S. L. Mathers (trans.),* The Greater Key of Solomon. *The DeLaurence Company, 1914, p. 81.*

as planetary day and hour. The *Greater Key* derives most of its power from the strict conditions observed in making the pentacles and the ritual conjurations of the planetary spirits that are needed to consecrate them. Here is an example of rain magic from the *Greater Key*: "The sixth and last pentacle of the Moon—This is wonderfully good and serveth excellently to excite and cause, heavy rains, if it be engraved on a plate of silver; and if it be placed underwater, as long as it remaineth there, there will be rain. It should be engraved, drawn or written in the day and hour of the Moon."[31]

I have made a paper Sixth Pentacle, and I make sure I keep it on my altar when I do rain work, but I have yet to try the silver plate and full Solomonic consecration. I find the idea of those elaborate rituals very appealing aesthetically, but in practice, the time, equipment, and a dedicated space involved would be difficult to work into my lifestyle and the constraints of the farming life.

The practice I have found to be effective in bringing much-needed rain on several occasions involves a two-step process of a ritual calling upon the angels and elementals governing the element of air, performed when the Moon is preferably waxing in an air sign (Gemini, Libra, or Aquarius), followed by petitioning the angels and elementals governing the element of water while the Moon is waxing in a water sign (Pisces or Cancer, but not Scorpio because the Moon is debilitated in Scorpio). In practice, I have often had to perform the ritual under less ideal lunar circumstances, but it has still been effective. The practice is a simplified form of ritual magic, and, as with the conjurations discussed above, it uses the essentially religious techniques of prayers and appeals to divine powers to call upon the angelic powers that control the operations of the elements in the sublunary sphere. This tradition has a very different take on the metaphysics underlying the natural world than that discussed above. In the view of the Hermetic Kabbalistic tradition, angels administer the cosmos at every level, acting out the will of God in the material world, which is the government of fate. The angels can be commanded in the name of God, circumventing the normal chain of command. This is a form of conjuration too but one that involves only "angelic" spirits rather than the demonic ones from the example above. This type of angelic magic was still forbidden in the religious climate of the Middle Ages, but it became more acceptable during the Renaissance and early modern periods.

I came across the ritual advertised on the blog of an astrologer and magician, Alex Scharaoschi,[32] and I was intrigued. I bookmarked it for later and remembered it during a prolonged dry spell last June. He charged a nominal fee for the text of the ritual and I figured it would be worth it for the peace of mind. I tried it, and the day after I had performed both rituals we got seven-tenths of an inch of rain and another inch a few days later. The summer continued to be a dry one, and each time I performed the ritual, within a few days there was rain. We again performed an emergency rain ritual on November 9, 2022, when extremely dry fall weather fueled an outbreak of fires in the forests of eastern Kentucky. The ritual was performed in tandem with some other magical friends, whose home was threatened by the fires. The following morning dawned to a steady rain which ended the fires and the drought. The consistent occurrence of favorable synchronicities constitutes the most reliable indicator of success in magical endeavors.

I have found it helpful to perform different ritual practices and prayers, before I get to the rain working proper, in order to get myself into the light trance state necessary to contact the angelic and elemental powers. Ritual framing, which serves to situate one in the magical cosmos, is helpful not only for getting the practitioner in the proper state of mind but also helps to create an energetic environment conducive to the spiritual forces with which the magician works, which are extensions of the mind of the Creator. The mind of the magician must reach outward and upward toward the world soul, by utilizing their inherent sympathy. What differentiates what we are doing from ordinary petitionary prayer is this merging of minds through the religious technology of ritual. The proper performance of ritual creates an altered state of consciousness, in which the mind of the operator expands—ascending, as it were—to the greater mental network and its intelligences. In my experience, to the extent petitionary prayer leaves the consciousness of the petitioner very much untouched, it is ineffective. Ritual magic engages the will as well as the imagination and leaves one feeling like they have done something, and this feeling is the key to its effectiveness in some way. The power to change reality still comes from the divine, however one conceives it, but the magician somehow participates in the flow of events, rather than remaining passive.

7

REIKI AND ENERGY TRANSFER

Applying the Eye of the Master to Promote the Well-Being of Animals and Plants

There is an old saying, "The eye of the master fattens cattle," which always meant to me that a farmer who pays close attention to his work is more successful. The saying's roots are more literal than that. The ancient Romans believed that the presence of the landowner on his farm had a beneficial effect on the whole farm and that his eyes projected a fertilizing virtue on the crops and animals, causing them to grow abundantly in response to his gaze. Like the evil eye, which was believed to wither and harm living things, the master's gaze was believed to be a projection of a force or substance that caused things to grow and prosper. The Roman agronomist Lucius Junius Moderatus Columella said of the visits of the landowner to his estate, "For the head of the household comes down the more willingly to feast his eyes upon his wealth in proportion to its splendour; and, as the poet says of the sacred deity, wheresoever the god has turned his goodly head, truly, wherever the person and eyes of the master are frequent visitors, there the fruit abounds in greater measure."[1] Scholar Britta K. Ager in her dissertation entitled "Roman Agricultural Magic" says, "Columella describes the eyes of the owner as something almost separate from the man himself. The master's gaze, as it sweeps across the plants, impels them to flourish and grow more vigorously."[2]

Perceptual and Morphic Fields

The idea that the eyes can project energies is, incidentally, common to all cultures—except materialist scientific ones. This idea can be verified easily merely by focusing on the backs of people's heads in a crowd or when passing them in a vehicle. Most people will turn around and make eye contact. And they will probably be wondering why you are staring at them. Visionary biologist Rubert Sheldrake calls this "the sense of being stared at."

Sheldrake mentions both the positive and negative powers associated with the human gaze. He proposes the theory that perception involves "perceptual fields extending out beyond the brain, connecting the seeing animal with that which is seen." He says that, "Minds . . . extend beyond brains through fields,"[3] comparing the perceptual field to well-known phenomena like magnetism.

Sheldrake has proposed elsewhere the idea of what he calls morphic fields, which guide the organization of developing organisms and regulate the activity of the nervous system and are comparable to the concept of the spirit body or vital soul. He says that the sense of being stared at is caused by the interaction of one person's perceptual field with that of another person. He states, "Some people may prefer not to use the word field, but instead talk in terms of vibrations, energy flow, chi, or nonlocal quantum effects that link observers to what they observe. Whatever theory is preferred, the sense of being stared at must depend on an influence at a distance of the looker on the person looked at, a projection of influences outward. This sense reveals that through the power of attention the mind is connected to the world beyond the body."[4]

It seems I have been unwittingly practicing ancient Roman agricultural magic my whole career by walking around the farm, evening being my favorite time for such walks, and admiring the beauty of the garden and its abundance, as well as assessing where work needs to be done. This is the instinctive magic of the relationship of mutuality that we humans can have with the plant kingdom. Most people have heard of the idea that plants like to be talked to or that they are in some way responsive to our thoughts and intentions. These ideas were popularized in the 1973 book *The Secret Life of Plants* by Peter Tompkins and Christopher Bird,

◆ *The author working with a scythe to gather fresh fodder for his cattle. Photo by Esmee McKee.*

and in the documentary of the same name.[5] We will return to this topic later in the chapter. Suffice it to say that talking to and singing to plants is just something that seems natural to me, as it does to many other "plant people."

PRACTICING THE ART OF ENERGY HEALING

In recent years, however, I have endeavored to raise this practice beyond the instinctual by deliberately trying to "send energy" to the plants via an energy healing modality originally from Japan known as Reiki. Reiki derives from the two Japanese words *rei* and *ki*. Rei literally means "universal" but has esoteric connotations and is understood as "all pervasive" in a spiritual sense. Ki is the universal life force that is known in Chinese as qi and is known to esoteric systems all over the world. So Reiki therefore means spiritually directed life energy.* In many ways Reiki resembles medical qigong, which was undoubtedly an influence on the founders of Reiki, including Mikao Usui, who discovered Reiki during a mystical experience as a student of Zen Buddhism in 1922.[6]

---

*There is a specific practice of projecting Reiki through the eyes, known as Gyoshi-Ho, which involves staring unfocused at a part of a client's body that needs healing and intending that healing energy flow there.

A Universal Life Force

The idea of a life force that flows through nature and enlivens all living things is common to all cultures. It was known to the Stoic school of philosophy in ancient Greece as the *pneuma*, "an all-pervasive force which . . . holds all things together—from the universe as a totality to an individual body within it."[7] The various philosophical schools of India refer to this force as *prana*, the Romans called it *spiritus*, and in Hebrew it is known as *ruach*. All of these are equated with wind and the breath of life. Indeed, the Taoist sage Zhuangzhe said, "The universe exhales its qi and it is named wind."[8] In North America, the Lakota people call the force which gives life to everything that lives *wóniya*.[9] A bit of research can furnish endless examples of that force that we are working with when we practice Reiki.

What makes Reiki different from other energy healing practices is that the ki in Reiki is understood to be directed by a divine or spiritual intelligence. The healer doesn't utilize their own life force to affect the healing, as in some other systems, but acts as a channel for the Reiki energy, such that we merely have to place our hands over the client to heal. The healer visualizes celestial light from the Creator entering into the crown chakra and flowing out from their hands into the body of the recipient. The healing force is then directed to where it is needed by the Reiki, which is seen as an intelligent spiritual force. I think of it in Western terms as an angel or a divine being—and this is how it came to me during my Master initiation—as Christ consciousness.

I got into Reiki when I heard an interview with author Peter Mark Adams about his book *The Power of the Healing Field*, in which he discusses Reiki and other healing practices. At the time, my father was terminally ill and in the last stages of kidney disease, although we still thought that dialysis and proper diet might give him a few years. I was very intrigued by the idea that I might be able to help him, as well as other members of my family and any friends that might need it, so I bought the book and signed up for a level one Reiki class. The experience of the previous two years—the pandemic years—had left me feeling a definite need to pick up my interest in healing again, after having paused my study of herbalism twenty years earlier to be a full-time farmer.

Illness in the Home

During the COVID pandemic, my family and actually myself in particular had gotten the disease before it was even officially in our state, in February of 2020. I thought it was the flu at first, but the infection settled deep in my lungs, and after six days I just wasn't recovering, which is very unusual for me. I decided to go to the emergency room, thinking I might need some antibiotics and that I would be sent home. I was surprised to learn that I had double pneumonia and was actually quite ill and had practically no white blood cells. I stayed in the hospital for six days, praying and meditating, but I was actually quite scared because I had never been seriously ill or hospitalized in my life. Because it was so early, I was not diagnosed with COVID, but my doctor compared my sickness with the virus that was spreading across China. I was released and recovered steadily at home, and as the panic of the pandemic erupted across the nation and the world, my family took it all very seriously. We quarantined at home and wore masks religiously for several years, and, because of my experience, I got all of the recommended vaccinations.

The fates had more tricks in store for us that year, because in June of 2020, my wife, Esmee, became seriously ill with an unknown illness. She had a low-grade fever, aches and pains, a strange rash over most of her body, and no appetite. We took her to the ER and they released her with some antibiotics. She got steadily worse. We took her back to the ER a few days later and they admitted her and began treating her for a strep infection. I suspected that it was a tick-borne illness and so did her doctor, and he ordered a test. Unfortunately, there is only one lab in my state that does the analysis on blood tests for tick-borne diseases and they had a waiting time of five days. Meanwhile, her condition was deteriorating. We were all terrified. My kids hadn't ever been away from their mom, it was the height of our growing season, and our garlic crop needed to be harvested.

She was transferred to a larger hospital, where the doctors took seriously the differential diagnosis of Rocky Mountain spotted fever and put her on the proper antibiotics. The test results confirmed the diagnosis of Rocky Mountain spotted fever, a rare disease carried by ticks and transmitted by a bite. Placed on the correct antibiotic, she began to steadily improve and was allowed to return home to recover after six days. During the course of a three-week illness, Esmee lost thirty pounds and had to walk with the aid of a walker. She began a long recovery and by the fall had regained her strength, but neither Esmee, nor I, nor my two little boys will ever forget that year, when illness came to our home as it did to the

◆ *"Angelic" cloud formation photographed over the author's farm.*

homes of so many. Many weren't as fortunate as we were to fully recover. We feel truly blessed to have come through the whole experience relatively unscathed.

Since then, it has become an interest of mine to figure out ways of navigating this life and the vagaries of fate that involve harnessing the hidden forces and energies that underlie the apparent solidity of our everyday world, including magic, astrology, and energy healing. Healing practices in particular seem to be an urgent concern, especially in light of my personal history and the collective trauma that we all went through in the pandemic years.

Angel Magic

When I began to study Reiki, I had already been practicing energy healing, using what I will call "angel magic," trying to call upon angelic healing powers to help my mom and dad when they had taken bad turns, as well as a childhood friend who had brain cancer. I was doing a series of eleven-day rituals in which I would pray for the three of them with the aid of the angel Mahasiah, whose specialty is healing. I was using a ritual from the pop magic book called *The 72 Angels of*

Magick, by Damon Brand,[10] as a template. I have since learned that these books, known as the Gallery of Magic series, have a very bad reputation for being simplistic among "serious" magic practitioners, to whom I would say, "You get what you put in." I agree that they can oversimplify the subject matter, they don't cite their sources, and they make extravagant claims for their methods. But the fact remains: I have had an impressive degree of success with some of Brand's books.

His book gives a shortcut to calling upon the seventy-two angels associated with the Hebrew phrase Shem Ha-Mephorash (which means "the ineffable name," actually a misnomer because that word refers properly to the Tetragrammaton, which we discussed in the last chapter, but that is what tradition calls these angelic beings). They are more properly called the angels of the Shem Vayisa Vayavo Vayet, which are the first three words in Hebrew of three verses from the Book of Exodus, chapter 14:19–21, each of which in Hebrew has seventy-two letters. There is a tradition in the Kabbalist school of Jewish mysticism that the entire Torah, or the first five books of the Old Testament (the whole of Jewish scripture), is one giant name of God, and that the individual verses and words as discussed in the last chapter are seen as spirit intelligences or angelic beings, encoded within the sacred text.[11] So in short, the seventy-two angels are seventy-two three-letter names encoded within the text and given by tradition certain "offices" or powers, one of which, Mahasiah, derived from the three letters MAH, is especially potent for healing.[12] The "Shem" angels have astrological significations, being associated with seventy-two five-degree segments of the zodiac, during which they are particularly potent. This system I was using did not reference any kind of astrological timing and still appears to be remarkably effective. I recently read over the twenty-nine angelic petitions I have made to various Shem angels over the last several years, and I found that most of my goals or targets have been realized—no small accomplishment for a system of practical magic.

Braucherei Healing

While I was calling upon the angel Mahasiah, I would picture myself (as Mahasiah) present with the targets and projecting light from myself to them. My parents and friend all improved (coincidence?), but the work took a great deal of concentration and left me tired and depressed, especially when I was doing three of these operations back-to-back. A friend told me that the healer is not supposed to use their own energy, but to use divine energy. Folk healers of all traditions say that it is the power of God that does the healing, but nevertheless,

✦ *Renaissance Kabbalist Athanasius Kircher's diagram, from* Oedipus Aegyptiacus,
c. 1653 CE, showing the seventy-two names of God. Image by Profesorb.

many traditions describe some transference of the disease from the client to the healer. I had come across this idea while researching the American Pennsylvania German Christian folk healing and magical tradition known as Braucherei or Powwow (yes, that's what it's actually called by the practitioners).

In his excellent handbook on the subject of Powwow traditions, *The Red Church*, author Christopher Bilardi says, "Many, many Pow-Wowers over the years have emphasized the demanding and sometimes stressful nature of Braucherei—for the Braucher. This is due to the above-mentioned 'energy' coming off of the patient and transferring (theoretically at least) into the healer. God then removes that 'disease' energy from the healer." Many Brauchers, as the practitioners are known in the Pennsylvania German dialect, have various decontamination rituals involving projecting the negative energy at a "contagion target," as Bilardi puts it, such as a tree, a stone, an animal, or even a mountain. Bilardi's teacher "decontaminates herself by way of shaking and wiping down her arms and hands—as if shaking off contagious water or dirt. However, despite this process of decontamination there are still some side effects. Fatigue can be one of these; or it is also common for Brauchers to feel 'phantom' pains of the illness that they removed from the patient a little while after the treatment. *But, to be frank, some Brauchers do get genuinely ill after treating people* [italics original]."[13]

Protecting the Healer Doing Energy Work

According to medical qigong practitioner Jerry Alan Johnson, if the healer is in a weakened state, it is possible for them to absorb disease from the patient, causing "turbid qi." There are various preventative exercises that must be done to protect the healer from absorbing negative energy from the patient. Johnson says that once the healer has absorbed toxic qi from the patient, they must turn the turbid qi back to clear qi by using "divine healing energy."[14]

In the Reiki tradition, disease is said to be caused by *byoki* or unhealthy ki,[15] which is similar if not identical to the idea of "turbid qi" in the qigong system. The healer sends Reiki, or divinely directed ki, into the affected area, removing blockages and allowing healthy ki to flow freely into the aura or energy body. (This concept, that of the energy body, is also attested to worldwide and is even mentioned in the New Testament. We will get back to it later.) Unlike Braucherei, Reiki practice is not believed to be draining to the healer. On the contrary, being a channel for Reiki is supposed to be healing to the practitioner, and that is my experience. Sometimes, when I have a backache or some other minor pain, I will actually perform Reiki on someone and I will find that both of us feel better afterward. Or, alternately, I will walk around the garden in the evening and project Reiki on the crops and feel my mood lift and the aches and pain fade away. All the same, however, the Reiki system does recommend a practice known as *kenyoku*, or dry bathing, which very much resembles the practice of Bilardi's teacher wherein the healer brushes off their aura after a session.

I came to Reiki as part of my interest in cunning craft, or being a spiritual worker for my family, friends, and community. It is, as I've said, an initiatic tradition, but unlike some, membership is open to all. I am interested in Braucherei as well and am actually partly Pennsylvania German, on my father's side. A distant relative of mine was included in the appendix to Bilardi's book, *The Red Church*. Braucherei is a closed practice, however, and you can only learn the practice from a living teacher of the opposite sex. I see Braucherei as very similar to Reiki, because it involves a practice of energy healing whereby the healer makes energy passes while praying, reciting charms, moving their hands over the "patient," and directing the divine energy.

Using *Historiolae* for Healing

There is an allied practice in the folk magical traditions of the world of using charms or *historiolae*, little rhymes that are believed to have a curative virtue when repeated by the healer during the healing. There is a grimoire associated with the Braucherei tradition, called *The Long Lost Friend*, which is a combination of recipes for herbal home remedies and charms that are to be recited to stop bleeding or cure warts or rashes. This type of medical magic is ancient and known from all over the world. I think it is tied to energy healing and perhaps serves to induce a calm suggestible state in the patient or perhaps a trance in the healer by occupying the conscious mind with the odd imagery and the sacred names. Here is an example of a charm to stop bleeding. I have recited this one while applying pressure to my son's nose when he had a nosebleed (my kids are used to my weirdness!):

> I walk through a green forest;
> There I find three wells, cool and cold;
> The first is called courage,
> The second is called good,
> And the third is called, stop the blood.
>
> + + +

The + symbols indicate that the healer is to make the sign of the cross three times over the bleeding wound.[16]

USING REIKI ON THE FARM

In this book, I have mostly written about "Western" esoteric traditions. But the Western esoteric system has a broken chain of transmission due to loss of the mystery schools and centuries of persecution by the hostile forces of the exoteric versions of the Christian faith. This leads to gaps in our system that need to be filled to be able to resurrect the mystery tradition and create a viable whole. Energy healing is one place especially where the West has lost its tradition (one exception being Braucherei; I'm sure there are others) and generally embraced the materialistic model of medicine that has prevailed since the nineteenth century. That system has brought us many advances that I would not want to do without. But I am not talking about giving up Western scientific medicine and all of its lifesaving advances. I am talking about taking back some power and agency for our health by using time-honored and time-tested techniques from around the world. So if you desire to be a community magic practitioner and healer in the twenty-first century, or even just a cunning farmer, you would do well to avail yourself of the best techniques that are on offer. Sensei Usui generously bestowed upon us in the West the miraculous healing practice known as Reiki as the selfless act of the loving-kindness of a bodhisattva.

I have used Reiki extensively with my family. For ordinary aches and pains, it is outstanding. It has reduced my personal use of ibuprofen, which I use to dull the aches and pains incidental to a life of farm labor, to practically none. A few moments with Reiki hands on a headache or a backache usually clears it up. I have sent distance Reiki to friends and family and those who have reported back report improvement, some even miraculous. I used my budding Reiki skills on my father as he was dying from kidney failure. I wasn't able to save my dad's life, but I was able to help him open up emotionally. That man, who his whole life would stiffen up when you hugged him and didn't like to be touched, allowed me to lay my hands on his withered body and he got so relaxed that he just fell asleep. He passed away several months later peacefully in his sleep after having said all of his goodbyes. He even made a joke in one of the last things he said to me. I give Reiki much of the credit for his peaceful death.

Reiki for Plants

I have used Reiki techniques extensively on the crops, both while they are in the fields and on seedlings in the greenhouse. While not a controlled experiment, I noticed increased yields, even during drought conditions, and decreased pest problems, such as not having to use my organic sprays for cabbage worms until June, which is unheard of. It is an enjoyable practice to walk up and down the rows of healthy growing crops, broadcasting Reiki from my hands while chanting a mantra.

Plants are spiritual beings without a doubt. They have been experimentally shown to respond to the thoughts and feelings of those who tend them. Peter Tompkins and Christopher Bird's book *The Secret Life of Plants* details several remarkable experiments in which plants were shown (using sensitive electronic equipment) to respond to the positive and negative intentions of their cultivators, flourishing with praise and withering with scorn.[17] So when we go out on the land, in among the plants we tend, and try to project love and positivity, they respond. One of my favorite things to do has been to sing to my plants or hum to them during times when they are stressed. My friend Salvador, whom I've mentioned several times, would call out to the plants, saying *"Animo, matitas!"* (Courage, little plants!) during a drought. We once had a patch of zucchini and assorted other squash that went for a month with no rain, yet continually produced baskets of squash, which we picked every day, praising the plants and offering them encouragement the whole time.

Salvador told me that his father taught him to be very "sentimental" with the plants. He told me a story about when he and his father were weeding the corn patch and it started to get dark, so it was almost time to quit for the day. Earlier in the day his father had told him that the corn plants were frightened of the weeds and that it was their job to help them. Well, at quitting time, little Salvador began to cry. His father asked him why he was crying, and he replied that he didn't want to leave the plants because they were frightened. His father told him not to worry, because they would come right back in the morning to help them first thing.

In Salvador's garden, many plants would volunteer from seed dropped the previous year, especially tomatillos, whose husk-covered tomato-like fruits are used to make the delicious *salsa verde* and other mouthwatering salsas used in Mexican cuisine. Every season his little garden would be absolutely overrun with the weedy tomatillo vines. He said it pained him to weed them out because, upon going to pull one out, it was as if he could hear them say, "Don't pull me out, look what I'm going to give you!" And seeing the little tomatillos forming on the vines, he wouldn't be able to pull them up.

That level of sensitivity toward our little green brothers and sisters is definitely noticed and rewarded by them. But unfortunately, it is not a practice encouraged in today's agricultural scene, where high yields, massive machinery, toxic chemicals, and biological experimentation with genetic modification is the norm, rather than the slow and patient work of listening to the voice of the plant spirits and responding with love and gentleness.

I hope I've made a case for the plausibility of energy healing techniques for not only human health but also for agricultural application, as well as for promoting the well-being of plants and contributing to the production of healthy foods, which are a fundamental part of maintaining human health. We have a relationship of mutuality with the natural world that, when honored and cultivated, leads to greater health and well-being, but when ignored leads to disease and disharmony and many of the problems we see in the culture around us. The only way to really get an understanding of these energetic practices is to dive in and try it yourself. A Reiki level one class is a good place to start, as is simply working with intent on the green things in your care.

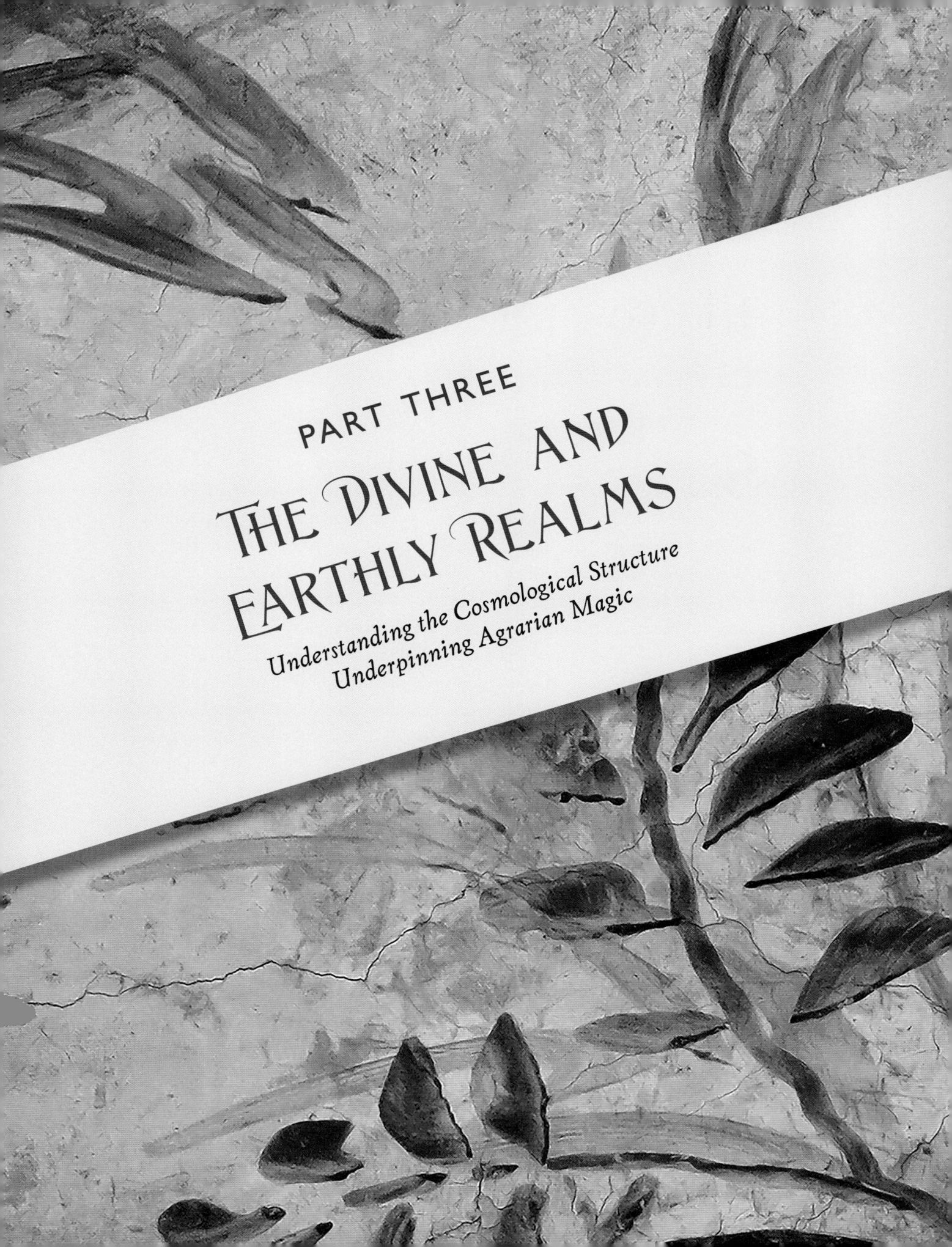

PART THREE

THE DIVINE AND EARTHLY REALMS

Understanding the Cosmological Structure
Underpinning Agrarian Magic

8

THE COSMOLOGY OF RITUAL

Shamanic cultures, including many of the Indigenous peoples of the world, have understood the cosmos to have a sort of tiered structure, usually composed of three main layers: the celestial world, also known as the upper world or the world of the divine beings; the terrestrial world, or "this" world of humans and animals; and the infernal world, also known as the underworld, the realm of the dead and other spirits of the underworld. The medieval/Renaissance cosmology of Henry Cornelius Agrippa also has a three-tiered structure, but in this case we have the supercelestial realm of God and the supercelestial angels, beyond the stars, transcending the physical universe; the celestial realm of the stars and the celestial angels and diamones; and the sublunar realm of the elements and the Earth, as well as the underworld, which is not a separate realm, but situated inside the Earth.

There are many variations on this theme in global mythology. A common thread is that the various worlds are connected by a "world tree" or cosmic axis, also known as the axis mundi, that runs through the center of all the worlds. As discussed in part one, the shaman or magician accesses these imaginal worlds by ascending or descending the axis mundi in visionary trance states and rituals. This enables the shaman to achieve contact with the gods and other divine beings in the upper world, the ancestors and other underworld powers in the lower realm, and the spirits of plants and animals in the archetypal version of the middle realm. As Mircea Eliade put it in his monumental work *Shamanism: Archaic Techniques of Ecstasy*, "The pre-eminently shamanic technique is the passage from one cosmic region to another—from earth to the sky or from the earth to the underworld. The shaman knows the mystery of the break-through in plane. This communication is made possible by the very structure of the

♦ Yggdrasil, *the world tree and cosmic axis of Old Norse cosmology. Illustration by Oluf Bagge in* Northern Antiquities, *1847.*

universe." Movement along the axis is possible, as Eliade states, because the axis "passes through an 'opening,' a 'hole'; and it is through this hole that the gods descend to earth and the dead to the subterranean regions; it is through this same hole that the soul of the shaman in ecstasy can fly up or down in the course of his celestial or infernal journeys."[1]

THE SACRED CENTER

In the sacred cosmology of the shaman, the place where the breakthrough between worlds was possible was the center. According to Eliade, "In the beginning, 'Center,' or the site of a possible break-through in plane, was applied to any sacred space, that is, any space that had been the scene of a hierophany and so manifested realities . . . that were not of our world, that came from elsewhere and primarily from the sky. The idea of a 'Center' followed from the experience of a sacred space, impregnated by a transhuman presence: at this particular point something from above (or from below) had manifested itself."[2] As the shamanic tradition developed into the magical and religious traditions of the world, the idea of a particular place where a hierophany (a revelation of the sacred) had occurred being a sacred center continued throughout history. One need only think of all the sacred sites of Jerusalem, for instance, as examples.

A parallel tradition, however, is the idea that the cosmic center is not localized in any particular place but that it can be invoked, that a space can be consecrated as a center, either more permanently as with a church or a temple, or temporarily as with the magic circles of shamans and magicians. The sacred space so consecrated is to be a region where the breakthrough between the worlds is made possible, a liminal space, a boundary between the worlds where the cosmic center is made manifest—an opening or hole between the worlds, in other words. The consecration of a temple, a church, or even a magic circle recapitulates the creation of the cosmos in miniature. Once the center is established, the work of traversing the worlds can begin.

The center, in addition to being situated between the upper and lower worlds, is also situated in the center point, where the rays of the four cardinal directions meet. It is thus the mean between the extremes. There is a rich symbolism in the traditions of the world regarding the directions and the cosmic and elemental powers linked to them. Wherever you are is the center in a sense, and by ritually situating oneself in the center, between the cardinal directions, beneath the upper world and above the lower world, one becomes the cosmic axis and embodies the ability to move between worlds. Traditions around the world have myriad ways of expressing the four-directional and multitiered cosmos in ritual as well as in sacred buildings.

My focus is not to attempt to appropriate modes of ritual from living cultures I have no ownership in but to dig into the shamanic roots of my own Western European, Judeo-Christian (as well as Islamic, by way of medieval magic), and classical Greek and Latin culture that preceded and continues to inform it, and peasant folk practice from the (also Western European) lands from which my ancestors came. So to that end, let us endeavor to begin to map the magical cosmos in which we operate. Let us also remember, as has been stated, that the Western tradition is a broken one, lacking in some major areas of esoteric practice due to a history of persecution by ecclesiastical authorities. So, therefore, a judicious and respectful borrowing of some practices from other traditions is occasionally warranted and necessary.

A FOURFOLD MAGICAL COSMOLOGY

Agrippa says that there is "a threefold world, elementary, celestial, and intellectual,"[3] by which he means, as mentioned, the physical world of the elements; the celestial world of the stars, planets, and signs; and the intellectual world, beyond space and time, of the angels and the creator. Agrippa announces his intention, as the thesis of the *Three Books of Occult Philosophy*, to present a system of magic appropriate to each of the worlds. It is one of the goals of this book to provide an introduction to all of these systems, and to start out I would like to present this helpful concept of a type of magic appropriate to each of the three worlds.

I would also like to postulate, in order to complete our sacred cosmology, a fourth world, by reintroducing the concept from shamanism but neglected by Agrippa, of the lower world, the land of the dead, of faeries and other chthonic spirits who dwell "within" the earth. Agrippa, who died while fleeing the Inquisition under a sentence of death as a heretic, had nevertheless attempted to avoid arousing the suspicion of the authorities by writing openly of such matters.

So we have, then, a fourfold world composed of the chthonic realm, the elementary or physical realm, the celestial realm, and the supercelestial realm. And we have four systems of magic, with a certain amount of overlap between the four: *goetia* and *necromancy* for working with chthonic spirits and spirits of the dead; natural or elemental magic for working with the forces of the world of nature and its spirits; astral or astrological magic for working with celestial spirits and lower angels; and theurgy for calling upon the aid of the divine, the hypercosmic

♦ *Byrhtferð's cosmological diagram, showing the four directions, the four elements, the four winds, the four seasons, and the four ages of man within the circle of the zodiac. From an English medieval manuscript from the late tenth century.*

✦ *A witch inscribes a magic circle in the dirt. Painting by John Williams Waterhouse, 1886.*

gods, and the archangels. Magicians situate themselves, like shamans, on the axis mundi, able to move between the worlds and to work with the various denizens of each. To access the center for magical work, we have to situate ourselves in it imaginally by means of ritual and vision.

In addition to the four worlds, we also have four cardinal directions: north, south, east, and west. The four directions are reflected in the four worlds as well, meaning that each level has its own four directions and its own spirits presiding over them. The four directions are linked to four elements, the four winds, the four archangels, the four Evangelists, the four living creatures, the four suits of the tarot, the four cardinal virtues, and the four humors. The list of quaternities is a long one.[4]

We can envision each plane as a circle, or more properly a sphere, or maybe a series of nested spheres. We can picture an axis running through the center of them all with which we can ascend or descend as we wish. The underworld is an extradimensional interior, lying within the terrestrial, but vast, not bounded by the dimensionality of the terrestrial sphere. The celestial sphere is unbounded and perhaps infinite in extension in space, comprising the physical cosmos and also its imaginal counterpart, the spiritual world of the planets and stars. The supercelestial realm is transcendent, existing outside of time and space, as the eternally present creation, the source that moment to moment emanates its being and light, holding us and all that exists in being. Imagine if a deep space telescope could look back fourteen billion years to be able to see the Big Bang; beyond that is the Source, yet eternally present and immanent in the heart of all existence.

USING RITUAL TO TRAVERSE THE REALMS

One of our vehicles to travel to these realms and our portal to allow communication with the denizens of the imaginal realms is ritual. Through the practice of ritual we physically bring the imaginal world into being in the physical. So when we want to talk to the gods and spirits, we set ourselves apart from the mundane by announcing our intention and saying, as they did in the Eleusinian Mysteries of ancient Greece, *"Hekas, hekas, este bebeloi,"* or "Begone, begone, ye profane." We are marking off our space and this time as sacred, as special and set apart. We create a boundary with our intentions, our actions, and principally our imaginations. We then seek to create, with the psychic, imaginal energy of will and intention, an image of the cosmos, a mandala in which to work. Many practitioners, as discussed elsewhere, actually make a circle on the ground with chalk or paint. This circle, commonly thought to be for protection from hostile spirits, is also a model of the cosmos in miniature and delineates the space where the ritual is to be performed.

Ritual of all types is a species of make-believe, of playacting for grown-ups. It is like drama (which indeed is a cognate of the Greek word *dromenon*, meaning "things done," which referred to ritual actions in the Eleusinian Mysteries),

✦ *John Dee and Edward Kelley summon spirits of the dead in a churchyard. Illustration by Ebenezer Sibly, c. 1790.*

where the practitioner assumes a role, that of priest(ess), or magician, or witch, or some other person capable of speaking to the primal and elemental forces of the cosmos, and one assumes (pretends, at first) that they can hear and respond. It is audacious to call out, as did the Hellenistic Egyptian magicians, into the firelit darkness and smoke of smoldering *kyphi* incense, "I summon you Headless One! You who created Earth and Heaven! You who created Night and Day! You who created Light and Darkness; You are Osoronnophris! Whom no man has ever seen!" The ritualist holds much of the drama in their imagination, and the circle, the place of working, the sacred space, like the dramatis personae, exists in a liminal space, physical yet imaginal. We may be standing on the floor in a spare bedroom, a basement, a home office, our friend's living room, a clearing in the forest, or even (as in conventional religious rituals) a church or temple, but when the ritual commences, we are between the worlds!

As Gemma Gary says of the witches' circle in her excellent book *Traditional Witchcraft*:

> The practice of marking out a circular area to delineate a hallowed space for the performing of rites, the working of magic and to contain raised forces is a very ancient one. However the purpose of the true witches' circle, ring, or "Compass Round" runs much deeper than mere delineation and containment. The most important function of the circle is that of access, for it is a place created and set aside for the ingress of virtues, powers, spirits, atavistic wisdom, and the manifestation of divine forces into the Craft of those who work within its boundary. Within the witches' circle may be found a map to the worlds that are drawn upon or traversed. The spirits, powers, and virtues of the crossroads are conjured into the circle's midst, through which runs the great axial road or "world tree" conjoining the depths, the quarter ways, and the heights. Within such a circle are the paths of access opened to the cross-quarter ways, the planetary, solar, and lunar forces and virtues of the starry height of Nevek, and the chthonic waters of creation, death, atavistic memory and wisdom within the underworld realm of Annown. Via the axial road is the chthonic fire; the serpent of the land, drawn forth from the depths to the heights.[5]

The cunning farmer may not actually have a designated space in which to work and may have to make do with the farm kitchen or hearth, or in my case, my study. So painting the floor with arcane symbols may not be possible. Delineating the space with symbols projected in the imagination of the operator on the floor or in the air, at the points of the cardinal directions for example, may be all that is possible.

In the highly informative appendix of *The Sworn and Secret Grimoire*, the recently deceased mage and scholar Jake Stratton-Kent, *requiescat in pace*, in discussing the stages and elements common to magical ritual, says that the purpose of a banishing ritual is "to invoke protective powers to guard and protect a ritual space or Circle, or indeed the instruments or operator. In the context of the pentagram [an ancient magical symbol, the familiar five pointed star drawn with lines, symbol of the microcosm, and the human being, among other things], names of God and angels are invoked, although the format has precedents involving pagan gods associated with the directions."[6] Stratton-Kent says that the drawing of symbols in the air and the use of imaginary circles in lieu of ones drawn on the floor is not without precedent in the magical tradition. As an example of this kind of consecration, here is a ceremony from a sixteenth-century anti-magical polemical work that, ironically, became an essential grimoire for succeeding generations of cunning folk, Reginald Scot's *Discoverie of Witchcraft*:

Of Magical Circles, and the Reason of their Institution

Magitians, and the more learned sort of conjurors, make use of circles in various manners, and to various intentions. First, when convenience serves not, as to time and place that a real Circle should be delineated, they frame an imaginary Circle, by means of Incantations and Consecrations, without either Knife, Pensil, or Compasses, circumscribing nine foot of ground around them, which they pretend to sanctifie with words and Ceremonies, spattering their Holy Water all about so far as the said Limit extendeth; and with a form of Consecration following, do alter the property of the ground, that from common (as they say) it becomes sanctifi'd, and made fit for Magicall uses.

How to Consecrate an Imaginary Circle

Let the exorcist, being cloathed in a black Garment, reaching to his knee, and under that a robe of fine Linnen that falls unto his ankles, fix himself in the midst of that place where he intends to perform his Conjurations: And throwing his old Shoos about ten yards from the place, let him put on his consecrated shoos of russet Leather with a Cross cut on the top of each shoe. Then with his Magical Wand, which must be a new hazel-stick about two yards of length, he must stretch out his arm to all the four Windes thrice, turning himself round at every Winde, and saying all that while with fervency:

"I who am the servant of the Highest, do by the virtue of his Holy Name Immanuel, sanctfie unto my self the circumference of nine foot round about me, +++.* From the East, *Glaurah*; From the West, *Garron*; From the North, *Carbon*; From the South, *Berith*; which ground I take for my proper defence from all malignant spirits, that they may have no power over my soul of body, nor come beyond these Limitations, but answer truly being summoned, without daring to transgress their bounds: *Worrah. Worrah harcot. Gambalon.* +++

Which Ceremonies being performed, the place so sanctified is eqivalent to any real Circle whatsoever."[7]

---

*The plus sign indicates making the sign of the cross in the air, and while the original text only has three, it would make sense to me to make four, one at each cardinal direction.

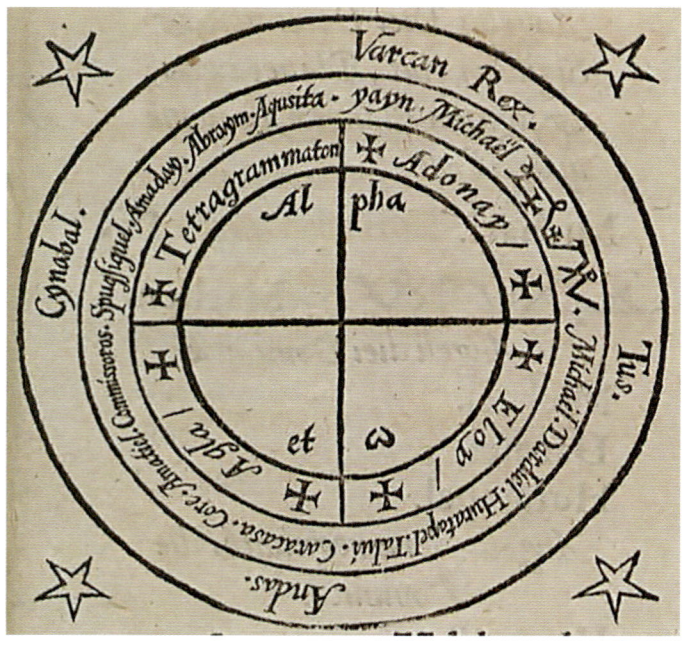

◆ *Magic circle from Pseudo–Peter de Abano's 1565 grimoire,* Heptameron.

Purification in Magical Ritual

Many magical practitioners, following ancient custom, commence every ceremony with rituals of purification and banishing. As Aleister Crowley wrote, "The Magician must therefore take the utmost care in the matter of purification, firstly, of himself, secondly, of his instruments, thirdly, of his place of working."[8] Certainly in a home or other place that is regularly used for mundane purposes, even in a temple that is used for many different kinds of magical work, this is

♦ *Banishing Earth pentagram.*

good practice. In a natural setting, when one is working with the spiritual powers resident in that place, banishing is probably bad manners and should be replaced with an invocation of the spiritual forces of the directions and elements of the place. But there are several opinions on the matter. "The purification of the operator, however, should not be neglected," writes Crowley.

For basic purification, many people use the Lesser Banishing Ritual of the Pentagram, also known as the LBRP, which is easy to memorize and energetically purifies the operator and the immediate space around them. The ritual calls upon the archangels of the four cardinal directions who rule the four elements, as well as four Hebrew names of God. Crowley says:

> It is usually sufficient to perform a general banishing, and to rely upon the aid of the guardians invoked. Let the banishing be short, but in no wise slurred—for it is useful as it tends to produce the proper attitude in mind for the invocations. "The Banishing Ritual of the Pentagram" is the best to use. Only the four elements are specifically mentioned, but these four elements contain the planets and the signs—the four elements are Tetragrammaton; and Tetragrammaton is the Universe. This special precaution is, however, necessary: make exceedingly sure the ceremony of banishing is effective! Be alert and on your guard! Watch before you pray! The feeling of success in banishing, once acquired, is unmistakable.[9]

Crowley tells us that a successful banishing produces a feeling of cleanliness. Here is my interpretation of the text of the Lesser Banishing Ritual of the Pentagram:

❦ The Lesser Banishing Ritual of the Pentagram (LBRP)

The first step in performing the ritual involves the performance of the Kabbalistic Cross in which the operator situates themselves between heaven and earth and calls down divine influence from heaven. It has the effect of grounding excessive or inharmonious psychophysical energy.

Begin by standing facing east, feet together, arms to the side. Visualize yourself standing on the earth and stretching out toward the heavens, growing taller until your head is among the stars in interstellar space. Visualize a sphere of brilliant white light descending from infinite white light and coming to rest right above your head, at the level of the crown chakra. With your closed right hand reach up to the sphere and vibrate the word. Vibrating sacred words and formulae involves chanting them with a tone and pouring all of the energy of your breath into the words.

Vibrate: "ATOH" (which means "Thine" or "Unto thee").

Draw the white light down in a straight line across the body until you reach the level of your groin, the sacral chakra, and vibrate: "MALKUTH" (which means "the Kingdom").

Visualize a line of white light extending down to the center of the earth.

Bring your closed hand from your groin up to the center of your chest, your heart chakra, drawing the white light up with it. Move your hand out to your right shoulder and vibrate: "VE-GEBURAH" (which means "the Power"), while visualizing the light extending out into infinite space.

Move your hand to your left shoulder, drawing the white light with it and vibrate: "VE-GEDULAH" (which means "the Glory"), while visualizing the light extending into infinite space as before.

Bring both hands to the center of your chest in prayer position and say: "LE OLAM, AMEN" (which means "forever" or "unto the ages"), visualizing the cross of light you have just formed with a rose of light blooming in in the center of your chest at your heart chakra.

Still facing east, with your right hand, form the sword fingers mudra with your index and middle fingers extended and your thumb holding down your ring and little finger. Extend your hands out in front of you and form the shape of the Banishing Earth Pentagram as follows:

Starting at the lower left corner of the five-pointed star, draw a line, visualizing the line in blue flame to the top of the star and then down the other side, up to the left arm, horizontally to the right arm, and down to the left leg, meeting where you started. Make the pentagrams large, about five feet from top to bottom.

Draw in your breath; make the sword fingers with both hands pointed up on either side of your head. Step forward with your left foot and point them at the center of the star. This gesture is called "the Sign of the Enterer." Vibrate: "EE-AH-OH-EH." Some people say "YOD HEH VAV HEH," preferring to spell out the Tetragrammaton. My personal practice is to pronounce it, as discussed in chapter 6.

With your right hand extended in the sword mudra, draw a line in blue light from the center of the pentagram, moving to the southern quarter of your working area, tracing the pentagram as before in blue flame. Make the Sign of the Enterer. Vibrate: "AH DOH NYE."

Draw a line in blue flame with the sword mudra as before, move to the western quarter, and trace the pentagram in blue light. Make the Sign of the Enterer. Vibrate: "EHIEH."

Draw a line in blue flame as before, move to the northern quarter, and trace the pentagram. Make the Sign of the Enterer. Vibrate: "AH GAH LA."

Draw a line from the center of the last pentagram to the center of the first one, moving back to the eastern quarter, completing the circle. Stand in the center of the space, visualizing all of the pentagrams shining in blue light in each of the quarters.

Visualize a towering luminous angelic figure, in yellow, holding a sword standing before you in the east. Say: "Before me RAH PHAY EL," vibrating the angelic name.

Visualize a towering angel of light in blue, holding a golden chalice, standing behind you in the west. Say: "Behind me GAH BRI EL," vibrating the name as before.

Visualize a towering luminous angel in red, in a gleaming bronze breastplate and holding a spear, on your right in the south. Say: "At my right hand MEE KIE EL."

Visualize a towering luminous angel in green holding a lamp in one hand and a shield in the other, standing before an ancient tree. Say: "At my left hand AUR EE EL."

Visualize the angels, the pentagrams, and a brilliant star of light shining far above you. Say: "For about me flame the pentagrams and above me shines the six-rayed star."

Still facing east, repeat the Kabbalistic Cross.

Proceed with the ritual work.[10]

There are many ways in various traditions of dedicating a space to a magical or spiritual work. One of my favorites is the First Degree Opening and Closing Ritual from John Michael Greer's excellent and highly recommended book on the Golden Dawn system of ritual magic, *Circles of Power*. The ritual was originally created by the Hermetic Order of the Golden Dawn and was used as a basic form to open a magical working. In Greer's words, this is a "ritual structure that includes the entire magical image of the universe with all its powers." This approach, he continues, "provides the framework for a wide range of symbolism in ceremonial work in just one ritual. It also, importantly, helps bring the invoked energies through in balance with the wider context of the energies of the universe."

You will need a chalice of pure water (preferably blessed, but it's not strictly necessary), as well as a thurible or a stick of incense.

Instructions, in my words where possible, for performing this simple ritual are as follows:

⚘ Basic Ritual for Opening and Closing

Begin by standing in the eastern quarter of your space, or west of your altar facing east if your altar is in the center of your space.

Proclaim with authority: "Hekas, hekas, este bebeloi," or "Begone, begone all you impure ones," as previously mentioned.

Perform the LBRP as described above.

Return to your place, facing east. Take the chalice of water in your right hand, hold it out before you, and say: "And so therefore first that priest who governeth the works of fire must sprinkle with the lustral waters of the loud resounding sea." Draw an equal-armed cross at the same level, visualizing it in vibrant blue light. Though this is not in the Golden Dawn version of the opening ritual, at this point I often say the versicle, from Psalm 51: "Purge me with hyssop, and I shall be clean; wash me and I shall be whiter than snow." (Actually, I usually say it in Latin: "Asperges me, hyssopo, et mundabor. Lavabis me, et super nivem dealbabor." This gives this lodge magic ritual a bit more of a medieval flavor, as well as reflecting the use of psalms in folk magic.)

Dip your fingers into the chalice and flick water in the direction you are facing three times. Repeat in the south, west, and north. Return to the center, facing east. Say: "I purify with water."

Raise up the incense or thurible. Say: "And when after all of the phantoms have been banished, thou shalt see that holy formless Fire which darts and flashes at the hidden depths of the universe, and hear thou the voice of the fire!" Go to the eastern quarter of the space. Make an equal-armed cross, visualizing it in red with the smoking incense stick or thurible. At this point, I like to add (again it is not part of the original ritual and can be omitted): "O Lord, be my tower of strength against all evil," or in Latin, "Domine Deus, esto mihi *tueris fortitudinis contra omnium malignorum."* As before, repeat in each of the four quarters, saying the versicle if desired.

Return to your place, facing east. Say: "I consecrate with fire." I like to add, "Let my prayer be guided to your sight, O Lord, like incense; the lifting up of our hands like the evening sacrifice," or in Latin, *"Dirigatur Domine Deus, ad te oratio mea, sicut incensum in conspecto tuo, elevatio manuum nostrarum sacri-ficium vespertinum."* As before, the versicle can be omitted.

Return to your place, facing east. Make the Sign of the Enterer. Circumambulate the space clockwise three times, making the Sign of the Enterer each time you pass the eastern quarter.

Visualize yourself ascending as if on a spiral staircase, rising above the level of the crosses and pentagrams. When finished, return to your place in the center, facing east. Make the Sign of the Enterer, and say: "Holy art thou, Lord of the Universe!" Make the Sign of the Enterer again and say: "Holy Art Thou, whom nature has not formed!"

Make the Sign of the Enterer yet again and say: "Holy Art Thou, the Vast and the Mighty One, Lord of the Light and of the Darkness!"

Then make the Sign of Silence by placing the index finger of your right hand on your lips. The space is then open for any work to proceed. When the work is finished, perform the closing as follows:

Repeat the Purification with Water and the Consecration by Fire in each quarter exactly as before.

Beginning in the east, circumambulate the space three times, making the Sign of Silence each time you pass the eastern quarter, visualizing yourself descending, as if on a spiral staircase, back to the level of the pentagrams and crosses. Facing east, extend your arms in the form of a cross, saying: "In the great name of strength through sacrifice YHShVH YHVShH (pronounced "Yeheshua, Yehowashah"), I hereby release any spirit that may have been imprisoned by this ceremony, depart in peace unto your abodes, and let there be peace between us, now and forever." Perform the Lesser Banishing Ritual of the Pentagram. Your circle is now closed.[11]

I highly recommend that the reader consult *Circles of Power* for the text of the original ritual, as well as other useful rituals. The book is invaluable to any-one interested in learning more about ceremonial magic. It covers some of the same territory as Donald Michael Kraig's classic *Modern Magick* but is, in my opinion, more comprehensive.

I also follow up this opening with the Lesser Invoking Ritual of the Hexagram from *Circles of Power*. These two practices are intended to draw in macrocosmic and celestial power to your circle for the accomplishment of whatever practical work you may want to do. Daily practice of these techniques is essential for mastery. The idea is that you prepare yourself for practical work with constant practice, and when you want to, say, invoke Jupiterian spirits because you need an infusion of cash, you have the ritual forms committed to memory and you don't lose focus and ceremonial momentum by fumbling with papers in the dark! (Incidentally, copying rituals and prayers you haven't memorized onto the Notes app of your smartphone and using the Night Shift setting can be really helpful.)

Another strand of witchcraft practice from the British Isles is represented by Gemma Gary's classic book *Traditional Witchcraft*. In it, the author details a ritual known as the Compass Rite, which is earthy and primal, and which involves purifying the space with a broom, conjuring the compass, and calling the spirits of the four directions. The operator next calls the Bucca, the dark, spiritual patron of the Cornish witches, and then offers a rhythmic chant known as "turning the mill," which is said while walking to the beat of the chant around a central fire.[12]

ADAPTING RITUALS TO YOUR ENVIRONMENT

I have often felt that the Golden Dawn–style opening rituals were not exactly at home in natural settings, where the practitioner is engaging directly with nature spirits, elementals, and land beings—and the more celestial and supercelestial types of beings called upon as guardian transmitters of ritual power could be overwhelming to the down-home spirits of my Kentucky forests and fields. It feels almost disrespectful to be bringing in gods and angels from the ancient Near East to banish the local spiritual fauna. Because of this, I am working on a hybrid ritual form that engages the energies of all of the cosmic levels in order to sanctify and invoke power, to protect the space without nuking it astrally. The form combines traditional ceremonial magic symbolism with elements of Trad Craft (Traditional Witchcraft) rituals like the Compass Rite as well as folk magic. It is something that would be at home indoors as well as out.

One modern author, Cat Heath, in her excellent book *Elves, Witches*

& *Gods*, recommends giving a small offering of cream, butter, cornmeal, or tobacco before engaging with magical work on the land to avoid offending the local land spirits.[13] We touched on the practice of offerings at the beginning of this book. I fully agree with this practice, and often carry a small amount of cornmeal or tobacco for exactly this purpose. This is a gesture of respect and friendship, which calls for a very different mindset from that which often accompanies ceremonial banishings, which can be a sort of ceremonial bulldozing of all of the other-than-human folk that occupy the space where a ritual is to be performed. This is not respectful or conducive to a good relationship, and it's likely to offend. The folklore of the fairy faith in many nations gives us many warnings of the dangers that can befall those who offend the invisible inhabitants of the land. Banish and ward your space, by all means, but start out with gifts and diplomacy, especially if working in a natural or unfamiliar setting.

At this point, you have a concept of how to get started in ritual magic and how to consecrate a space and perform a banishing and opening ritual. The next task is to find an appropriate place. Basically, you need a quiet space where you can focus and not be disturbed. As I stated earlier, the cunning farmer is a blue-collar magician who doesn't necessarily have a temple or dedicated space for the performance of their rituals, so a spare room in the house, a barn, a shed, or outdoors in good weather is what we have to work with.

In my spiritual work, sometimes I practice my rituals inside and sometimes outside. I do inside work in a home office, where I have a desk in a corner that I use as an altar. Inside work is done when the weather is bad outside, when I need an elaborate setup for whatever I'm doing, or when I need books. Outside work I either do in my backyard, in a little spot I use as a circle, or on a large flat rock at the edge of a steep hill in an oak grove near my house. I live in a very rural setting, and it is unlikely anyone will disturb me or even hear me if I get up to praying or chanting too loud. Folks in the city or more populated areas will have to be more discreet. It is unlikely that the casual interloper will assume that you are just performing beneficial magic to help your friends and neighbors! Outdoor work for me usually consists of regular daily devotions to the planetary angels and spirits of the day, devotions to the Full and New Moon, and offerings to the Earth and other nature spirit work.

Working outdoors can be very powerful, especially in a natural setting. The gods send us epiphanies through the forces of nature and animals. Many, many times, my invocations have been followed by a chorus of coyotes, or a sudden gust of wind whipping up the dry leaves in a whirlwind, a brilliant shooting star, a sudden flash of light in the sky, the hoot of a screech owl, or the clouds suddenly parting and bathing the scene in a brilliant beam from a Full Moon. Epiphanies are especially frequent when one calls on Hekate near a Full Moon, at Moon hour,* preferably, if possible, when the Moon culminates in the sky and the tidal forces are at their peak. The coyotes almost always call back, which is appropriate to her symbolism.

*This refers to the planetary hour of the Moon. See chapter 10 for a full discussion of planetary hours.

9

THEURGY

The Magic of the Supercelestial Realm

In the last chapter, we introduced the idea of the four worlds of the magical cosmos and the concept of the magician as the mediator between the realms, and discussed the creation of the sacred space as a liminal zone for travel between the realms. In this and the following chapters, we will be looking into the theory, practice, and history of the ways magicians and other spiritual workers have drawn on the powers of each specific cosmic realm. In this chapter, we will focus particularly on what we will call the supercelestial realm or the divine realm, the transcendent reality that exists outside of space and time, the zone of pure being and limitless light from which the whole created cosmos emanates.

In Western Neoplatonic magic, as practiced by the cunning farmer, even when our aims are mundane we often bring energy from the top of the hierarchy down, calling upon the Divine by the names appropriate to our goal, and then the archangel of the appropriate sphere, then the angel, and finally the elemental or chthonic spirit given charge over the aspect of nature we wish to influence. We do this whether the goal of the operation is to call the rain, petition for profit in business, or seek some other earthly goal. The supercelestial realm, the realm of the transcendent divine, is thus the starting point for operations of what we may call high magic, as distinguished from operations involving only elemental beings, nature spirits, and chthonic entities, which is traditionally called low magic. So, therefore, when practicing this type of magic, we aim to draw down divine power from the highest sphere in our cosmology. An understanding of the philosophy and lore of the angelic realm and the transcendent is fundamental to the practice of magic as we understand it.

IN FORMA DUNQUE DI CANDIDA ROSA
MI SI MOSTRAVA LA MILIZIA SANTA,
CHE NEL SUO SANGUE CRISTO FECE SPOSA.
PARADISO, c. XXXI, v. 1-3.

◆ *Dante's vision of the Empyrean Heaven. Engraving by Gustave Doré for the* Divine Comedy *(nineteenth century).*

DANTE'S EMPYREAN HEAVEN

The highest sphere has been known by the mystics of various traditions by so many names, we will only be able to name a few. It has been called Keter or Ain Soph by the kabbalistic traditions, the Pleroma by the Gnostic Christians, the Source by the Hermetic mystics, the One or the Good by the Platonist and Pythagorean traditions, the Godhead by Christian mystics, or simply God or Heaven. We are not describing a place or a person but rather a transcendent reality beyond being in which all of our conceptual categories break down into absurdity, a state of being that is as much No-thing as it is the All. But despite being utterly paradoxical, this fundamental level is considered to be the ultimate reality. Everything else is conditional or dependent on something else for its existence. It alone is self-existent, and from It proceed all things, from unity into multiplicity, the One into the Many.

In the astrological model we have been using as a map to the spiritual cosmos, this realm is known as the Empyrean, the Heaven of Fire, traditionally considered to be the abode of God and the angels. The medieval Italian poet Dante spoke of the Empyrean Heaven, saying, "O splendor of God, through which I saw the high triumph of the true kingdom, give me the power to tell how I saw it!"[1] Indeed, his descriptions of his vision are difficult to conceptualize, as are those of many mystics, but he begins to describe it thus: "Light is thereabove which makes the Creator visible to that creature which has its peace only in seeing him."[2] He likens it to the shape of a vast white rose made of light, composed of a circular throng of angels, vast in size and emanating pure joy. The angels "had their faces all of living flame, and their wings of gold, and the rest so white that no snow reaches that limit."[3]

This is obviously the language of mystical experience, the words used by a traveler to this imaginal realm to convey his ineffable experience to the reader. The modern mind is capable of accepting mystical experiences of the exalted realms of God and the angels as long as they remain within the domain of subjective hallucinatory experiences or the merely imaginary states of mind experienced by poets and artists in their reveries. But when we say that Dante and those like him were experiencing reality, the reality of the transcendent, the modern skeptical mind balks. We have done well to get away from much of the literalism that accompanies the religious systems of the past, but we have lost much in the bargain, including the ability to tap into the reservoir of pure consciousness that is the raw beating heart of the cosmos and the true source of all.

The Empyrean Heaven, which was known as First Heaven, was believed by St. Basil of Caesarea (329–379 CE) to have "existed already before the creation in the form of incorporeal light." There was, he declared, "a certain mundane condition, older than the birth of the world and proper to the supermundane powers, one beyond time, everlasting, without beginning or end. In it the Creator and Producer of all things perfected the world of His art, a spiritual light."[4] This is interesting, because we have an everlasting realm of pure spiritual light existing eternally outside of space and time, from which the Creator sends forth creation. I see parallels with the big bang of modern cosmology in which suddenly a massive flash of light erupts into space and time, being itself the beginning of space and time. Where did this sudden eruption of light and energy come from? Who knows. But it is certain that it came from outside of space and time.

✦ *Aion, the god of time. Roman mosaic, 200–250 CE.*

Platonic Influences

The ancient Platonic tradition, which has been extremely influential on all subsequent esoteric movements, in common with its Pythagorean predecessors, believed that all things had their origin in the One, which was also called the Monad, the Good, Being, Existence, the All. It was equated with the primordial god Aion, who engendered the father and mother of the gods, Kronos and Rhea, who, in the words of modern Pythagorean teacher John Opsopaus, "rule all of the principal dualities of the universe, including light and dark, form and substance, mind and matter, being and becoming, identity and diversity. These dualities are all manifestations of Unity (the Monad) and Plurality (the Indefinite Dyad)."[5]

Opsopaus says, "According to the myths, Aion is bisexual and by self-fertilization gives birth to the Father and Mother of the Gods. But since Kronos and Rhea together create ordinary (determinate) time and space, Aion's creation of Them is outside of time, an eternal Emanation."[6]

Cosmology and Literalism

It's important to remind oneself, when discussing these abstract-sounding cosmological schemes, that they have their origin in ineffable states of mystical ecstasy and are an attempt to put the insights gained in those exalted states into everyday language. The mystical and religious concepts of the mystics are not to be taken literally, as if they were referring to things in this concrete world of cause and effect. This is the pitfall modern people continually fall into with the various spiritual documents that they come into contact with. The Bible, for instance, is full of mystical and visionary language, visions of prophets, seers, and mystics who had face-to-face encounters with the sacred. Yet most people today see only two ways of understanding it: interpreting it literally or rejecting it as a bunch of fairy tales. Similarly, the Pythagoreans weren't imagining Aion literally giving birth to Kronos and Rhea. These are the primordial forces of Creation, similar to the fundamental forces of modern cosmology but on a spiritual level.

Notice the inclusion of goddesses and the divine feminine in the emanation systems of the schools of thought influenced by Platonism. Besides Rhea, the *Chaldean Oracles* honor Hekate as the personification of the world soul and the feminine principle. Gnostic systems, inspired by Philo of Alexandria, name their primordial feminine principle Sophia or Wisdom, who is also the biblical Chokmah of the Wisdom of Solomon and the Books of Sirach and Proverbs, herself a direct survival of the Ugaritic goddess Asherah, wife of YHWH. In the older, pre-Christian systems, there are goddesses in the highest realm who are the primordial mothers of creation and from whose fecund wombs issued forth all creation.

Many of these Pythagorean-influenced systems envisioned a primal triad, which took different forms in the different schools. The basic structure was a primal monad which emanated a dyad, usually a gendered pair, either seen as mother and son with the monad being the father or a bisexual monad emanating a primal couple who were parents to all subsequent emanations. As the *Chaldean Oracles** state, "For in every world shines a Triad ruled by a Monad," and, "For in the Womb of this Triad all things are sown."[7] We can see that this might have had a lasting influence in helping shape the Christian concept of the Trinity.

The Pythagorean and Neoplatonic systems influenced all of the monotheistic faiths as well as Gnosticism. The later faiths emphasize the idea of unity in divinity and tend to ignore the idea of diversity completely. The gods and goddesses of the Pythagoreans and Platonists are demoted to the angels and demons of the Islamic, Jewish, and Christian faiths, and the Aeons and archons of the Gnostics. But to the Pagan philosophers, all of the gods and goddesses were aspects and emanations of the One, the most exalted aspects of which were seen to dwell outside the cosmos in the highest sphere. These were called by the Neoplatonist philosopher Iamblichus the "hypercosmic Gods," the gods beyond the cosmos, dwelling in the intelligible realm, a nonphysical realm of pure consciousness, another word for the Empyrean, which is a later term. Iamblichus said these nonphysical Gods, whom he also called intelligible Gods, directed the planets and stars, which he called the visible images of the gods:[8] "If they are mounted on the heavenly spheres as incorporeal and intelligible and unified entities, they have their originating principles in the intelligible, and it is by thinking their own divine forms that they direct the totality

---

*The *Chaldean Oracles* are a fascinating document consisting of fragments of a lost work on theurgy of the same name dating from the second century CE and assembled from quotations in works that have survived from late antiquity. The surviving fragments, which are cryptic and highly symbolic statements relating to mystical philosophy and visionary states, have been arranged in a traditional order by successive generations of scholars.

of the heavens through a single infinite act."[9] So we have a system of emanations, whereby a transcendent unity emanates a duality, then a multiplicity of gods, still nonphysical and purely intelligible—consciousness, in modern terms, whose thoughts direct the cosmos as archetypes.

Angelic Hierarchies

It is at this point that we have to introduce and unpack a term that may be problematic to modern sensibilities: *hierarchy*. This word was coined by a fifth-century Christian Neoplatonist who wrote under the name of Dionysius the Areopagite, a first-century Athenian convert of St. Paul whose true identity has been lost to history. The word *hierarchy*, meaning "sacred rule," was coined by St. Dionysius to describe the ordered system of emanation of spiritual beings, which he viewed as angels (rather than gods, in contrast to the philosopher Proclus, from whose work he borrowed liberally), which poured forth from God into the physical cosmos. This is the Neoplatonic conception of procession, or *prohodos*.

When the mystic Dionysius contemplated the heavens, he saw an "outpouring of Light which is so primal . . . and which comes from that source of divinity."[10] This divine emanation took the form of ranks of angels, arranged in nine ranks. The ranks, whose names were derived from scripture, are, in descending order:

✦ Assumption of the Virgin, *featuring choirs of angels,*
as envisioned by Francesco Botticini, 1475–76.

Thrones, Cherubim, Seraphim, Authorities, Dominions, Powers, Principalities, Archangels, and Angels. This comprises what he calls the "heavenly hierarchy." St. Dionysius's and Dante's visions both shaped the way subsequent generations of Christians envisioned the imaginal realms above the heavens. This itself is a Christianization of Proclus's ranks of divine emanations, known as *henads*. Hierarchies, says Dionysius, "are a certain perfect arrangement, an image of the beauty of God, which sacredly works out the mysteries of enlightenment in [its] orders and levels." A hierarchy "bears in itself the mark of God" and passes the divine light generously down to lower members of the cosmic scheme, eventually even to the earthly realm and to humans.[11]

Herein lies the problem. The concept of the divine hierarchy was used to justify the ecclesiastical hierarchy of the pope, cardinals, archbishops, bishops, priests, deacons, minor clergy, emperor, nobility, and finally, the common people. It is in this political sense of hierarchy that we use the word today, and it has been used since the time of St. Dionysius to justify first the Imperial Roman power structure and later the medieval power structure and the divine right of kings. The idea of divinely ordered cosmic structure reflected in the power structure of human society is an ancient one and was followed in ancient India and Mesopotamia, and indeed over much of the world. The idea is rightfully distasteful to modern democratic sentiments; it constitutes a misuse of the geography of the imaginal for the purposes of political propaganda. So, should we abandon all ideas of a structured cosmos, arranged in ontological levels from pure light and perfect being to dark fallen matter and all of the world-hating, flesh- and body-denying metaphysics that it entails? There is much negative discourse regarding the Neoplatonic school these days because of this supposedly world-denying antiphysical stance. If all goodness flows from the divine realm through successive layers, ultimately settling "down" into our shadow-filled fallen world, then what should motivate us to love this world and our fellow humans and work for the benefit of all?

It is important to remember, again, that these theories are metaphors and maps of the imaginal cosmos and are to be followed to the extent that they are helpful and set aside for better models in other situations where they don't serve the project at hand. The concept of an angelic hierarchy might be helpful when one wants to contemplate the structure of the spiritual realm in order to connect with higher intelligences, gods, and angels but perhaps not helpful when one wants to set up a just and equitable system for governing human societies. That is exactly what Plato was trying to figure out when he introduced the idea of

✦ *Hildegard of Bingen's vision of the angelic hierarchy, twelfth century.*

the ultimate transcendent reality, the One, in his dialogue *The Republic*. Direct knowledge of the transcendent archetypal realm he considered essential for anyone invested with the awesome responsibility of governing.

As I have pointed out elsewhere in other chapters, spatial metaphors—like higher and lower, ascending and descending, and above and below—aren't to be taken literally. There is much talk in the tradition that would lead people to believe that the divine is remote and uncaring, up there in the ninth sphere outside the cosmos, surrounded by equally distant angels. In truth, however, the divine realm is as intimate as one's own breath, and there is no space anywhere that is not filled with the presence of the source. It is what animates us and holds us in being and life. What keeps us distant from the source is our own impaired perception. In the apocryphal Gospel of Thomas, when asked by his disciples, "When will the Kingdom come," Jesus replied, "It will not come by waiting for it. It is not a matter of saying 'here it is' or 'there it is.' Rather the Kingdom of the Father is spread out upon the Earth and men do not see it."[12] Here is another saying from the Gospel of Thomas that gives us a sense of ideas of divine immanence current in this period: "Jesus said, 'It is I who am the light which is from above them all. It is I who am the all. From me did all come forth, and unto me did the all extend. Split a piece of wood, and I am there. Lift up the stone and you will find me there.'"[13]

UNITING HEAVEN AND EARTH

A fellow Kentuckian, esoteric author, and friend, Normandi Ellis, in her recent and highly informative work on angelology, *The Ancient Tradition of Angels*, writes:

> Angelic hierarchies operate on separate realms on a continuum that describes the density of the energy pattern as it coalesces into form. Hierarchies are not about degrees of power, but rather degrees of energetic operation. Looking at the light codes sent out by our sun and the unseen impulses of the exploding big bang, we observe that form follows energy. All spiritual evolution, whether falling from grace or ascending to heaven, is engaged in the same hierarchical structure of energy forms. From ferns to archangels, all unfold through process. At its deepest levels, Creation is about the divine process of unfoldment.[14]

The later Neoplatonists, incorporating ideas from the influential mystical writings known as the *Chaldean Oracles* (one of the founding documents of theurgic mysticism, which has unfortunately been lost to history and is now available in fragmentary form in quotations in other works), founded a system of spiritual practice known as theurgy, in which the practitioner, using ritual, invokes divine powers into the human sphere for the purposes of mystical knowledge and union, as well as what we would consider today to be magic. For the theurgist, this cosmos, although fallen and distant from the source, is shot through and thoroughly animated with the divine, and the gods are present here with us in the material realm. The divine contains everything within itself, as St. Paul was reputed to have said of his version of God: "In him we live, move, and have our being."[15]

The human soul, too, is divine, and by means of theurgic prayer and ritual, that divine within can be unified with the greater divine. As Iamblichus states, "For that element which is divine and intellectual and one—or if you so wish to term it, intelligible—is aroused, then, clearly in prayer, and when aroused, strives primarily towards what is like itself, and joins itself to essential perfection."[16] Moreover, he writes, "By virtue of the sacred liturgy . . . the divine is literally united with itself."[17] This theurgic prayer is a far cry from the petitions known in conventional religion, whereby lowly supplicants beg favors from a divine sovereign, infinitely removed from them. It is a very far cry indeed from the "total depravity" of Calvinist Christianity. Iamblichus also says, while we have a divine element in our souls, "We are inferior to the gods in power and purity and in all respects, it is eminently suitable that we entreat them to the greatest degree possible. The consciousness of our own nothingness . . . makes us naturally turn to supplications. And by the practice of supplication . . . we gradually take on the perfection of the divine."[18]

The theurgists used ritual for the purpose of raising the soul to the level of the gods and achieving mystical states of union by calling the gods, angels, daimons, and other spiritual entities into the material realm for the purposes of attaining knowledge, for divination, and in order to inhabit material objects, such as statues and talismans. They were the ancestors of today's ceremonial magicians, and indeed many of the historical practices of theurgists described in the classical sources continue to be practiced to this day. The idea of bringing together harmonious material objects associated with a particular god—such as flowers, perfumes, incense, colored fabric, candles, and other materials, in a ritual setting, in which the right words and names, the *nomina barbara*, are used at a propitious astrological time, or *kairos*—is still the essential practice of ceremonial magical ritual invocations.

Ruth Majercik, in her extremely informative introduction to her translation of the fragments of the *Chaldean Oracles*, lists the various types of magical operations performed by theurgists in late antiquity. The list would be familiar to any practicing magician of today. One important practice was known as *systasis*, or "conjunction," known today as "invocation," which means, according to Aleister Crowley, "to call in," and "evocation," which means "to call forth" or to summon. The term refers to contact or communication with a particular god or spirit. Says Majercik:

> The principal means of effecting this contact was through the use of various invocations: the adept "called upon" the god by uttering his divine names, which amounted to a lengthy recitation of unintelligible vowel and consonant sounds. Such *nomina barbara*, or *voces mysticae*, are found throughout the magical papyri as well as in certain Gnostic and Hermetic texts and are a staple of late Antique magical practice. Although a cursory glance at these lists of sounds reveals what appears to be a random selection of so much vocal gibberish, closer scrutiny shows that there are definite patterns not only to the arrangement of vowels and consonants but also in terms of numerical equations, all of which had potent magical properties. By rhythmically chanting these sounds (which equalled the "hidden" or divine name of the god), the adept was able to effect the proper conjunction with the deity.[19]

Another theurgic practice was the "animation" of statues, by placing inside them magical substances, such as stones or herbs, in a ritual context, and at a propitious time—similar, perhaps, to the creation of astrological talismans. Majercik writes: "The operative principle behind these procedures is that of sympatheia, a notion which assumes a direct correspondence between a given deity and his or her symbolic representative in the animal, mineral, or vegetable worlds. Thus by properly fashioning and consecrating the god's "material image" (and then placing it in the god's statue) he or she could generally be persuaded to appear (generally in the form of light) and answer the questions put to him by the theurgist."[20] This is more akin to the evocation of the modern mage, in which the spirit is called into something, such as a triangle of art, a crystal, a bottle, or a brass vessel. I refer the interested reader to Proclus's brief treatise, "On the Priestly Art," in which he discusses the principle of *sympatheia* in theurgic ritual. We will discuss sympatheia in more depth when we discuss astrological talismans in the next chapter.

Other theurgic ritual practices included what Majercik terms "binding and loosing," which appear to be akin to what we call today channeling and trance possession, in which the theurgist merges their consciousness with that of the god, speaking and acting as the god and exhibiting miraculous phenomena such as levitation or manifesting luminous apparitions, which were "seen entering or leaving the medium's body."[21] Practitioners also engaged in *anagoge*, or what we might term astral travel or out-of-body work, or celestial ascent in a ritual context. An interesting example of a theurgic ritual of cosmic ascent may be seen in the text of the so-called Mithras Liturgy, which was probably not a Mithraic ritual at all, although the "great god Helios-Mithras" is mentioned as the revealer of the ritual. The text is a set of instructions for a ritual of what may be a guided meditation and a series of invocations, addressed to a female student. The instructions call for the ingestion of the juices of unspecified herbs and spices as well as *pranayama*-like breathing exercises to aid the initiate on her ascent. The text makes for fascinating reading.

> Give ear to me and hearken to me, NN, whose mother is NN, O lord, you who have bound together with your breath the fiery bars of the fourfold root,
> O fire walker, PENTITEROUNI,
> Light-maker, SEMESILAM,*
> Fire-breather, PSYRINPHEU,
> Light-breather, OAI,
> Fire-delighter, ELOURE,
> Beautiful light, AZAI,
> Aion, ACHBA,
> Light-master, PEPPER PREPEMPIPI,
> Fire-body, PHNOUENIOCH,
> Light-giver, . . .
> Fire-sower, AREI EIKITA,
> Fire-driver, GALLABALBA,
> Light-forcer, AIO,
> Fire-whirler, PYRICHIBOOSEIA,
> Light-mover, SANCHEROB,
> Thunder-shaker, IE OE IOEIO,
> Glory-light, BEEGENETE,
> Light-increaser, SOUSINEPHIEN,
> Fire-light-maintainer, SOUSINEPHI ARENBARZEI MARMARENTEU
> Star-tamer, . . .
> Open for me.[22]

---

*This is Hebrew: *shemesh olam*, or "eternal sun."

HISTORY OF THEURGY

After the closing of the Platonic Academy by an edict of the Christian Emperor Justinian (482–565 CE) and the exile of its last official head, Damascius, in 529 CE, the final destruction of the traditional Pagan religions was nearly complete. With the suppression of the "heretical" alternative mystery traditions of the Gnostics and others by imperial Christianity in the fifth and sixth century, victorious Christianity became the container, for better or worse, for the Western mystery tradition. Many of the teachings of the theurgists and mystics were forced into the "clerical underground" of the medieval magicians and into the elaborate symbolism of the alchemists, as well as, some say, into the esoteric teaching of groups such as the Knights Templar and the heretical Christian Cathars.

Another extremely important locus for the preservation of the Hermetic and Platonic teachings of late antiquity were the esoteric traditions of the Islamic faith, groups such as the Sufis and the Brethren of Purity. Their esoteric writings began to trickle into Europe in the eleventh century, and as noted scholar of medieval magic Richard Kieckhefer says, "In the twelfth century well over a hundred were translated from Arabic into Latin, or else written in Latin specifically as paraphrases of Arabic learning."[23] This included works on astronomy, astrology (such as the works of Ptolemy), and astral magic, such as the infamous *Picatrix*. These works all shaped how European magicians envisioned the myriad spirits who filled the various layers of the multitiered cosmos.

Jewish magic and mysticism was also a major contributor to the understanding of the structure and dynamics of the heavens and their denizens. Because Judaism and Christianity both shared the holy texts that became known to Christians as the Old Testament but which were known to Jews as the Tanakh, they shared many images and understandings of the divine realms, however differently these shared roots worked out in practice—from the various Books of Enoch, which depict the various levels of the heavens and, ultimately, the throne of God, to the Hekhalot literature that depicts a descent to the throne of God (yes, you read correctly, as discussed, there is no reason the throne of God shouldn't be within as opposed to above), to the magical texts, *Sefer Ha-Razim* and *Sefer Raziel*, which describe magical techniques for invoking angels. These two texts were extremely influential, especially the latter, whose structure of angels presiding over the seasons and times, as well the various signs of the zodiac and how and when to call them, had a long afterlife in the Christian angel magic of the Middle Ages and Renaissance.

Prohet heſekiel am erſten capitel

Im dreyſſigſten zar am funfften tage des virrden monden da ich war vnder den gefangen am waſſer Chebor thet ſich der
hiemel auff vnd got zaige mir geſicht der ſelb funffte tage des monden war eben im funfften zar nach dem Joachim der
koning Juda gefangen war weg gefurt da geſchach des herren wort zu heſkiel dem ſonn Buſi des prieſters im lande dre
Chaldeer am waſſer Chebar daſelbſt kam des herren haund vber in.

♦ *The vision of Ezekiel as interpreted in a sixteenth-century German text.*
Illustration by Heinrich Vogtherr the Younger.

The Jewish mystics of antiquity practiced a variety of mystical exercises aimed at freeing the consciousness from the confines of the body in order to ascend or descend to the highest sphere, the throne room of God. This was also conceptualized as the Merkabah, the chariot, as seen in the vision of Ezekiel. In the same tradition are the apocryphal Books of Enoch, which detail the celestial ascension and visions of the various heavens of the biblical patriarch Enoch, who was taken to heaven by God, prefiguring the Christian story of the ascension. According to Gershom Scholem, the famous scholar of Jewish mysticism, "The dangers of the ascent through the palaces of the Merkabah sphere are great, particularly for those who undertake the journey without the necessary preparation, let alone for those who are unworthy of its object." This was nothing other than seeing God face to face! "Angels and archons storm the traveler 'in order to drive him out'; a fire which proceeds from his own body threatens to devour him,"[24] writes Scholem.

Fearsome guardians stand guard at the gates of the seventh palace, the throne-chariot of God. "And at the door of the seventh palace stand angry all of the heroes, warlike, strong, harsh, fearful, terrific, taller than mountains and sharper than their peaks. Their bows are strung and stand before them; their swords are sharpened and in their hands. And lightnings flow and issue forth from the balls of their eyes, and spider webs of fire from their nostrils and torches of fiery coals from their mouths. And they are equipped with helmets and with coats of mail, and javelins and spears are hung from their thews,"[25] says the *Hekhalot Rabbati*, *The Book of the Greater Palaces*, which details the ascent of the mystic Rabbi Ishmael. If you think the guardians sound fearsome, you should see their horses! The text says, "All those who descend to the Merkabah and ascend are not harmed, but they behold all this trouble and descend safely and come to stand and bear witness and tell the fearful and terrific vision." The vision of God on his throne, as Rabbi Ishmael says, looks like:

> *A lovely face, a majestic face*
> *A face of beauty, a face of flame*
> *Is the face of . . . the Lord God of Israel*
> *When He sitteth upon the throne of His glory*
> *And His dignity is established in the dwelling of His majesty.*
> *His beauty is more lovely than the beauty of the powers,*
> *His majesty surpasseth the majesty of the bridegrooms*
> *And brides in their bride-chambers.*
> *He that beholdeth Him is at once torn to pieces,*
> *He that glimpseth His beauty at once poureth himself out as a*
> *vessel.*[26]

In the Middle Ages in Europe, a hybrid Christian theurgy was developed, composed of the esoteric traditions of the Hellenistic world (the Greco-Egyptian), transmitted through the Byzantine Empire and the Arab world, and the Jewish tradition. All these traditions had met and blended in the past in the cultural melting pot of Hellenistic Alexandria, but they were to be reintroduced in the cloistered monastic libraries and centers of learning in Western Europe, baptized in a new Christian guise. Gone were the Pagan gods of the Empyrean realm—Aion, Kronos, and Rhea—replaced with the Father, Son, and Holy

Spirit, and also myriads of angels filling all the spheres. Pagan Neoplatonic theurgy was a major influence on the Orthodox Christian church in many ways, philosophically and practically. St. Dionysius, the Christian student of Proclus, the Pagan theurgist and magician, in his treatise, *The Ecclesiastical Hierarchy*, explicitly compares the mass to theurgy. Many Christian ritual elements in the liturgical churches retain a distinctly Neoplatonic flavor. One of my favorite liturgical prayers from the Episcopal church, the Eucharistic Prayer D, says, "You alone are God, living and true, dwelling in light inaccessible from before time and for ever. Fountain of life and source of all goodness, you created all things and fill them with your blessing."[27] Pure Neoplatonism, straight from Pseudo-Dionysius.

Enochian Magic

Perhaps the most advanced system of angelic magic in the West was (and is) the system developed by the sixteenth-century English intellectual, polymath, and mystic John Dee. Dee's system, which came to be known as Enochian, began as a series of magical experiments in summoning angels, undertaken by Dee and his scryer (a medium who was adept in communicating with the angels, what we would today call channeling), Edward Kelley. Dee, in the words of contemporary magician Aaron Leitch in his excellent book on the subject, *The Essential Enochian Grimoire*, says, "Before long . . . Dee found himself holding daily conversations with archangels such as Michael, Gabriel, Raphael, and Uriel. These entities introduced the men to an entire host of previously unknown angels who would eventually reveal an entirely new system of magick."[28] Drawing on the medieval Solomonic tradition of angel summoning, which he learned by studying books like the *Sworn Book* and other works, such as the *Ars Notoria*, the *Ars Paulina*, the *Heptameron*, and Agrippa's works, Dee, with the help of the angels, managed to create a complex and powerful system of Renaissance theurgy.

Following in the footsteps of the Jewish Enochic tradition of celestial ascent we have spoken of above, the Merkabah mystics, and the Hekhalot tradition, Dee and Kelley asked the angels to reveal to them the true Book of Enoch, to which they agreed. According to Leitch, "The angels revealed nothing less than the mysterious 'book' Enoch had read and copied from in heaven—the Book of Life itself, the ultimate record of all things in the universe." The angels called this book the *Holy Book of Loagaeth*, and "it contained the words God had used to create the world, entrances into the forty-eight gates of heaven, and it recorded the destinies of all things in creation—past, present, and future."[29] In common with the *Hekhalot Rabati*, in which are revealed the passwords to enable the visionary to get past the celestial guardians at the gates of the spheres of the heavens, the primary purpose of the *Book of Loagaeth*, says Leitch, "was to open the gates of heaven and either call out the angels or spiritually enter the gates—as Enoch had done—to visit the celestial realms directly."[30]

Another fascinating revelation of Dee and Kelley's angelic informants was an entire angelic language, believed by Dee to be the original Adamic language spoken before the "confusion of tongues" in Genesis 11:9 (when, according to the myth, YHWH confused the speech of all humankind, who had up to that point spoken one language). Dee believed his revealed language to be the sacred words used by God to speak all things into existence, as in the Kabbalah of the medieval Jewish mystics. This myth of a primal divine language of creation is an extremely appealing and long-lived one. Indeed, the divine Iamblichus also believed that the ancient languages of the Egyptians and the Chaldeans were closest to the speech of the gods and thus more appropriate and potent for magic and theurgy. The angelic or Enochian language, as it has come to be called, does have a particularly sonorous ring to it. As magical languages go, it certainly makes for some powerful-feeling *voces magicae*.

"Abrahamic" Religious Symbolism

The magic of the Ninth Sphere is the magic of Angels, Gods, and the Divine Names. It is the raising of the divine elements of the soul "up" to the level of the Gods and bringing the divine into the human realm by means of ritual practice in ritual space. Prayer, ritual, and contemplation are the three main practical means of achieving contact with the upper realm. There are many schools of practice, and in this chapter, we have been mainly focused on Neoplatonic and Hermetic theurgy and have just begun to dip our toes into the deep waters of the tradition of angel magic in the Medieval period and the Renaissance.

In practical terms, however, when you work with divine and angelic forces of the transcendent sphere in your life, it is important to remember that the Source is infinite generosity and infinite love. It fills all space with being and life, and to work with it you only need to ask. I'm not saying that you're going to get everything that you pray for. We all live under the government of Fate, but we don't know what's possible until we try, either. This isn't *Pollyanna*. Spiritual growth takes work and practice.

Many of us have had a seriously flawed introduction into the methods of interacting with that Higher Consciousness commonly referred to as the Divine. Many of us were brought up with problematic patriarchal and authoritarian notions of God as divine dictator and angry "sky daddy" who is ready to curse us to eternal damnation if we step out of line. Some have rightly rejected these notions and embraced a more secular or atheistic worldview. In my opinion, this counts as spiritual growth! If this type of God is the thing you think of when someone says "divinity" or anything spiritual or religious, then you must destroy this concept and its hold over you by any means necessary. And I fully support you in your efforts.

The God presented by most of the religious traditions of the Western world, at least in the popular understanding, though not in the understanding of the mystics of those traditions, is what the Platonists and the Gnostics referred to as the Demiurge, a limited creator being distinctly further down the ontological "hierarchy" from the true divinity, the One or the Source, we have been discussing. The One is not a person (although it will take that form for you, and when it does it is no longer the One but a God or an Angel). The One does not condemn or curse or hate; it gives, endlessly, every moment, generously, of its infinite being. It is you in your deepest inner nature and as it flows out into all natures endlessly, it draws us back to it, which we experience as a longing and desire for spiritual wholeness. Procession and return. As the Hindu scriptures say, "*Tat Tvam Asi*" or "That art Thou." You can call it God, but that word is problematic for many. You can call it Goddess, if you prefer. It has no gender or qualities of any kind that can be defined by language. Those come later. The One is primal, the eternal beginning.

DAILY THEURGY

Contemporary ceremonial magician Lon Milo DuQuette, in his very helpful book on the Enochian magical tradition, *Enochian Vision Magick*, has this advice to the reader regarding prayer, which is integral to magical work:

In magick, prayer has nothing to do with religion. It does, however, have everything to do with the magician's grasp of the facts of life in the magical world in which he or she intends to operate. Part of that reality is the mystical fact that existence itself is an expression of consciousness, and there are levels of consciousness higher and lower than the one we are currently attuned to. By the simple act of uttering a prayer, our apparently limited consciousness reaches out and makes contact with a higher level of consciousness, and by doing so, we insert ourselves into the divine circuitry of the cosmos. And as we do, part of us (a part that we might not be aware of) wakes up, as it were, and recognizes our rightful place in this circuitry.[31]

So a daily practice of prayer, to a deity of your choice, is a good place to start. This type of thing is intensely personal. Let your taste and your spiritual instincts be your guide. I prefer to use prayers from the Hermetic and Gnostic tradition or the Orphic Hymns in my daily practice. I think the loftier the language, the better. The hymns of Synesius, authored by the Neoplatonic philosopher Synesius of Cyrene, who was pressed into service as a Christian bishop, are an excellent example of a theurgic prayer that invokes the power of the Divine Source.

Another helpful practice from the Golden Dawn tradition of Western magic is the Middle Pillar Ritual, which is an exercise intended to draw spiritual power down the central column of the subtle body from the heavens through the crown chakra, identified with the kabbalistic Sephirah, Keter, down through the body and into the earth. The energy can be circulated a number of times using several different visualizations. It is an excellent way to raise power for practical magic or healing purposes. It is easy to learn, only takes a few minutes, and can be practiced daily.

⚘ The Middle Pillar Ritual

Stand in your sacred space, facing east. Close your eyes. Take several deep, full breaths, releasing all bodily tension in preparation to call upon the highest divine forces and to bring them down into your personal sphere.

Visualize yourself standing on the sphere of the Earth in the center of the universe. Take a moment to formulate this picture fully. Next, imagine a ray of brilliant white light descending from the infinite distance far above your head which descends to form a sphere of the same brilliant light about eight inches in diameter immediately above your head, at the location of the crown chakra.

Inhale deeply, breathing the light into the sphere. While slowly and completely exhaling, vibrate (chant) the divine name Ehieh (Eh-Hee-Yeh). Inhale. Repeat three times.

Visualize a beam of white light descending from the sphere of light above your head down the central axis of your body until it reaches the throat, the location of the throat chakra, where it expands into an eight-inch sphere of brilliant pale blue or lavender light.

Inhale deeply into the sphere, as before. Exhale slowly while vibrating YHWH Elohim (EE-AH-OH-EH El-Oh-HEEM), Inhale. Repeat three times.

Visualize the beam of brilliant white light as before, descending from the sphere at your throat down to the level of your solar plexus, the space where the ribs meet in the center of your chest, where it expands into an eight-inch sphere of brilliant golden light, the color of sunshine.

Inhale deeply into the sphere. While exhaling, vibrate YHWH Eloah Va-Daath (EE-AH-OH-EH ELOAH VA-DAAT), Inhale slowly. Repeat three times.

Visualize the white beam of light as before, descending to the level of the lower abdomen or groin area, where it expands into an eight-inch sphere of brilliant purple light.

Inhale deeply into the sphere. While exhaling, vibrate the divine name Shaddai El Chai (ShADDAI EL CHY). Inhale once again and repeat three times.

Visualize the beam of brilliant white light once again, this time descending into the ground beneath your feet and expanding into a black sphere, grounding the descending energy. Inhale into the sphere. Exhale, vibrating Adonai Ha-Aretz (ADONAY HA ARETZ). Inhale deeply. Repeat three times.

Revisualize the five colored spheres. Inhale deeply and slowly while drawing in the energy from the sphere above your head down the back of your spine into the sphere below your feet. Exhale and draw the energy up the front of your body into the white light sphere above your head. Circulate this energy six or more times while visualizing your energy body glowing brighter at each pass. Beware of getting light-headed. Slowly return your breath to normal and end the visualization.

It is good to combine the Lesser Banishing Ritual of the Pentagram with the Middle Pillar ritual. These practices, like their theurgic forebears, use divine and angelic names, *nomina barbara*, to call upon celestial forces and bring them into the sphere of the ritualist.

Linking your consciousness to the highest energies of the divine source and calling them down into your personal sphere has a cleansing and purifying effect on the spiritual atmosphere, driving away malevolent forces and miasma that may have built up. This is why prayers like in the Book of Psalms or the Lord's Prayer are believed to drive out evil influences—because they do!

In closing, I want to emphasize the practical nature of angelic work. There are many ways of engaging with the angelic energies of the supercelestial world for magical work of a very practical and earthly nature. It's not all lofty contemplation of transcendent powers. Let this chapter serve as an introduction to a complex subject. Speaking for myself and my particular brand of practice, when I do magical work for real effects in my physical and embodied life, with goals such as healing, financial success, or working with elemental or planetary energies, I often start with the Divine Source and its angelic mediators who bring the divine energies into our physical sphere. Let us remember the fundamental and eternal power of the Infinite Source, the One, the Good, the True, and the Beautiful, which pours out its love and light each instant, holding all things in being, and let our spiritual longing lead us back to our divine Source.

10
ASTROLOGICAL MAGIC
The Magic of the Celestial Realm

In the last chapter, we discussed magical and spiritual practices intended to make contact with the spiritual forces of what we called the supercelestial realm, the transcendent divine powers "above" or "beyond" the cosmos that are also immanent and permeate all of physical existence. Again, this is part of the body of practices that the ancients referred to as theurgy, or literally "god-work." Our topic for the present chapter also falls in the category of theurgy, since it seeks to work with the divine powers that animate the stars and planets.

The magic of the celestial realm is also known as astrological magic. We will be exploring what astrological magic is and how it has been understood by some of the cultures that have practiced it throughout history. Astral or astrological magic (the terms are used interchangeably) is defined as the practice of using astrological timing techniques, symbolic images, and ritual to produce an alteration in reality by "drawing down" the celestial or astral forces into the human sphere. These forces have been understood at different points in history as spiritual entities such as gods, daimons, or angels, or as impersonal rays emitted by celestial objects such as stars or planets. Both of these theoretical models have been important to the tradition, and they are by no means exclusive. Many thinkers have subscribed to both.

While the relevance of astrological magic to the very down-to-earth practicalities of the work of the cunning farmer may not be immediately apparent to the modern reader, astrological talismans were considered by the ancient practitioners of the art to be among the most effective methods of swaying the very earthy spirits of the sublunar realm, and foundational texts on astrological magic such

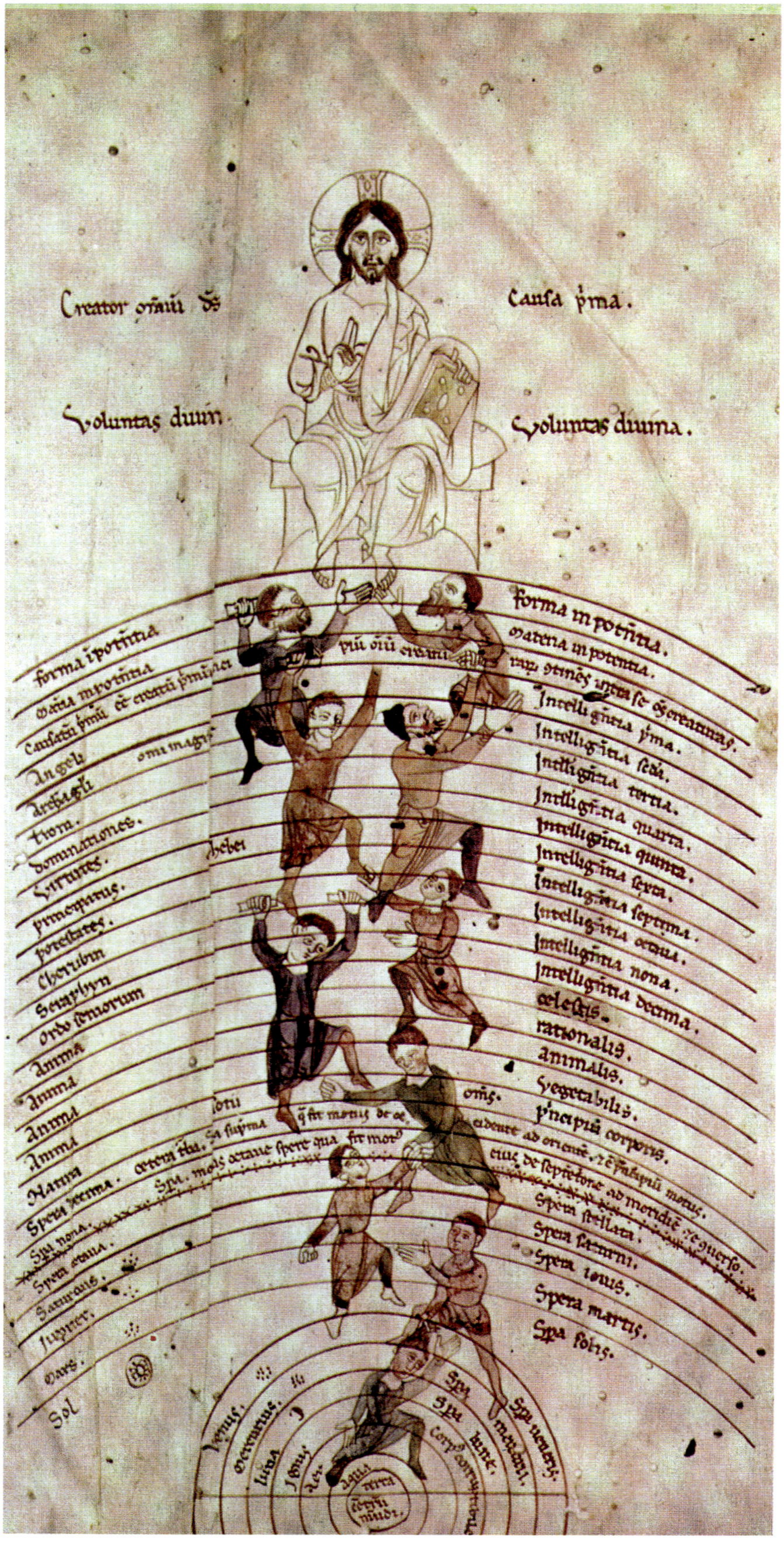

♦ *A Christianized version of the Hermetic ascent of the soul through the spheres toward union with the Divine, from a twelfth-century manuscript.*

as the *Picatrix* and one of its source texts, the aptly named *Nabatean Agriculture* of Ibn Washiyyah, furnish many examples of celestial magic used for earthly and agricultural purposes. Ancient people knew they lived and died by the success of the crops upon which all depended, thus, agricultural magic was of paramount importance to all.

EARLY ASTROLOGICAL MAGIC

It must be remembered that in the traditional cosmology the planets and stars govern or at least influence all things here on Earth, and that methods to channel their influence or to aid in the production of crops and livestock have been practiced since the days of ancient Mesopotamia. Astrological magic has been employed for practices like the control of noxious pests such as rats and scorpions, as well as for encouraging rain and increasing the yield of crops and the fertility of the land. These practices and others were considered to fall into the category of natural magic, which is the practice of manipulating the hidden sympathies inherent in Nature, her occult virtues, for human benefit, and in this occult science we see the beginnings of modern disenchanted science. Therefore, in this chapter we will explore the theoretical background of astrological magic and offer some simple suggestions for getting started using the power of the planets in a variety of magical operations.

Star beings have a presence in mythologies worldwide, from the time of the earliest humans, who looked up at the myriads of stars that populated the unpolluted skies. Nearly every culture on Earth has mapped its mythology against the backdrop of the heavens. Shamans and sages have predicted the rise and fall of dynasties, plagues, and wars, as well as the fortunes of agricultural production, be it feast or famine, by tracking regular movements of the planets against the seemingly unchanging "fixed" stars arranged in constellations. The stars were understood to be both the markers of time and its rulers, divine beings, eternally looking down on the inhabitants of Earth, more or less indifferent to human fortunes.

With the development of what we would recognize as astrology in Mesopotamia, in the sixth or seventh centuries BCE, an empirical system of mapping the correlations of planetary motions, angles, and combinations against the historical events in the kingdom of Babylon and its neighbors was developed. Babylonian and Assyrian astronomers kept meticulous records of celestial observations and coincident events. In an example from one of the cuneiform tablets

✦ *The Neo-Assyrian Venus Tablet of Ammisaduqa, a record of astronomical observations of the planet Venus. From Nineveh, first millennium BCE. Image by Fæ.*

from a translation by Assyriologist Reginald Campbell Thompson, "When Jupiter goes with Venus, the prayer of the land will reach the heart of the gods, Merdoach [Marduk] and Sarpanitum [a Mesopotamian goddess, the spouse of Marduk, here equated with Venus] will hear the prayer of thy people and will have mercy on thy people."[1] This quote illustrates the link between the efficacy of prayers and the position and movements of celestial bodies that came to be at the heart of the tradition of astrological magic.

It is not enough to pray to the stars. They have to be in a place where they are disposed to hear the petitions offered. If they are in an unfavorable aspect, the prayer will go unheard. These are powerful, but not omnipotent, beings. They have times, seasons, and moods when they can be more or less helpful or harmful. The example above didn't list the sign in which the planets had their conjunction, but when Marduk (Jupiter) is with his wife (Venus) and they are in the sign of Pisces, that is an extremely favorable time for both planets and for making contact with the beneficial energies of both.

The idea that the stars are intelligent beings, having souls, which were known as the "intelligences" of the planets, is one with a long history. As Plato said (or more likely his student Philip of Opus), "We must call them the divine host of stars—endowed with the fairest of bodies and the happiest and best of souls."[2] If the stars—vast, intelligent, ancient beings—have, as Plato said, "a sufficient term of life, covering many ages," is it not possible in a conscious universe, in which all things are in sympathetic connection, that they might actually exert an influence on the lives of conscious beings here on Earth? Is it unthinkable that a huge and complex celestial object like our Sun has an internal conscious life? British biologist Rupert Sheldrake thinks it is possible, or actually likely, and has written an interesting paper on the subject, arguing on the basis of panpsychism and the complexities of the Sun's magnetic field that it is entirely possible that not only the Sun but other stars, solar systems, black holes, galaxies, and the universe as a whole may be conscious on some level. He writes, "The Sun is influenced by the electromagnetic patterns of activity within the galaxy as a whole, which could in turn be closely connected with a galactic mind, perhaps centered in or around the supermassive black hole at the galactic centre. The galactic mind could influence what happens here on earth through its effects on the sun and the solar wind."[3] I take it as axiomatic that the universe is conscious and animate, as well as everything in it on some level, actually being "made" of mind stuff, so Sheldrake's considered conclusion is not too shocking.

In a mental universe with conscious stars, and one in which conscious biological entities on a smaller scale can and do influence reality with their minds, it would seem that incredibly vast, inconceivably ancient minds—like those of stars, may be godlike in their scope—could almost certainly affect the fabric of reality. (You've gone with me this far, so let's just descend a little bit further down the rabbit hole!) Perhaps ancient people, skilled in the arts of consciousness alteration and ritual practice, combined with the science of astronomy, discovered methods with which they could achieve contact with the celestial minds of the stars, and these techniques became the basis of the Hermetic art of astrological magic. In the Hellenistic melting pot of cities like Alexandria, the scientific and astrological knowledge of the Babylonian magi was blended with the magical and ritual techniques of the ancient Egyptian priesthood to achieve a hybrid system of astral theurgy capable of making contact with the very souls of the stars, for the purpose of gaining wisdom as well as spiritual and temporal power.

THE *PICATRIX*

With the rise of imperial Christianity and the suppression of the ancient Gnostic and Pagan religious systems that preserved these arcane techniques, the remaining initiates went underground. Many retreated beyond the reach of the Christian authorities of the Byzantine Empire into the Mesopotamian frontier town of Harran, an ancient city located in what is now Turkey. There, the Pagan traditions of astral religion, Hermeticism, and Neoplatonism persisted and were tolerated by the Muslim authorities, who gained control of the city in 640 CE. The Sabians, as the natives of Harran were known, gained tolerance as "people of the book" and monotheists, having as their prophet Hermes Trismegistus, whom the Islamic conquerors identified with the biblical prophet Enoch, known in Arabic as Idris.[4] The eighth-century Sabian author Thabit ibn Qurra, greatest of the Harranian mages, master of mathematics, geometry, astronomy, astrology, and magic, practiced the ancient art of astrological magic, the science of talismanic images, and transmitted it to the Islamic world, where it was taken up by Muslim Neoplatonists such as the Brethren of Purity.[5]

As mentioned in the introduction, ideas were reintroduced to the European world in a book written in tenth-century Andalusia by a Muslim polymath known as Maslama al-Qurtubi, known as the *Picatrix* in its Latin translation. The text was originally given the Arabic title *Ghayat al-Hakim*, meaning "Goal of the Wise." It is an anthology that the author, "one wise philosopher," according to the mysterious author himself, compiled "from over two hundred works of philosophy,"[6] write Dan Attrell and David Porreca in their translation. A compendium of philosophy and folklore from a variety of traditions, including Hellenistic, Hermetic, Sabian, Indian, and Nabatean, to name just a few, the *Picatrix* mainly deals with the theory and practice of astrological magic. It also has been referred to as a medieval "black ops" manual, cutting-edge tenth-century science in the service of the wealthy and powerful of the day to allow them to gain an advantage over their rivals by occult means.

The *Picatrix* is a window into another time, and there are many aspects of this fascinating book that don't translate well in the twenty-first century. The text has recipes for drugs, deadly poisons, grisly and bizarre experiments that use human and animal parts, and petitions to various astral powers that call for animal sacrifice. There are recipes that would be right at home in the cookbooks of the witches from *Macbeth*. None of this material is what draws me to the *Picatrix*; I mention it to warn the sensitive that there is certainly some shocking content. The tenth century was a very different time, when people lived closer to death and violence than most of us do in the West these days. Just keep that in mind when you read primary magical source texts from this period.

The magical techniques described in the text may be understood as making contact with celestial powers when those powers are enhanced by their position in the heavens, by electing with painstaking care an auspicious moment to prepare an image that is symbolic of the power one wants to invoke, by suffumigating with a harmonious incense, and by performing a ritual with prayers and petitions to the correct heavenly powers. This requires skill in astrology, as well as artistic or artisanal capabilities and the patience to wait for the opportune moments when the stars are in the precise condition for their powers to be drawn down.

The Picatrix's View of the Cosmos

Worldview is fundamental to magic. In order for magic to be effective, the magician must live in an enchanted cosmos, the kind of world where magic can happen. That means fundamentally reexamining one's assumptions about how reality is constructed. We have already discussed cosmology at length, but it is crucial to the understanding of the astral magic of the *Picatrix* and related texts to know something about the kind of thought-world it inhabits. The text was composed some seven hundred years before Galileo and six hundred years before Copernicus and so therefore shares with all of traditional astrology the view that the Earth is the center of the cosmos. Although today we know that this is not literally true, this is still the perspective one observes when looking out at the stars from our terrestrial home. This view, articulated in ancient times by Plato and then developed by Aristotle and systematized by later thinkers like Ptolemy, saw the *cosmos* (the Greek word for "order") as composed of a series of moving concentric spheres with the Earth at the center. As stated in book one of the *Picatrix*:

> All things in the world are ordered in grades according to the things that govern them. The first of all things of this world, nobler, higher, and more perfect than anything found in the world, is God Himself, who is the shaper and creator of all. There follow in order . . . after consciousness, spirit; and after spirit, matter; and this matter [aether or quintessence] is immobile [and] unalterable. After this comes the sphere of nature, which is called the prime mover of all things moved, and is the source of all generation and corruption that happens in the world. Next comes the sphere of the fixed stars, and below this are found the other sphere[s] in order down to the sphere of the Moon."[7]

We have discussed the basic Platonic and Ptolemaic cosmology in chapter 3, so there is no need to repeat the particulars. Here we want to emphasize the idea that communication between the spheres and ontological levels of creation is possible under certain conditions because this cosmos is endowed with life and intelligence at all levels. In this view the prayers and petitions of the magician move up the hierarchy, as it were, as the influences of the stars and planets, enacting the will and providence of the creator, move down.

Plato believed that the creator created the soul of the universe and "formed within her the corporeal universe" and united the two. He wrote, "The soul, interfused everywhere from the center to the circumference of heaven, of which she is the external development, herself turning in herself, began a divine beginning of never-ceasing and rational life enduring throughout all time. The body of heaven is visible, but the soul is invisible and partakes of reason and harmony."[8]

An eternal harmony and order prevails throughout the entire creation, which the ancients believed to be divinely ordained and named *pronoia*, or providence. All of the parts from the highest to the lowest were believed to be in sympathy or invisibly connected by the divine soul of the cosmos, such that all parts were in communication. The philosopher Plotinus stated it like this: "The motion which directs the universe and the revolution of the stars arranges each thing in accordance with their relationship to each other brought about by their aspects and risings, settings and conjunctions."[9]

·SATVRNVS·

◆ *A depiction of the planetary intelligence of Saturn, from the fifteenth-century*
astrological treatise Sphaerae coelestis et planetarum descriptio.

In addition to the fundamental principle of sympathy or correlation between the movements of the heavens and events in the terrestrial sphere, in the magical world of which the *Picatrix* partakes, it is postulated that by communication with the celestial intelligences it is possible to change how the manifestation of these forces plays out in the sublunar realm in which we live. Indeed the *Picatrix* states that the "attraction of celestial spirits" through astrology is fundamental to magic. Plotinus states that "this universe is 'one living being encompassing all the living beings inside itself,' having one soul extending to all its parts. . . . This unified universe is in a condition of sympathy, and is one in the manner of a living being, and the distant parts are actually close together."[10] So the task of the magician, then, is to use the web of cosmic sympathies to draw down influences from one part of the universe to another.

Asking, "How are we to explain the operations of magic?" Plotinus answers, "They are due to the operation of sympathies, and because there is by nature a concord of things that are the same and an opposition of the things that are not the same, and by the variegation of the many powers that contribute to the one living being. Indeed many things are attracted and bewitched without any other person contriving it; and in fact the real magic is the 'Love' in the universe and the 'Strife' which accompanies it."[11] He goes on to say that the magicians "use figures endowed with powers, and by working themselves into certain figures, effortlessly bring those powers to themselves, being part, as they are, of one living being."[12]

Plotinus states that, "When one prays to a heavenly being, something emanates from that being to him or something else." Astral magic, as practiced in the *Picatrix*, postulates the mediation of what are referred to variously as celestial spirits, intelligences, daimones, or angels. The universe of the magician is home to a myriad of spiritual beings on all levels mediating between all parts of the cosmos. These are the chains of sympathy that connect the distant parts of the one living being that is the universe. Plato says:

> It is, of course, the stars and the bodies we can perceive existing along them that must be named first as the visible gods [the planets], and the greatest, and most worshipful, and clear sighted of them all; after them and below them, come in order the daemons and creatures of the air, who hold the third and midmost rank, doing the office of interpreters, and should be particularly honored in our prayers, that they may transmit comfortable messages. Both sorts of creature, those of the ether and those of the air . . . are wholly transparent; however close

· IVPITER ·

✦ *A depiction of the planetary intelligence of Jupiter, from the fifteenth-century astrological treatise* Sphaerae coelestis et planetarum descriptio.

they are, they go undiscerned. . . . The universe being thus full throughout of living creatures, they all, so we shall say, act as interpreters . . . of all things, to one another and to the highest gods, seeing that the middle ranks of creatures can flit so lightly over the earth and the whole universe."[13]

As we have seen, in the ancient Platonist cosmology, the stars and planets were seen as visible gods, the physical bodies whose souls were the intelligible gods, dwelling above the spheres, outside the physical cosmos, tasked by the creator with the regulation of time and administration of providence in the lower cosmos. Appearing to mark the time with their regular and predictable movements, they were believed to be the dispensers of fate, as archaeologist and historian Franz Cumont states: "Undoubtedly Destiny holds sovereign sway over the whole world, and the celestial orbs by their combined movements are the authors of all that was, and is, and is to come."[14] Cumont goes on to say the following of the role of the stars in the astral religion of the ancient Greeks: "Each of the constellations, each star which glittered in the cosmic vault, was equally divine."[15] The divine power of the creator, which creates and maintains the universe in existence, emanates from the creator outside the cosmos, beyond the ninth sphere, through the fixed stars. It is imprinted by the gods of the zodiac and the decans, which govern time, fate, and becoming, through each of the whirling spheres of the planets, each adding their influence, down through the elements to bring about creation, destruction, and change on Earth.

Each of the visible gods had a quality or dynamic that they imparted to the material world: Saturn, cold, distant, and slow moving, imparts the qualities of permanence, coldness, age, and wisdom to the spheres below; Jupiter imparts generosity, abundance, rulership, and benevolence; Mars imparts violence and bellicosity, as well as energy and vigor; the Sun imparts rulership, wealth, warmth, and light; Venus imparts beauty, friendship, love, and harmony; Mercury imparts swiftness, intelligence, cunning, stealth, and occult power; the Moon imparts change, growth, mystery, and emotion. The angles and interplay between these bodies creates the qualities of manifestation here below.

The stars and planets, however, are living spiritual forces. *Picatrix* scholar Liana Saif states, "Despite talking of the stars and planets as causes and signs, we must not understand the celestial dynamics to be mechanical. Vital and volitional agents transmit astral influences to the world below, which are called rūḥāniyyāt, spiritual powers."[16] The *rūḥāniyyāt* are the angels that govern the heavenly spheres. The intellectual model of astral magic expressed in the *Picatrix*,

· VENVS ·

◆ *A depiction of the planetary intelligence of Venus, from the fifteenth-century astrological treatise* Sphaerae coelestis et planetarum descriptio.

Saif writes, "combines the Neoplatonic anima with the Stoic view that the cosmos is a network of causal channels through which the vivifying power of the pneuma flows."[17]

The mysterious group of Muslim intellectuals known as the Brotherhood of Purity were Neoplatonist scholars and mystics who were contemporaries of Maslama al-Qurtubi, author of the *Picatrix*. "Their treatment of celestial mediation greatly influenced" his understanding of the workings of celestial causation, states Saif. In their body of works, known as *The Epistles of the Brethren of Purity* or the *Rasa'il*, Saif says, "They explain that the celestial bodies are the causes for the stability of the forms of species and genera in their bodies. They experience change and transformation due to the motion of the planets and stars. The celestial bodies are defined as 'the instruments' of the Universal Soul which animates the world, propels the spheres and produces the forms themselves."[18] These spiritual agencies are vital for the working of the cosmic system. Saif explains, "These souls are the manifestations and multiplications of the power of the Universal Soul. In their *Rasail*, the *Ihwan al-Safa* explain that 'the first power which flows from the universal soul towards the world is in the noble luminous entities that are the fixed planets [the fixed stars], then after them into the moving planets.' They use the term rūhāniyyāt to describe the localizations of this power in the planets, being the agency whereby planets influence the world below."[19] Magic can only be practiced "when the planets and their *rūhāniyyāt* are in beneficial states,"[20] when the celestial bodies are well dignified and well aspected, which requires precise knowledge of astrology.

COSMOLOGY IN OTHER TRADITIONS

Spiritual powers governing the planets and stars, and through them administrating and influencing fate in the terrestrial sphere, are part of many diverse traditions. The Babylonians worshipped the stars, which were seen as the personifications of their gods. As Cumont tells us:

Among the stars the most important were conceived to be the moon and the sun . . . then the five planets, which were . . . identified [with] the principal divinities of mythology. To them was given the name of Interpreters, because, being endowed with a particular movement not possessed by the fixed stars, which are subject to a motion of their own, they above all others make manifest the purposes of the gods. But worship was also bestowed on all the constellations of the firmament, as the revealers of the will of Heaven, and in particular on the twelve signs of the zodiac, and the thirty-six decans, which were called the Counsellor Gods; then, outside the zodiac, on twenty-four stars, twelve in the northern, and twelve in the southern hemisphere, which, being sometimes visible, sometimes invisible, became the Judges of the living and the dead. All these heavenly bodies . . . announced not only hurricanes, rains, and scorching heats, but the good or evil fortune of countries, nations, kings and even of mere individuals.[21]

It is interesting that here the decans are mentioned in connection with Babylon, when they originated in ancient Egypt as gods connected with the heliacal rising of certain stars and asterisms, associated with ten-day increments of time and ten-degree segments of the ecliptic, and later the zodiac. These gods were rulers of time, or *chronocrators*, believed to be responsible for the government of the cosmos and of fate. The oldest Egyptian texts regarding the decans are Coffin Texts from the Ninth and Tenth Dynasty in ancient Asyut. Dorian Greenbaum, in *The Daimon in Hellenistic Astrology*, states, "Since the decans are stars of constellations, the deities linked to them thus became a part of religious ritual and astral divination. . . . As representatives of decan stars, sub-beings very much like daimons represent and report to superior gods."[22] In another ancient Egyptian context, Greenbaum writes, "Decans are associated with arrows of protection sent by gods of heaven, earth, and the underworld." These star powers, "associated with the decans and their gods could both attack and protect, punish and guard, those with whom they came in contact. They could both cause illness and be propitiated to prevent it. Because the decans rise and culminate at 10-day intervals, their influence is activated at these particular periods. Thus the Egyptian tradition perceives these decanal beings, like daimons, as both destructive and protective."[23]

The Hermetic tradition of Hellenistic Egypt further developed the concept of decans as world-administering, daimonic entities who act as "careful guardians and overseers of the universe." According to the sixth Hermetic fragment of Stobaeus, which is framed as a philosophical dialog between a teacher to his student:

The force which works in all events that befall men collectively comes from the Decans; for instance, overthrows of kingdoms, revolts of cities, famines, pestilences, overthrowings of the sea, earthquakes—none of these things, my son,

◆ *Devils from the astrological and magical manuscript* Kitab al-Bulhan *or* Book of Wonders, *composed in fourteenth- and fifteenth-century Baghdad.*

take place without the workings of the Decans. . . . For if the Decans rule over the seven planets, and we are subject to the planets, do you not see that the force that is set in motion by the Decans reaches us also, whether it is worked by the Decans themselves or by means of the planets?[24]

The Hermetic writings present a hierarchically structured universe of spiritual entities that dispense fate and fortune to mortals in the sublunar realm. In the words of Book 16 of the *Corpus Hermeticum*, "The intelligible cosmos, then, depends from god and the sensible cosmos from the intelligible, but the sun, through the intelligible cosmos and the sensible as well, is supplied by god with the influx of the good. . . . Around the sun are the eight spheres that depend on it: the sphere of the fixed stars, the six of the planets, and the one that surrounds the Earth. From these spheres depend the demons [daimones] and then, from the demons, humans. Thus all things and all persons are dependent from god."[25] These daimones, although this text doesn't use the word *decans*, play the same cosmic function of the decans in the Stobaeus Hermetic fragment above, as "they follow the orders of a particular star, and they are good and evil, according to their nature." In nearly identical wording, we are told that they "oversee human activity" and that "when the gods enjoin them they effect through torrents, hurricanes, thunderstorms, fiery alterations and earthquakes; with famines and wars, moreover, they repay irreverence," which "is mankind's greatest wrong against the gods."[26]

Over and over, we find references in the ancient literature to cosmic beings administering fate and dispensing both fortune and misfortune to lowly humans in the material world. We seem to be at the mercy of these lofty astral powers, without, it seems, recourse to change our fate. Astrology can help us learn our fate, perhaps reconcile us to it, but in ancient times it was the task of priests to intercede with these cosmic powers and for magicians to compel them. The distinction in practice, however, is not so well defined; many priests in ancient times performed what we may call magic, and many magicians were also priests. Indeed, our word *magician* derives from the Persian word *magos*, meaning a Zoroastrian priest. Hans Dieter Betz, the great papyrologist, explains the situation like this: "Humanity is essentially at the whim of the forces of the universe. Religion is nothing but taking seriously this dependency on the forces of the universe." The magician, Betz says, "claimed to know and understand the traditions of various religions. While other people could no longer make sense of the old religions, he was able to. He knew the code words to communicate with the gods, the demons, the dead. He could tap, regulate and manipulate the invisible energies."[27]

♦ *Detail of the relief depicting the zodiac on the ceiling of the Temple of Hathor in Dendera, Egypt. The decans can be seen represented as small figures around the circumference of the inner circle.*

The Gnostic sects of Hellenistic and late antiquity incorporated the astrological cosmology into their strongly dualistic systems, believing that the planets, stars, and decans were malevolent demons, known as archons (Greek for "ruler"), who ruled over the cosmos. According to Greenbaum, "The seven planetary archons are considered to be demons ruling over the world of corruptible matter. The creator of the planetary realm is a demiourgos often called Yaldabaoth. Thus in Gnosticism, the planetary gods of the Greeks are transformed into archontic demons, each ruling a material world which the soul must pass through and transcend in its ascent to heaven."[28] Whether the planets are angels or demons, the Gnostics probably practiced some form of what we now call astrological magic. In the writings of Irenaeus, the Christian bishop of Lyon and notorious heresy hunter who recorded the doctrines of the Gnostic sects he opposed, the early Gnostic teachers Saturninus and Basilides are reported to "practice magic; and use images, incantations, invocations, and every other kind of curious art. Coining names as if they were those of angels, they proclaim some of these as belonging to the first, and others to the second heaven; and then they strive to set forth the names, principles and powers of the three hundred sixty five imagined heavens."[29] It is very likely that the Gnostics and their baroque cosmologies of heavens populated by angels and archons with strange, barbarous-sounding names contributed to the traditions that eventually came to be Western magic. In the intellectual melting pot of Hellenistic antiquity, the various magical, cosmological, and religious systems of the eastern Mediterranean blended together into multiple local syncretisms. Zoroastrianism from Persia, Mesopotamian astral religion, Judaism, and Egyptian and Hellenic polytheism all influenced one another to a greater or lesser degree. Astrological magic in general, and the *Picatrix* in particular, are products of this syncretism.

Judaism absorbed the angelology of the Persians and the astrology of the Babylonians during the Babylonian captivity and the subsequent domination of Palestine by the Persians. The gods presiding over the planets, the decans, and the zodiac itself were absorbed into the spiritual hierarchy of "monotheistic" Judaism. No longer gods but angels, these beings still administered fate and the cosmos, but at the behest of the Most High, Yahweh.

The Enochic literature, the various apocalyptic Books of Enoch, as we have seen, deals with the heavenly ascent of the biblical patriarch Enoch, who is taken bodily through the seven planetary spheres, here referred to as the seven heavens. Then he's taken through the eighth heaven, according to 2 Enoch, "called in the Hebrew tongue the *Muzaloth*, changer of seasons, of drought, and of wet, and the twelve constellations of the circle of the firmament,"[30] next the ninth and all the way to the tenth heaven, the very throne of God. Along the way, he encounters the various archangels and their throngs of subordinate angels who execute the will of God in the cosmos. They make the weather and make the Earth bear fruit, punish the sinners, and bless the righteous. The names and descriptions of these angels are given, and they were called upon in the Jewish magical tradition to come to the aid of the magician. The much later kabbalistic tradition holds, in the words of Aryeh Kaplan's commentary on the earliest text of that tradition, the *Sefer Yetzirah*, that "the stars also form an important link in God's providence over the physical world. Between God and man, there are many levels of interaction, the lowest being those of the angels and the stars. The Midrash thus teaches, 'there is no blade of grass that does not have a constellation (*Mazal*) over it telling it to grow.'" He continues, "God's providence works through angels, but these angels, in turn, work through the stars and planets. As some authorities put it, the angels are in a sense, like souls to the stars."[31] Kaplan goes on to say "that stars and planets are the 'bodies' of these angels. As such, each star serves as a focus for a particular angel, maintaining it as an integrated whole, even though it may have many different tasks. . . . There is, therefore, a one-to-one relationship between the stars and their angels. Each star has its particular angel and each angel its particular star."[32]

AGRIPPA'S COSMOLOGY

The relevance of achieving effective communications with the spiritual intellences that govern the weather, the growth of plants, and everything in between to the practice of the cunning farmer is obvious. If we can accept the notion of the old cosmology being a spiritual analogue for the metaphysical processes governing the playing out of archetypal forces in the material world, then we can use this system as a magical paradigm for communicating with those forces, the celestial spirits.

In order to better understand what celestial spirits are in the magical tradition and how the magician is to interact with them, we will have to jump forward to the Renaissance and to the sixteenth-century thinker Henry Cornelius Agrippa and his compendium of occult lore, *Three Books of Occult Philosophy*, which we have discussed in other chapters. Agrippa, following Platonic tradition, divides the cosmos into three realms: the elementary, or realm of the elements; the celestial, or realm of the stars and planets; and the intellectual, the transcendent realm of the divine mind. Each realm gives rise to its own particular branch of magic. By using the sympathies of the cosmic soul, we may align our operations with the celestial bodies and use their influence to work changes in the elementary realm of matter. As Agrippa states, "The celestial souls send forth their virtues to celestial bodies, which transmit them to this sensible world. For the virtues of the terrene orb [the Earth] proceed from no other cause than the celestial. Hence the magician that will work with them, useth a cunning invocation of superiors, with mysterious words, and a certain kind of ingenious speech, drawing the one to the other, yet by a natural force through a certain natural agreement betwixt them, whereby things follow of their own accord, or are sometimes drawn unwillingly."[33] He continues, quoting a pseudepigraphic work attributed to Aristotle (which sounds suspiciously like one of our quotations from Plotinus above), "When anyone by binding or bewitching doth call upon the Sun or other stars, praying them to be helpful to the work desired, the Sun and the other stars do not hear his words, but are moved by a certain conjunction, and mutual series, whereby the parts of the world are mutually subordinate the one to the other, and have a mutual consent, by means of their great union."[34]

In the great chain of being, inferior things can move superior things by sympathetic resonance, "so when anyone moves any part of the world, other parts are moved by the perceiving of the motion of that." The influence goes from "lower" to "higher" and back in the system of resonance and sympathy that is the living ensouled cosmos of the astral magician:

> Celestial bodies therefore move the body of the elementary world, compounded, generable, sensible, from the circumference to the center, by superior, perpetual, and spiritual essences, depending on the primary intellect, which is the acting intellect: but upon the virtue put in by the word of God . . . the cause of causes, because from it are produced all beings . . . . The word therefore is the image of God, the acting intellect is the image of the word; the soul is the image of this intellect; and our word is the image of the soul, by which it acts upon natural things naturally, because nature is the work thereof.[35]

In this chain of images, causation can work upward as well as downward, which is how the magician uses words and symbols here in the space and time of the elemental world to attract the celestial souls to work wonders.

Agrippa counsels the magus not to mistake the physical star for the celestial spirit that ensouls it and to not forget that magic, for the astral magician, is a spiritual labor by which one raises one's consciousness to the contemplation of the celestial souls with which he or she is engaging, being transformed thereby: "Our soul therefore, if it will work any wonderful thing in these inferiors, must have respect to their beginning. . . . Therefore we must be more diligent in contemplating the souls of the stars than their bodies, and the supercelestial, and intellectual world, than the celestial corporeal, because that is more noble, although this be excellent [the physical stars, planets, and constellations], and the way to that; without which medium the influence of the superior cannot be attained to."[36] In other words, the way of the astral mage is the way to the higher divine realm. No one can hope to bring down the celestial powers without first raising themselves up to them. Giving the example of the Sun, which receives its true light, its being from the intellectual world above the stars, he says:

> He that desires to attract the influence of the Sun, must contemplate upon the Sun, not only by the speculation of the exterior light, but also the interior. And this no man can do unless he return to the soul of the Sun and become like to it, and comprehend the intellectual light with an intellectual sight as well as with a corporeal eye. . . . And when he hath received the light of the supreme degree, then his soul shall come to perfection, and be made like to the spirits of the Sun . . . if he hath attained faith in the first author."[37]

In other words, to obtain success in the art of astrological magic, it is necessary to become celestial, to raise one's soul to the level of the celestial spirits one is seeking to invoke.

Agrippa is worth quoting at length here, because he sums up the whole theory of astral magic very neatly. The celestial intelligences are appointed "for the government of every heaven and star, whence they are divided into so many orders, as there are heavens in the world, and as there are stars in the heavens." There are angels who rule each of the planets, "angels, who might rule the signs, triplicities, decans, quinaries, degrees and stars. . . . Twelve princes of the angels who rule the twelve signs of the Zodiac, and thirty-six who may rule the decans, and seventy-two which may rule the quinaries of heaven, and the tongues of men and nations, and four which rule the triplicities and elements, and seven governors of the whole world, according to the seven planets." The astrologers have given them "names and seals, which they call characters, and used them in their invocations, incantations, and carvings, describing them in instruments of their operations, images, glasses, rings, papers, wax lights, and such like; and if any time they did operate for the Sun, they did invoke by the name of the Sun, and by the names of solar angels, and so of the rest." By means of these techniques, the astral magicians call down the assistance of the celestial powers, who "being visible to none, do direct our journeys and all our businesses, and are oft present at battles, and by secret helps do give the desired success to their friends, for they are said, that at their pleasures they can procure prosperity and inflict adversity." There is, Agrippa says, "no part of the world destitute of the proper assistance of these angels."[38] One only needs to know the proper times, names, and symbols with which to communicate with them.

◆ *Copper astrological talisman with an inscription in Arabic, from the Mamluk dynasty of Egypt and Syria (1250–1517). Image by Sailko.*

WORKING WITH CELESTIAL POWERS

This is the task of the magician, to know how to communicate with the hidden forces that govern manifestation in the material world and by interacting with them to raise his or her consciousness to the divine. As Plato said, "And the motions which are naturally akin to the divine principle within us are the thoughts and revolutions of the universe. These each man should follow, and by learning the harmonies and revolutions of the universe, should correct the courses of the head which were corrupted at our birth, and should renew the thinking being to the thought, renewing his original nature, so that having assimilated them he may attain to that best life which the gods have set before mankind, both for the present and the future."[39]

Practically, then, how do we work with the celestial powers for magic? How do we bring their influence to bear in the sphere of our lives? The ancients, following Plato, believed, because form is the organizing principle in the universe, that if one were to craft an artistic image congenial to the spirit of the planet, star, constellation, sign, or decan, at the precise time that that celestial entity was in a potent astrological state, in a condition of positive dignity—linking it also by harmonious aspect to the Moon, which as we have seen conducts the "higher" forces into the lower realm, as she is the mediatrix of celestial power to the elemental realm in which we live—then the form would become manifest in the material world. As Renaissance astrologer and magician Marsilio Ficino wrote in his classic work, *Three Books on Life*, "When we fashion images rightly, our spirit, if it has been intent upon the work and upon the stars through imagination and emotion, is joined together with the very spirit of the world and with the rays of the stars through which the world-spirit acts. And when our spirit has been so joined . . . a certain vital power, is poured into the image."[40] According to Ficino, the mind, imagination, and emotion of the operator are as critical to the success of the magic as the timing and the material. This is a subset of alchemy, our astral magic, in which the mind of the operator is the vessel in which the reaction is accomplished.

If you want to do astrological magic—as with any magic—you have to decide what your aim is. If you want love, for example, you may work with Venus. If it's money you want, Jupiter or the Sun. If you want protection from harmful spiritual forces, you might choose the fixed star Algol, or maybe you could make a lunar mansion talisman. You may find, however, that especially with some of the slower-moving planets, you may have to wait years for Jupiter or Saturn to get in the condition of astrological power required to make an effective talisman. There are other magical methods to accomplish the same goals, without such exacting electional requirements, such as making a talisman for a fixed star or lunar mansion with a similar quality or effect.

Another, simpler way to use the powers of the planets for magical effect is to use ancient systems of planetary days and hours. In ancient Greece, the days of the week were assigned to the rulership of the seven planets, according to the order of the planets in the Chaldean system, filtered through the principles of sacred musical harmony. Each planetary intelligence was a *chronocrator*, a time-ruler, ruling over the days and hours according to an orderly and mathematical division of the days of the week as well as the hours of the day. The concept of the planetary day spread throughout the Roman Empire and was picked up by the so-called barbarian nations and is reflected in our modern English names for the days, each day having been assigned to the rulership of one of the gods of the Anglo-Saxon people:

Sunday is the Sun's day.

Monday is the Moon's day.

Tuesday (Tew's day) is the day of Mars, because Tew was the Germanic god of War.

Wednesday (Woden or Odin's day, considered equivalent to Mercury) is Mercury's day.

Thursday (Thor's day, the Germanic Jupiter) is Jupiter's day.

Friday (Freya's day) is Venus's day, as Freya was considered the Germanic version of Venus.

Saturday (Saturn's day) is for Saturn, of course.

Planetary hours and days are periods of time in which the magical influence of the planet ruling that day or hour is more strongly felt. The seven days of the week we have discussed, but similarly, the twelve hours of the day and the twelve hours of the night are similarly divided between the seven traditional planets, reflecting both an orderly and precise mathematical divison of both time and the geometrical space of the zodiacal sphere of 360 degrees. A planetary day starts at sunrise, and the first hour is assigned to the ruler of that day. The other planets follow *in reverse Chaldean order*, which is the traditional order of the planets in astrology, based on the speed of their motion. For instance: On Sunday, the first hour belongs to the Sun, the second, Venus, the third, Mercury, the fourth, the Moon, followed by Saturn, Jupiter, and Mars. Each planet is likewise assigned an archangel who governs the affairs of that sphere. Following Agrippa and Francis Barret, whose table is reproduced on page 182 for your convenience, Michael is assigned to the Sun; Anael, or Haniel, is assigned to Venus; Raphael to Mercury; Gabriel to the Moon; Cassiel (or more properly Tzaphkiel) to Saturn; Sachiel (Tzadkiel) to Jupiter; and Samael to Mars.

To use planetary hours for magical pursuits, it is merely necessary to time one's magical operations for the day and/or hour of the planet that rules those operations. For example, for aggressive magic, Mars day and hour would be appropriate; for love magic, Venus day and hour would be best; for communication magic, Mercury day and hour; and so forth. Even mundane pursuits—including those of an agricultural nature—can be timed in this way. Thus working soil can be begun on Saturday, during the hour of Saturn. Planetary days and hours are

I shall here set down the Table of the names of Spirits and Planets governing the hours; so thou shalt easily know by inspection, what Spirit and Planet governs every Hour of the Day and Night in the Week.

Hours Day.	Angels and Planets ruling SUNDAY.	Angels and Planets ruling MONDAY.	Angels and Planets ruling TUESDAY.	Angels and Planets ruling WEDNESDAY	Angels and Planets ruling THURSDAY.	Angels and Planets ruling FRIDAY.	Angels and Planets ruling SATURDAY.
	Day.	*Day.*	*Day.*	*Day.*	*Day.*	*Day.*	*Day.*
1	☉ Michael	☽ Gabriel	♂ Samael	☿ Raphael	♃ Sachiel	♀ Anael	♄ Cassiel
2	♀ Anael	♄ Cassiel	☉ Michael	☽ Gabriel	♂ Samael	☿ Raphael	♃ Sachiel
3	☿ Raphael	♃ Sachiel	♀ Anael	♄ Cassiel	☉ Michael	☽ Gahriel	♂ Samael
4	☽ Gahriel	♂ Samael	☿ Raphael	♃ Sachiel	♀ Anael	♄ Cassiel	☉ Michael
5	♄ Cassiel	☉ Michael	☽ Gahriel	♂ Samael	☿ Raphael	♃ Sachiel	♀ Anael
6	♃ Sachiel	♀ Anael	♄ Cassiel	☉ Michael	☽ Gahriel	♂ Samael	☿ Raphael
7	♂ Samael	☿ Raphael	♃ Sachiel	♀ Anael	♄ Cassiel	☉ Michael	☽ Gahriel
8	☉ Michael	☽ Gabriel	♂ Samael	☿ Raphael	♃ Sachiel	♀ Anael	♄ Cassiel
9	♀ Anael	♄ Cassiel	☉ Michael	☽ Gabriel	♂ Samael	☿ Raphael	♃ Sachiel
10	☿ Raphael	♃ Sachiel	♀ Anael	♄ Cassiel	☉ Michael	☽ Gahriel	♂ Samael
11	☽ Gabriel	♂ Samael	☿ Raphael	♃ Sachael	♀ Anael	♄ Cassiel	☉ Michael
12	♄ Cassiel	☉ Michael	☽ Gabriel	♂ Samael	☿ Raphael	♃ Sachiel	♀ Anael
Hours Night.	*Night.*	*Night.*	*Night.*	*Night.*	*Night.*	*Night.*	*Night.*
1	♃ Sachael	♀ Anael	♄ Cassiel	☉ Michael	☽ Gahriel	♂ Samael	☿ Raphael
2	♂ Samiel	☿ Raphael	♃ Sachiel	♀ Anael	♄ Cassiel	☉ Michael	☽ Gahriel
3	☉ Michael	☽ Gabriel	♂ Samael	☿ Raphael	♃ Sachiel	♀ Anael	♄ Cassiel
4	♀ Anael	♄ Cassiel	☉ Michael	☽ Gahriel	♂ Samael	☿ Raphael	♃ Sachiel
5	☿ Raphael	♃ Sachiel	♀ Anael	♄ Cassiel	☉ Michael	☽ Gabriel	♂ Samael
6	☽ Gabriel	♂ Samael	☿ Raphael	♃ Sachiel	♀ Anael	♄ Cassiel	☉ Michael
7	♄ Cassiel	☉ Michael	☽ Gahriel	♂ Samael	☿ Raphael	♃ Sachiel	♀ Anael
8	♃ Sachiel	♀ Anael	♄ Cassiel	☉ Michael	☽ Gahriel	♂ Samael	☿ Raphael
9	♂ Samael	☿ Raphael	♃ Sachiel	♀ Anael	♄ Cassiel	☉ Michael	☽ Gahriel
10	☉ Michael	☽ Gabriel	♂ Samael	☿ Raphael	♃ Sachiel	♀ Anael	♄ Cassiel
11	♀ Anael	♄ Cassiel	☉ Michael	☽ Gabriel	♂ Samael	☿ Raphael	♃ Sachiel
12	☿ Raphael	♃ Sachiel	♀ Anael	♄ Cassiel	☉ Michael	☽ Gahriel	♂ Samael

Note, The day is divided into twelve equal parts, called Planetary Hours, reckoning from sun-rise to sun-set, and, again, from the setting to the rising; and to find the planetary hour, you need but to divide the natural hours by twelve, and the quotient gives the length of the planetary hours and odd minuets, which shews you how long a spirit bears rule in that day; as Michael governs the first and the eighth hour on Sunday, as does the ☉. After you have the length of the first hour, you have only to look in the Table, as if it be the fourth hour, on Sunday, you see in the Table that the ☽ and Gabriel rules: and so for the rest it being so plain and easy you cannot err.

✦ *Table of planetary hours and spirits from Francis Barrett's* The Magus *(1801).*

not a substitute for good astrological elections, but can be used to enhance good elections, or to get some planetary power when electional timing is not possible.

Incorporating Planetary Magic into Rituals

The first step of getting started with planetary magic is entering into a relationship with the planets as spiritual forces.

Daily planetary practice is something I started doing when I took a class on astrological magic with astrologer Christopher Warnock several years ago. He recommends a practice that consists of doing a daily devotion to the planetary spirits and/or angels of the planet that presides over that day in order to develop a relationship with the planetary spirits. As discussed earlier in this chapter, each day is governed by one of the seven traditional planets, and communication with the spirits of those planets is more effective on those days. Each planetary day starts at sunrise. The period between sunrise and sunset and the period between sunset and sunrise are each divided into twelve equal planetary hours, starting with the hour of the planet of the day and proceeding in reverse Chaldean order through the planets, repeating until sunrise the next day.

This division of day and hours under the governance of the planets reflects the fact that to ancient people the planets were not seen as rocks in space but rather as gods who ruled over time itself. Indeed, as time passed, planetary hours and days came to be seen as more important for magical purposes than the astronomical movements of the planets themselves. In the tradition I follow, we try to use both whenever we do very important work. For daily devotions, following Warnock, planetary day is sufficient. (Interested readers should visit Warnock's very helpful and informative website, renaissanceastrology.com, for more information about planetary practice.)

For basic planetary devotions, you do not need a circle or a banishing or a purification, just a prayer or one of the *Orphic* or *Homeric Hymns* to the planet or planetary god in question and maybe a colored candle or incense appropriate to the day. Traditional medieval magic sources such as the *Picatrix*, the *Hygromanteia of Solomon*, or the *Heptameron* also have prayers to planetary spirits and angels that you may want to incorporate into your practice. In the *Heptameron* the planetary prayers are listed as the Conjuration of Monday (and the other days of the week), and are in both seventeenth-century English and Latin. The full text of the *Heptameron* can be found in Agrippa's *Fourth Book of Occult Philosophy*, as well as Joseph H. Peterson's *Elucidation of Necromancy*. See the illustration on page 182 for a chart of the planetary hours and their angelic governors.

The seven planets of traditional astrology are considered to be celestial intelligences through which the world soul transmits its power to the sublunar sphere, our material world. These celestial influences govern events in the material sphere in an archetypal manner that in most cases transcends the realm of physical causation. Notable exceptions are the Sun and the Moon, whose influences on the elemental world are very physical and felt by all living things.

The actual manner of influence these celestial objects exert, and the means by which it is exerted, is and will remain mysterious. There has been much debate throughout history about whether the correspondences described by astrology are signs or causes. For our purposes here, the debate is not relevant. Our working hypothesis should be that the celestial intelligences are vast, living spiritual forces, agents of the creator and Source, with whom we interact and who shape our lives, all that lives, and reality itself.

In all the various branches of the Platonic tradition, the cosmos is seen as a continuous process of emanation, whereby the worlds are constantly bathed in Being and Life issuing from a transcendent source beyond space and time but eternally present. The signs and planets of astrology are living, moving symbols of the qualities that the energies issuing from the cosmic Source take as they manifest in the material realm.

Our bodies and psyches are imprinted with the energies present at the moment of our birth. At any given time, those energies are expressed more or less imperfectly. We have all been stamped by fate with certain deficiencies and strengths. The dignity of the planets expressed in our natal chart is a symbol of the gifts and challenges that we have been handed. The cultivation of relationships with the celestial intelligences can be seen as a way of remedying these imperfections. Devotional practices that engage with these spiritual forces, who are variously seen as gods or angels and administer the power of the Creator, is an important practice for anyone who intends to call upon these powers as allies in their work. Nothing less than linking the planetary powers imperfectly expressed in our own souls to the perfect archetypes of the divine mind is the goal of this practice. This practice is an excellent preparation, both spiritually and in terms of developing a working knowledge for other works of planetary magic, such as talismans and petitions, which we will discuss in future chapters.

I can personally attest to the palpable experience of the power of invoking a planetary archetype into one's life after a successful ritual, particularly a benefic like Venus or Jupiter. It can be experienced as a sort of emotional high, complete with favorable synchronicities.

To sum up, I'd like to outline a series of simple rituals for conducting planetary magic. Let's say its focus will be the lunar spirits of Monday. On Monday, Moon hour usually comes around 10:30 or 11:00 p.m., and my house is usually quiet enough to do magical work, so that is when I would start.

At that time I would prepare the altar with purple candles and burn a lunar incense, such as the head of a dried frog or the eyes of a bull—yes, Agrippa recommends these, but I'm just kidding! Maybe camphor or copal would be better.

A Monday Planetary Ritual

1. Perform the Lesser Banishing Ritual of the Pentagram (LBRP) and the First Degree Ritual of Opening. (These rituals are included in chapter 8 of this book.)

2. Next, perform the Lesser Invoking Ritual of the Hexagram (if you really want to get advanced you can do the Greater Hexagram Ritual, which is specific to the planet you are invoking).

3. Perform the Middle Pillar Ritual. (For those who aren't familiar with them, the rituals mentioned in this exercise can be found in John Michael Greer's *Circles of Power* and other sources. The Middle Pillar ritual is also included in chapter 9 of this book.)

4. Perform the Conjuration of Monday from the *Heptameron*.

5. Perform the *Orphic Hymn* to Selene. (There are many translations of the *Orphic Hymns* available. Eighteenth-century classicist Thomas Taylor's verse translation is widely available and in the public domain. I advise the reader to pick a version they like and to get familiar with these beautiful hymns to the gods for all occasions.)

6. Meditate on the lunar influence, asking a petition for any of the things governed by the Moon: psychic skills, intuition, divination, or perhaps plant growth.

7. Perform the First Degree Ritual of Closing. (This ritual is included in chapter 8 of this book.)

When you learn this basic form, you can adapt it to any planetary or astrological magical operation or to schedule any practical magical work for a day when the planetary energy is conducive to that work. You can elaborate it as you wish. It's like a recipe. Or to simplify, the second or third step—or both—can be dropped and the LBRP can be substituted for the opening ritual. In fact, that's a good way to start and add steps as you learn them.

To embark on this path of planetary magic requires basic astrological knowledge and the ability to read charts and use an ephemeris, which is a reference book with the daily positions of the planets. It is not too difficult, but it is an excellent way to learn astrology. It requires patience, in order to wait for the proper *kairos*, or time, to work with the powers in question. What recommends astrological magic to me, as a practitioner, is the delicious sense of participation it gives me, the feeling of participating in the life of the cosmos as a whole. The effects of the talismans, when properly elected and executed, are unmistakable. When working with benefic planetary powers, Jupiter and Venus, for example, there is a high that follows the consecration of the talisman, an elation, a kind and loving atmosphere, as well as a series of favorable coincidences, in which the effects manifest. I have made Mars and Saturn talismans as wards to protect my home and farm from danger, and have had no ill effects, but odd coincidences let me know they are working. Some people who are very sensitive psychically might actually be a bit put off by the presence of a strong warding talisman. But mostly I think that troublemakers are uncomfortable around them.

I have noticed since practicing astrological magic a gradual improvement in many areas of my life, both materially as well as emotionally and spiritually. The goals of this practice are nothing less than the elevation of one's spirit and the attainment of wisdom, more than any purely material gain. I have heard some extremely extravagant claims for success by some of the veteran practitioners of the art. I am inclined to believe them, as they are masters in both the theory and practice of the science of images. I am not quite there yet, but I am very satisfied by both the process and the results. It's like standing in a powerful current of energy. When you call down a celestial spirit, the presence is palpable.

Astrological magic is not the most accessible magic for beginners, requiring as it does technical skill in astrology as well as skill in making the jewelry or metal plates that are the most powerful talismanic objects. Paper talismans are easier for beginners but are not considered as powerful or long lasting. But you have to start somewhere. It is important to remember that practical results are only part of the desired outcome. Spiritual practice as participation in the procession and return of divine energy, from being to becoming and back again, is the actual goal of the practice.

Although it is possible to learn the skills necessary for practicing astrological magic by diligent study of the *Picatrix* applied to an existing knowledge of astrology, I personally found it very helpful to enroll in the astrological magic course offered by astrologer Christopher Warnock, who was responsible for reviving the nearly lost tradition and was instrumental in making a good modern English translation of the *Picatrix* available in an affordable edition. The course starts out with the basics of astrology: signs, planets, houses, essential dignity, and aspects. It then moves on to the election and making of the different types of astrological talismans. There are talismans made with planets, fixed stars, constellations, decans, and lunar mansions, as well as custom-made talismanic elections designed to enhance the power of one of the houses of traditional astrology, which allow the magician to focus the intention on a particular area of life, career, business, relationships, or wealth. For those lacking the time or inclination to learn astrological magic but who nevertheless want to give it a try, Warnock also has a wide variety of astrological talismans available for sale on his website.

11
EARTH MAGIC

He too is happy who knows the country gods,
The sister nymphs, and Pan, and old Sylvanus[1]

VIRGIL, *THE SECOND GEORGIC*

In this chapter, we continue our exploration of the energies and inhabitants of the various realms of the spiritual cosmos in which the cunning farmer operates. We descend the world axis, the great cosmic tree known to various traditions, and we reach the ground, the world of the elements and their intelligences, the elementals and nature spirits. These are the forces of nature, which in our animist reenvisioning of the universe we learn to see as living things, as ensouled beings, as persons, part of a holistic network of animate beings, connected to one another in a web of sympathy and mutuality. Relating to our environment, to nonhuman nature in this way, as spiritual or even divine, requires a shift in our perception and our values, and the shift begins in both our heart and our imagination.

The first step is to take stock of where you find yourself—at home, in your yard, at your local park, or on the land on which you live if, like me, you are a country dweller. In our modern world, many of us live in towns, surrounded by concrete, chemical lawns, twenty-four-hour artificial lights, and climate control, stepping from our home to the car to the office and back, only spending minutes outside per day. That we are alienated from the world of nature is putting it mildly. But even so, if that is your situation and you want to have greater contact with the natural world, the first step is to go outside. That means braving the heat, the cold, the bugs, and the wild twenty-first-century weather, which

✦ *Mother Earth, from the seventeenth-century alchemical book*
Atalanta Fugiens, *by Michael Maier.*

requires proper outfitting and planning. Do it by yourself, if you can safely. And declare yourself, introduce yourself to the powers of the place, the winds, the trees, the plants and animals, and their guardian spirits. Get quiet, sit down, and wait to see what comes to you, what thoughts run through your mind. If you can, separate the monkey-mind chatter from the quiet whisperings of the almost timeless spirits of place.

But what are the powers and spirits of a place? Are they energies, or do they have personalities? This is the same question we asked about the stars: Are we dealing with rays or spirits? The answer now, as it was then, is a resounding yes to both. The Earth-based cultures of the world, including our pre-urban, preindustrial ancestors, experience the natural world as populated with a host of entities, every culture having a veritable zoology of nature spirits and elementals, each having a name and attributes. Our culture in the postmodern era has largely lost this rich daimonology of land spirits. Most people can no longer hear the whisperings of the spirits in the wind and rain or catch a fleeting glimpse of a half-imagined being disappearing into the brush in a dense wood.

We need to reinvigorate the mythic imagination that is our birthright as human beings. Our modern world seems to be empty of meaning and context, but it is my sincere belief that this is an illusion caused by taking what we experience on the surface of things to be the only truth, rather than looking deeper into the true nature of our remarkable existence. Our prevailing mode of life despoils the natural world because we long ago desacralized it. The forests, seas, oceans, deserts, and mountains, which archaic humanity called home, were believed by them to also be home to myriads of spiritual beings; they were the very landscapes in which the primal mythic drama of creation played out. As the great Irish poet William Butler Yeats put it:

> Once every people in the world believed that trees were divine, and could take human or grotesque shape and dance among the shadows; and that deer, and ravens and foxes, and wolves and bears, and clouds and pools, almost all things under the sun and moon . . . were not less divine and changeable. They saw in the rainbow the still bent bow of a god thrown down in his negligence; they heard in the thunder the sound of his beaten water-jar, or the tumult of his chariot wheels; and when a sudden flight of wild duck, or of crows, passed over their heads, they thought they were gazing at the dead hastening to their rest. . . .[2]

Contrasting the archaic worldview with our own, Yeats continues:

> Men who lived in a world where anything might flow and change, and become any other thing; and among great gods whose passions were in the flaming sunset, and in the thunder and the thunder-shower, had not our thoughts of weight and measure. They worshipped nature and the abundance of nature, and had always, as it seems, for a supreme ritual that tumultuous dance among the hills or in the depths of the woods, where unearthly ecstasy fell upon the dancers, until they seemed the gods or the godlike beasts, and felt their souls overtopping the moon. . . . They had imaginative passions because they did not live within our own strait limits, and were nearer to ancient chaos, every man's desire, and

had immortal models about them. The hare that ran by among the dew might have sat upon his haunches when the first man was made, and the poor bunch of rushes under their feet might have been a goddess laughing among the stars; and with but a little magic, a little waving of the hands, a little murmuring of the lips, they too could become a hare or a bunch of rushes, and know immortal love and immortal hatred.[3]

Yeats means imagination as an organ of spiritual perception, and it is the methods of cultivating this imaginative passion in our relationship to nature and the land that will be the main subject of this chapter.

In the following pages I will review some basic techniques of ritual practice that I have found helpful for establishing a relationship with the spirits of place, as well as report some of the experiences I have had with what I believe to be land spirits following these practices and throughout my life. When I look back at my life, which has had more than its share of wandering in the woods and fields, with some acquired wisdom of experience and a willingness, perhaps lacking in my youth, to suspend skepticism, I can reinterpret the strange and sudden moods and whims that came over me during those times as the experience of the presence of spiritual forces. And following some of my ritual practices, I have had some striking signs and synchronicities, epiphanies that made me believe I had been heard and that my gestures had been appreciated.

We will also be exploring some of the lore and theory regarding the spiritual denizens of the natural world in the various traditions from which the cunning farmer draws—namely, the Western esoteric tradition and its Hermetic and classical antecedents. There is a lot of material to sift through, though, so forgive me if I leave out something you consider essential. Because the Celtic fairies have a link to burial mounds and the dead, the folklore regarding the fey traditions of the Celtic nations will also be mostly covered in a future chapter on the infernal or underworld realm. In this chapter, we will be focusing primarily on the spirits of place, plants, animals, weather, and terrestrial waters (rivers and streams). There is truly a great deal of overlap with fairy lore, but not the space to do more than acknowledge the parallels in this chapter.

CONNECTING WITH THE SPIRITS OF PLACE

The cunning farmer, being primarily an agricultural magician, works within cultivated nature, on the edge of wild nature, and must call upon the spiritual

♦ *Demeter, Persephone, and Triptolemus. From the Great Eleusinian Relief, initially found in Eleusis, Greece, and created in the mid–fifth century BCE. Image by TimeTravelRome.*

powers of both the garden and farm, personified by classical deities such as Demeter and Persephone, as well as the deities of wild nature, such as Pan and Silvanus.* These deities are known as *chthonic* deities, from the ancient Greek word meaning "soil" or "earth." The word *chthonic* refers to the deities of agriculture or the underworld, linking the soil to the land of the dead and its deities.† The eminent scholar of ancient Greek religion, Walter Burkert, describes a fundamental opposition in the religious worldview of the ancient Greeks between the bright ouranian (heavenly) gods of the Olympian pantheon, per-

---

*I work with mostly Greek names for gods and goddesses in my more Pagan practices, because the mythology and personalities of these deities are so well developed.

†"*Chthonic* deities" may refer to both gods of the Earth and gods of the underworld.

taining to the celestial and supercelestial realms, and the chthonioi, the gods of the earth.[4] The gods of the earth were also associated with the dead and the underworld. Hades and Persephone, the god and goddess presiding over the land of the dead, were regarded with fear by the ancient Greeks, but Hades is also Pluto, the giver of wealth, and Persephone represents the cycle of vegetation. The well-known myth of her abduction and subsequent return, giving rise to the withering of plants in autumn and their renewal in spring, became the basis of one of the most profound and long-lived mystery religions of antiquity, the Eleusinian Mysteries.

The linking of the cycle of vegetation with the life and death of mortal humans is ancient as well, probably as old as agriculture itself. Joseph Campbell, the mythographer, believed that the transition of human societies from nomadic hunter-gatherers to settled agriculturalists prompted a major religious shift. Paraphrasing the German anthropologist Leo Frobenius, Campbell wrote: "Two contrasting attitudes toward death appear among the primitive peoples of the world. Among the hunting tribes, whose lifestyle is based on the art of killing, and who live in a world of animals who kill and are killed and hardly know the organic experience of a natural death, all death is a consequence of violence and is generally ascribed not to the natural destiny of temporal beings, but to magic."[5] According to Campbell and Frobenius, "For the planting folk of the fertile steppes and tropical jungles, on the other hand, death is a natural phase of life, comparable to the moment of planting the seed, for rebirth." In the new relationship with planting crops, "The vegetable world supplied not only the food, clothing and shelter of man . . . but also his model of the wonder of life—in its cycle of growth and decay, blossom and seed, wherein death and life appear as transformations of a single, super-ordinated, indestructible force."[6]

This brings us back to the universal life force, which is personified all over the world in the dying and reborn gods and goddesses of vegetation who represent the annual death and rebirth of vegetation and crops. Like the hapless Persephone, abducted by cruel Hades to the dark and terrifying land of the dead, they rise again in glory and splendor as the newly sown wheat and the flowers of springtime.[7] Many other deities from the ancient world take this pattern, explored at length by pioneering anthropologist James George Frazer in his monumental study *The Golden Bough*. Many of the religious systems he describes take the form of what he terms sympathetic magic. By reenacting the annual death and rebirth of the gods and goddesses of vegetation, the peoples of the ancient world sought to stimulate the natural processes they were imitating.

Death and agriculture are intimately associated. Any beginning farmer quickly learns that the craft of farming involves killing in order to allow favored plants and animals to thrive. After all, one of the most enduring symbols of death is the Grim Reaper, and reaping is farmwork. The animal husbandry that has always been a part of the life of the farm can't be done without culling the herd. Roman writer Cicero noted that "All things go back to earth and rise up from the earth."[8] Thus, the Earth Mother receives all the living back into the soil to be reborn anew in another form. As the poet Aeschylus wrote, "Yea, summon Earth, who brings all things to life / And rears and takes all things to her womb." Affirming the association of the chthonic goddesses of the Earth and agriculture with the cult of the dead and the underworld, Jane Harrison notes, "And so the Mother herself keeps ward in the metropolis of the dead, and therefore [quoting Plutarch] 'the Athenians of old called the dead "Demeter's people."' On the festival day of the dead, the *Nekusia* at Athens, they sacrificed to Earth. To a people who practised inhumation, such ritual and such symbolism were almost inevitable. When the Earth-Mother developed into the Corn-Mother, such symbolism gained new life and force from the processes of agriculture. Cicero records that in his day it was still the custom to sow the graves of the dead with corn."[9]

The gods and goddesses represent the personification of universal spiritual energies, archetypal forces linking the imaginal and material cosmos. These are vast powers that have taken a personal form, or perhaps have been given a personal form, so that humans can interact with them. In themselves, they are not what they appear to be in the stories and myths. It must be remembered that myths are teaching stories to give voice to ineffable spiritual and philosophical truths, not accounts of the deeds of superhero-like characters. These stories point out spiritual truths, but like a finger pointing at the Moon is not the Moon, they are not truth itself. As the Roman Neoplatonist philosopher Sallust said, myths "assert to all men that there are gods; but who they are, and of what kind, they alone manifest to those who are capable of such exalted knowledge."[10]

Gaia and the Earth Goddesses

The most important, to my mind, of the earthly deities is the titaness Gaia, Mother Earth herself, spouse of the titan Uranus, whose name means "heaven." This is the primal couple, the first divine beings to emerge from Chaos, who engendered all of the gods and everything else that exists. Gaia is the living Earth, soul and body, nurturer and mother of all life. Another goddess, Demeter, the goddess of grain whose name simply means "the mother," is probably a more specific form of Gaia, the first mother of all. This mythology makes a primal mother the source of all material life. As Jane Harrison states in *Prolegomena to the Study of Greek Religion*, "These primitive goddesses reflect another condition of things, a relationship traced through the mother, the state of society known by the awkward term matriarchal." The ancient Earth mother, according to Harrison, was also known by the epithet Karpophoros, meaning "fruit bearer" or "productive," and also "Lady of Wild Things."[11]

So the most important of the farmer's gods are chthonic goddesses: Earth herself, from which all things spring; Demeter, the goddess of agriculture; and Persephone, the goddess who represents the rebirth of vegetation itself, dwelling in the darkness with Hades, as the Queen of the Dead, half the year and then springing back to life as the Sun begins to journey to the south in the spring. Another chthonic goddess, important to me personally and one of my favorites, is Hekate, often associated with the lunar goddess Artemis, who presides over crossroads and magic, and in mythology leads Persephone back from the land of the dead. She is spoken of in exalted terms as the feminine principle itself in the theurgic work the *Chaldean Oracles* and was also known as the patroness of witchcraft.

Pan and the Devil

Of the gods associated with the natural world, the most famous is the rustic god Pan, which literally means "all." He is usually pictured in art as half goat and half man playing the syrinx (another name for the Pan flute), who keeps watch over flocks as patron of shepherds. He is sometimes equated with a sort of wild masculinity, an unbridled, uncivilized counterpoint to the civilized masculinity of Zeus or Apollo. In modern neo-Pagan practice Pan has become the Horned God, the consort to the Great Goddess, and the attributes of nature gods and goddesses of many cultures have been elided into this divine power couple. There is nothing wrong with this bit of modern syncretism; it is in the nature of the archetypes to take on forms according to the spiritual needs of the human cultures in which they find themselves. In the ancient Roman world, Pan was equated with Silvanus, the god of uncultivated nature and forests who likewise presided over flocks and herds. Silvanus was, according to the *Interpretatio Romana* (the Roman practice of Romanizing the gods of the cultures with whom they came in

◆ *A statue of Pan from Pompeii, Italy, likely from the late Hellenistic period.*

contact), associated with various Gaulish and Celtic deities, such as Callirius, the Horned God of the British tribe the Brigantes.[12]

As Europe became increasingly Christianized and the gods of the Pagan nations were forced into the dualistic narrative of the dominant Christian culture, the symbolic attributes of Pan and the goat-like minor nature deities (the satyrs) were adopted as the form of the Christian personification of the evil principle, variously known as Satan or the Devil. This character, who is a syncretic blend of various Pagan deities and poorly defined characters from the Old and New Testaments of the Bible, was increasingly equated with the gods of every pantheon the conquering faith came across, especially the more chthonic ones. Many Traditional Witchcraft practitioners have taken this on and now assert a difference between the "folkloric" devil and the theological one, saying their devil, the one the witches were accused of worshipping, is the former rather than the latter and the confusion between the two is largely due to the churchmen equating the native deities worshipped by the country people with their own religion's evil principle. Modern neo-Pagan Wiccans, in contrast to the Traditional Witchcraft movement, which tends to embrace the transgressive nature of the role of the witch, have generally downplayed any association between their Horned God, who is simply the masculine fertility principle, and Old Nick.

My own approach, as a dual-faith practitioner and a monist, is to avoid associating the nature spirits and gods to whom I pray with the evil principle of the Christian faith. My own Hermetic understanding of Christianity is non-dualist and has no ultimate evil principle, and although I may call the ultimate principle the Good, it is not "good" in a moral sense, which implies opposition to "evil," and there is no duality in the One. However, without getting into

✦ *Hermes Kriophoros. From Greece, fifth century BCE. Image by Tetraktys.*

✦ Roman relief of Hermes with caduceus (in the Metropolitan Museum of Art).
Created sometime from 27 BCE to 68 CE. Image by Foto Ad Meskens.

too much of a digression, there is some truth in equating the chthonic gods of the underworld with the destructive passions that come from expressing the telluric forces that arise from the Earth, or the powers of the underworld, often represented by serpents or dragons, or dark, dangerous gods like Hades. These forces can be destructive when unbalanced, but they are also the sources of biological fertility and are the fiery energy of life itself. The key to harnessing these forces is keeping the balance, walking the line between destruction and creation; it is central to the task of the magician.

The deeper symbolism of these animal gods, like Pan and his Christian alter ego, the Devil, is that they represent the vital soul, the "lowest" of the three souls that Plato postulated, and are often equated with the lower chakras in the yogic system, which contain our instincts for survival, dominance, aggression, and sexuality. Many of these natural drives, which exist in all of us, have been denied and repressed in most people by the ouranian, solar bias of Western civilization, which rightly sees them as dangerous. They have become subsumed into the shadow archetype of Western culture and despised and denigrated and personified as the Devil or Satan. The dragon must be made conscious, raised and brought into the light, and made to serve the magician, who raises the serpent and harnesses its energy for fertility and healing. I don't mean to sound naive; evil is a fact, and it is ugly, and its metaphysical ontology is a mystery. Let there be no doubt that it should be resisted. But evil is most destructive when it is unconscious and masquerading as good. I am not talking about giving free rein to everyone's lower drives and impulses but rather honestly acknowledging them and making them conscious. The Devil is in all of us. He becomes Pan, the life principle, when we honor and accept our animal nature and our kinship with all life, and he devolves into Satan when we deny him and project him onto other people and other groups of people. The phrase "shadow work" is used rather lightly these days, but it is serious business, the spiritual challenge of a lifetime.

Hermes and Mercurius

The father of Pan, Hermes, divine trickster and messenger of the gods, was originally a rustic deity. Born to the nymph Maia in the mountains of Kyllene in Arcadia, in ancient Greece, he was the patron of shepherds and cattle rustlers, according to Walter Burkert. He is often depicted as a fertility god, in ithyphallic form, famous for seducing nymphs and multiplying the fertility of herds. He is the pyschopomp, the leader of the souls to the underworld.[13] He was known to the Romans as Mercurius and is of great importance as the patron of magicians and all other practitioners of the Hermetic arts. It is important to introduce him here because of his continuing role in the later alchemical tradition and his contributions to the depth psychology of Carl Jung.

In one of his works exploring alchemical symbolism, *Alchemical Studies*, Jung introduces the fairy tale known as "The Spirit in the Bottle," which tells the story of a spirit trapped in a bottle hidden in the roots of an immense old oak tree. The son of a poor woodcutter finds the bottle and releases the spirit,

♦ *The alchemical Mercurius, from* The Twelve Keys of Basil Valentine. *Engraving by Matthäus Merian the Elder (1593–1650).*

who identifies himself as Mercurius, who—to make a long story short—gives them miraculous powers and wealth.[14] The fact that the spirit is trapped in a bottle concealed in the roots of the tree links him to the telluric powers we discussed above. To Jung and to the alchemists, Mercurius was a principle capable of transforming evil into good or base metal into gold, as a mediating principle that partakes of both evil and good. Jung describes him thus: "If Mercurius is not exactly the Evil One himself, he at least contains him—that is, he is morally neutral, good and evil, or as Khunrath [the alchemist] says: 'Good with the good, evil with the evil.' His nature is more exactly defined, however, if one conceives him as a process that begins with evil and ends with good."[15] Hermes is the god who connects us to our deeper nature, the roots of our tree, as it were, and brings our darkness to light, redeeming and transforming it into alchemical gold. He contains the evil dragon and the light. He is the Abraxas of the Gnostics, both Christ and the Devil, unifying them into one whole. He is the divine double, the one who leads the souls of the dead to the next world. To the righteous, he appears as Christ, but to the unregenerate as the Devil, and to the wise, he is one, himself.

Minor Gods and Goddesses

The minor nature gods suffered the same sort of demonization as Pan when they were forced by theologians into a place in the dualistic Christian mythology. As St. Augustine stated, "The gods of the nations are most impure demons, who desire to be thought gods, availing themselves of the names of certain defunct souls, or the appearance of mundane creatures."[16] Claude Lecouteux, in his highly informative study of nature spirits in medieval European history, *Demons and Spirits of the Land*, describes the gradual process in detail. The legend of the fallen angels in the Enochic literature and the few tantalizing verses regarding the Nephilim, giants, and demons born from the sexual liaisons between the sons of God and the daughters of men provided the proof text for fitting the nature spirits into the Christian mythos. Lecouteux quotes the churchmen of late antiquity, such as Martianus Capella, who had this to say: "The places inaccessible to men are inhabited by a host of *very ancient* creatures [emphasis in original] who haunt woods, glades, and groves, and lakes and springs, and brooks; whose names are Pans, Fauns, *Fontes*, Satyrs, Sylvans, Nymphs, *Fatui* or *Fantuae*, or even *Fanae*." Lecouteux also notes Martin of Braga, a sixth-century bishop from Portugal, equated the nature spirits explicitly with the angels who fell from heaven with Lucifer, complaining that the "ignorant folk worship them as gods and offer sacrifices."[17] Lecouteux explains that the Latin terms for the local spirits, used by the churchmen to condemn their local cults, cover and obscure the diversity of Indigenous spirit traditions. This is the way of all totalizing narratives, a familiar story that has repeated itself all over the world: to force everything into its place in the imperial myth.

Despite this, rural people of the ancient world (and some in the present as well) saw the natural world as populated with innumerable minor deities. Every hill, mountain, river, spring, and grove had its nymph or dryad. There were satyrs and pans and myriads of other invisible inhabitants. Skipping ahead to the Renaissance, our old friend Henry Cornelius Agrippa spoke of these natural spirits as "not so noxious, but most near to men, so that they are even affected by human passions, and many of these delight in man's society, and willingly dwell with them: some dote upon women, some upon children, some are delighted in the company of divers domestic and wild animals, some inhabit woods and parks, and some dwell about fountains and meadows." Agrippa continues, "So the fairies, and hobgoblins inhabit champion fields; the naiades, fountains; the potamides, rivers; the nymphs marshes, and ponds; the oreades mountains; the humedes meadows;

◆ *Nymphs, Hermes, and Pan. Relief from about 330–310 BCE, Athens, Greece (in the National Archaeological Museum of Athens). Image by Zde.*

the dryads and hamadryads the woods, which also satyrs and sylvani inhabit, the same also take delight in trees and brakes; as do the naptae, and agapae, flowers; the dodonae in acorns; the palae and feniliae in fodder and the country."[18]

As the medieval Neoplatonist Bernardus Silvestris wrote in his *Cosmographia*, "Wherever earth is most delightful, rejoicing in green hill, flowery mountainside, and river, or clothed in woodland greenery, there Sylvans, Pans, and *Nerei*, who know only innocence, draw out the term of their long life. Their bodies are of elemental purity: yet these too succumb at last in the season of their dissolution."[19] These spirits of the woodlands and fields, dwelling in places of natural beauty, are long lived but not eternal.

They can be glimpsed by the patient and quiet visitor to the fields or the forest, occasionally out of the corner of one's eye—a flash of light, a whirlwind lifting the leaves in the still morning, the sudden passing of a shadow in front of the Moon. They are easy to miss and easier still to disbelieve in, for skeptical modern people. Agrippa gives instructions for summoning them:

> He therefore that will call upon them, may easily do it in the places where their abode is, by alluring them with sweet fumes, with pleasant sounds . . . adding songs, verses, enchantments suitable to it, and that which is especially to be observed in this, the singleness of wit, innoncency of the mind, a firm credulity, and constant silence; wherefore do they often meet children, women, and poor and mean [humble] men. They are afraid of and fly from men of a constant, bold, and undaunted mind, being in no way offensive to good and pure men, but to wicked and impure, noxious.[20]

Many people in the Celtic nations leave the good people, as the rustic spirits of the countryside are called there, gifts of milk, cheese, and butter to gain their favor. I visit favored places in the woods near my home, or even in my yard outside my home on the farm, and leave gifts of food, cornbread, milk, beer, and sacred tobacco. I have heard that you should never approach empty-handed, so—as I mentioned near the beginning of this book—I always bring something for them, whatever they are

The spirits, fey folk, dryads, nymphs, elementals, or whatever they are called by the various nations and traditions can influence our consciousness with sudden and unaccounted-for shifts of mood and weird impulses. This is the origin of our word *panic*, a sudden irrational impulse to flee at the approach of the god Pan. A sudden sensation of fear, when one is alone in the woods, is not uncommon,

◆ Hylas and the Nymphs, *by John William Waterhouse, 1896.*

and, in parts of the world where there are large predators, is not unreasonable and could very well be a sign that one is being stalked by a very physical threat. But for many of us, sudden changes in consciousness when one is alone in nature could very well be a sign of the arrival of a spirit.

One story about nymphs and nature spirits and their effect on human consciousness comes from Plato in his dialogue *Phaedrus*. Plato's dialogues are philosophical dramas in which the setting, known as the frame narrative, is important to the meaning of the whole. In this dialogue, Socrates, the wise philosopher, and his student, named Phaedrus, decide to take a walk outside the walls of the city and out into the country for a wide-ranging discussion around the topic of eros—love or desire. They stroll down by the cool shady banks of the River Illissus and sit down in the shade of a plane tree. Socrates takes note that the spot where they are sitting has been consecrated to the water god Achelous and to some nymphs, and that it is a beautiful, fragrant spot. They launch into the discussion, with Phaedrus reading a treatise on love and Socrates responding to it. Socrates begins to wax eloquent and launches into verse, declaring himself to be possessed by a nymph, *numpholeptos*, and changes the direction of the conversation to a discussion of divine madness and inspiration, contrasting it with ordinary madness and artistic inspiration. Divine madness is a form of spirit possession, where one is possessed not by a nymph but by a god. He is speaking of the mystical ecstasy that raises the soul to the sublimest heights of spiritual union with the beloved, launching into a literally inspired discourse on love, divine longing, heaven, the soul, reincarnation, and the beatific vision of archetypal beauty. At the end of the dialogue, Socrates offers the following prayer: "Dear Pan, and all ye other gods that dwell in this place, grant that I may become fair within, and that such outward things as I have may not war against the spirit within me. May I count him rich only who is wise, and as for gold, may I have only so much of it as a temperate man might bear and carry with him."[21] Socrates, it must be remembered, was an adept and initiate of the highest caliber and likely only attracted spirits of the highest vibration. The rest of us, however, often have more humble experiences in our contact with nature spirits.

Scholar Jennifer Larson tells us that the setting of the dialogue described in the *Phaedrus* was a classic example of a *locus amoenus*, a place characterized by flowing cool water, shade, and lush vegetation, a place pleasing to the senses, a place of refreshment to take shelter from the midday Sun on a hot day, and a place that was certain to be haunted by nymphs. "Such a place is never without a divine presence, which accounts for the appeal of the landscape and its strong influence upon the susceptible observer," says Larson.[22] What follows is an example from my own life of a personal experience of nympholepsy, during which an incongruous and reckless mood came over me suddenly and unaccountably, causing me to make a very dangerous mistake.

Looking back on a lifetime of wandering in the forests and woods in the several places I have lived, I have had several experiences of odd mood shifts while in natural settings, one of which nearly cost me my life. One day I was swimming with some friends at an idyllic location in Fairmount Park in Philadelphia, below a weir that was built to hold back water in order to power a mill. The water was smooth and soft and deliciously cool, and the day was hot. The lush shade of the trees was a welcome respite from the August Sun; it was a true locus amoenus *in the classic sense. I knew, and in fact warned my friends, to stay away from the weir because of the dangerous current below it. No sooner had the warning left my mouth than I had a sudden impulse to throw myself into the same circulating current below the weir. I was instantly caught and pulled under, struggling to stay afloat and eventually becoming exhausted and inhaling water and sinking, fully conscious until the* lack of oxygen caused me to black out. I had been thinking the most banal thoughts and having regrets—one in particular, being a teenager, was that I was to die a virgin! With that being my last thought, I awoke to blinding light pouring into my eyes and oxygen flooding my starved brain. I had lost consciousness and simply floated out like a rag and was saved by one of my companions, who had given me mouth-to-mouth resuscitation. In reflecting on the incident, I could never account for my seeming death wish. I had read a bit of Jung and Freud by that point in my life, and I chalked it up to some unconscious urge to self-destruction, called by Freud thanatos, *the death instinct, which was not usually something I was prone to, being a fairly cautious young man. Perhaps some other force was at work. One writer tells us: "Nympholepsy is a dangerous and unwelcome mental state, connoting loss of self-restraint, abduction or even disappearance and death."[23] But whatever force was responsible for my rash behavior, I was shaken by it! Perhaps, I too was* numpholeptos, *lured to near drowning by the spirits of that place.*

✦ The Water Sprite, *by Theodor Kittelsen, 1904.*

The ancient Greeks said of children who drowned that they were "snatched by nymphs,"[24] their favorite prey being young men. "Nymphs incarnate a dangerous and anomalous femininity often associated with premature death; seductive and dangerous, they lie in wait for attractive boys or men, and then drown them."[25] Perhaps the nymph of that place took pity on me and released me from her watery grip (or maybe I just wasn't attractive enough for her).

That place, Magaree Dam, is near a notoriously dangerous swimming hole, known as Devil's Pool, where, as Lenni-Lenape Indian legend has it, the Great Spirit defeated the Dark Spirit and buried him in the pool.[26] Another lush, green *locus amoenus*, many drownings and accidents have happened there. In folklore, the presence of the word *devil* in a place-name is often a sign of the presence of a nature spirit, possibly a malevolent one. After all, not just Greek mythology but global folklore is full of water spirits that lure young men into the water and drown them, preferring the innocent. The nymphs of ancient Greece, the siren Lorelei of the Germans, and the rusalka of the Slavic peoples are of this type.

Years earlier, when I was maybe twelve years old, I had a weird accident in Fairmount Park near the same dam, caused again by an unaccountable urge to do something dangerous and thoughtless, in this case rolling a large rock down a hill that my mom and her friend were sitting at the bottom of. No one was hurt, but I was left wondering later, as were the women I had put in danger: Why did I do that? Maybe it was just a kid being stupid, but maybe there were other trickster spirits in that place, capable of putting destructive urges into the permeable minds of unshielded children.

Twentieth-century occult author and psychologist Dion Fortune, in her work *Psychic Self-Defense*, had this warning for those who choose to interact with nature spirits, which she refers to as the Deva Kingdom:

There are many people for whom the Deva Kingdom, as the sphere which the elementals share with the Nature Spirits is sometimes called, has a great fascination, and then they try by meditation and ritual to get in touch with it. In my opinion it is decidedly risky for a person who is not an initiate to attempt this work. It is extremely apt to lead to mental imbalance, if not actual obsession. Not that the nature contacts are evil, but they are profoundly disturbing to the human consciousness because they stir those atavistic depths which the psychoanalyst has laid bare by means of his technique.[27]

This was exactly what had occurred to me after my near drowning, except I never thought that my accident had anything to do with elementals or spirits, being something of a skeptic at the time. Fortune speaks of those who commit suicide due to a sudden urge to throw themselves from a height or those who swim too far out to sea and drown as willing sacrifices to the elemental gods. Based on my experience, they are probably not exactly willing! As one who has a great fascination with nature spirits, even spirits in general, I am probably one of those people Fortune was warning. While not exactly what she would have called an initiate, I do practice basic shielding and banishing, and I try to be respectful to my local spiritual fauna by leaving regular offerings and by not antagonizing them.

I have written that I believe our consciousness to be permeable and not confined to the inside of our skulls, that we are eddies in an ocean of consciousness, surrounded by other eddies of many kinds, some more subtle and some more dense. The more dense of these we may experience as physical beings and the subtle eddies we experience as spirits, or transient moods, flights of fancy, sudden ecstasies, or obsessions, which don't come from within but pass through the permeable membrane of our psyche, if we are open to them. Openness to these influences is not always a good thing. You want to let in the helpful influences, the muses and the kindly spirits that bring happiness (the Greek word for which is *eudaimonia*, or "having a good spirit"), but you want to keep out the harmful. This is why psychic shielding and banishing is absolutely essential to keep entities of a lower vibration out of your psychic space. The ritual described below contains a Druidic practice known as the Sphere of Protection and serves the purpose of psychic shielding nicely. I find it to be more harmonious and less intrusive in a natural setting than the Lesser Banishing Ritual of The Pentagram. The LBRP, which we discussed in chapter 8, is perfect for cleansing oneself of low-level psychic pollution and for cleansing a small space, but is not recommended when working with nature spirits in a natural setting.

Many of my more recent experiences with the spiritual realm have typically involved signs given in response to offerings left, and mostly good fortune in horticulture— good yields of healthy crops. Here is an example from my own practice:

I descend the hill that my house sits upon and enter the woods by a narrow path that leads into a thicket of cedars and wild plum trees, descending steeply into a dry creek bed that I cross. Then I begin the ascent of a hill forested mostly by chinkapin oak trees, which are tall, stately trees common in this part of Kentucky. At the top of the hill is a meadow that is cut for hay occasionally, but my sacred place is a small flat spot with a large flat piece of limestone and a rough perimeter of flat limestone rocks that ring the edge of the forest at the top of the slope. Once there, I assemble my gifts, calm myself, and clear my mind. I call the four elements in turn, facing the cardinal directions, as well as the spirits of sky and Earth. I call up the telluric current from the Earth into my body and call down the solar and lunar currents. I envision myself as the great cosmic tree, the world axis, poised between heaven and Earth. I circulate the currents of energy down from my branches and up from my roots with my breath. I raise the energy and focus it into a sphere of white light that I visualize extending all over my land and my neighborhood, a protective current of energy raised from the Earth and called down from the heavens. I speak aloud to my tutelary gods and to the spirits of the place and offer them my gifts. I pour out a libation and offer some tobacco and some grain, both from the harvest of our farm. I meditate in silence for a few minutes, and then I leave, thanking the spirits.

As I leave, I often say, "Send me a sign if my gift was pleasing to you," and wait a few minutes for the scream of a hawk or the snort of a deer, which I take to be an answer. A few times, I have been answered by silence, shrugged my shoulders, and walked back up the ridge. On one occasion, I walked up to the barn in the fading light of the gloaming to return my tractor to its parking space and, as I looked up to the sky, in a part of the sky where there were no bright stars, there appeared three bright lights, forming a perfect equilateral triangle, separated by ten degrees of sky. I was shocked and reached for my phone to take a photo, only to have them fade out one at a time. A few weeks later, I performed the same ritual, and again asked for a sign, again heard nothing, and walked back to the farm to close the chicken coop. I looked up to the sky in the fading but still-bright twilight, only to see a brilliant fireball streak overhead and explode in the sky above me.

I have different experiences on the farm across the road, part of which we lease for crop production and part of which is largely abandoned, except for the hay being cut and hunters being there during deer hunting season. It has the ruins of a once well-kept homestead, though the house burned long ago and the barns and outbuildings are falling into a state of advanced disrepair. It is a very sacred place with a powerful presence that I have really fallen in love with over the years, raising crops there and cutting firewood. It is definitely haunted, but I'm not sure if it is haunted by the former inhabitants of the homestead or something older and less human. I leave offerings there too, in an oak grove and at the barn near the outbuildings, at night, on the Full Moon. On one occasion, I left an offering of tobacco, whiskey, and water, and asked, "If there are any spirits here, knock three times." And I was immediately answered with three clear raps on a loose piece of metal roofing. The hair rose on my neck, but I stayed. I left my offering, and asked them to send me a sign if it was pleasing to them, and I instantly felt a sharp pain in my right side, in my ribs. I then remembered that the old man who used to take care of the farm, whose family had owned it, was a strict teetotaler. Maybe he didn't like the whiskey. As I walked away, I heard three clear knocks coming from the abandoned garage.

I went over there recently, when the Full Moon was in Aquarius. I walked onto the property and was immediately greeted by the shrill bark of a coyote, close by and directed at my nocturnal trespassing. It was insistent and wouldn't quiet down until I ultimately left, and I distinctly heard the rustling of several more quiet individuals in the underbrush nearby. I was going to ignore it and continue to the abandoned homestead to talk to the spirits when I remembered that the coyote is a bad omen in many Native American traditions, a trickster spirit with no particular love for humans. Even though I'm sure it was a very flesh-and-blood mammal, probably a mother with pups, it was definitely a clear warning sign. So I turned around and went home.

I'd like to report one more uncanny occurrence at the leased farm. I had gone back to the oak grove to leave an offering to the land spirits earlier in the day, and in response to my usual question I heard the cry of a hawk overheard. "A good sign," I thought, and went home for lunch. Later in the day, on a walk with my older son, I returned to the vicinity of the abandoned outbuildings. For some reason, I became aware of an incongruous clump of trees growing out of the meadow, not twenty feet from the country lane that leads to my farm. I have driven and walked by this particular clump of trees literally thousands of times and had never thought to investigate it or question why those trees had been allowed to grow up in the middle of the field. This day, however, I became seized by a peculiar curiosity and decided to take a look at that clump of rangy box elders to see what lay within. There, under the tree, were several large flat rocks that appeared to be covering up a hole in the ground. I got down on all fours and took out my phone to shine a light under the rock while peering underneath. The light illuminated a rather coffin-like stone-lined chamber about six feet long and two or three feet wide that was covered by two large flat pieces of limestone. The whole thing had a strikingly primitive look about it, reminding me of Neolithic burial chambers in Europe, but on a much smaller scale. The bottom of the *chamber was dirt, and it was impossible to tell how far down the chamber had originally gone. Previous inhabitants had obviously tried to cover it up and may have even filled it in with dirt partially. Its original purpose remains mysterious, although I have several rather mundane theories about it, ranging from the shaft of a lead mine to an ice house, a spring house, or even a root cellar. I keep having a feeling that my attention was directed there, and maybe for a reason. The mysterious chamber is on private property, right next to a public road, so I will not be investigating any time soon.*

I'm not sure what kind of presences I'm dealing with on the land here, whether they were once human or have never been human, whether they are fey, or ghosts, or some other class of nature spirit. They have stolen my cow (see chapter 1), poked me in the ribs, and pulled my elbow. I have actually seen flashes of silver light hovering close to the ground, well out of firefly season, and have had a whirlwind travel directly across a field and pass right between a farmhand and myself while we were working a few feet apart. I have seen shadows pass before the rising Moon and lights among the stars that blink and flash and disappear. This place is beautiful and strange and certainly haunted by animate presences.

The skeptic will say that all of these experiences are vague and probably coincidental. However, silencing the inner skeptic, or at least quieting it down a bit, is an essential part of the work. Those interested in coming to know nature spirits intimately would do well to heed Agrippa's advice, quoted above, to maintain "a singleness of wit and a firm credulity, and keep silent." In other words, keep an open mind and believe! A willed suspension of disbelief is the first step to cultivating imaginative perception. The constant reality testing of the skeptical mind—the habit of asking oneself, *Is this real?*, and telling oneself, *This is probably just a coincidence*—is a sure way to block imaginative perception and nascent psychic abilities. We are trying to recover some of the way of seeing that Yeats spoke of, by interacting with the living nature around us and seeing what is revealed when we treat the landscape, the plants and animals, the rocks and hills, as if they truly were the dwelling places of the gods and lesser gods. I think anyone who carries out the experiment long enough will find, as I have, that the more reverence is paid to the natural world around us, the more the divine within it is unveiled.

12

WORKING WITH EARTH ENERGIES

This chapter will build on the previous chapter by introducing what I have called geurgy, or the practice of working with Earth energy or Earth spirit force to bring about meaningful change on the farm or homestead. These are techniques of re-enchantment, practices for benediction, creating fertility in the land, and healing the damage done by careless humans on a spiritual level. We are living in a perilous time, a time of great and unprecedented change, when many of us feel unmoored and disconnected. Now more than ever, we need to find a way to draw up the power of the Earth and draw down the powers of heaven for reestablishing our broken connection to the cosmos. We will have to examine this topic from the point of view of several traditions before putting it into practice on our land to commune with the spirit forces of our home, wherever it may be. What will follow will be a discussion of theory as well as practice, an exploration of the topic as well as an outline for future experiments in areas where present expertise is lacking. I have not, at present, built any megaliths, but it is definitely on my to-do list!

In addition to the nonphysical creatures of the natural world, the animist cosmology also includes the concept of a universal life force, a conscious energy that permeates all of nature and gives life to living things. This force can be manipulated and concentrated by adepts for healing and other "magical" purposes. It arises from the depths of the Earth and flows down from the stars and other celestial bodies. It is thought by Taoist practitioners—who have made an esoteric science of this energy, called in their tradition, again, qi—to be polarized as yin energy, which flows from the Earth, and yang energy, which flows

✦ *Ouroboros engraving from Lambsprinck's* De Lapide Philosophico *published in Lucas Jennis's* Musaeum Hermeticum (1625). *A Latin caption appears on the original text; here is the English translation from Stanislas Klossowski De Rola's* The Golden Game[1]: *"The perfect hieroglyph of the* Prima Materia, *the Stone of the Philosophers and their Subject is the Dragon, because its scales, its volatility and its venomous nature are equivalent to that of the mineral Subject. From a virulent poison, the Stone of the Philosophers becomes the Medicine of the Wise."*
– *Lambsprinck.*

from the heavens. Other systems refer to these two basic currents as the telluric current and the celestial current. The essential energy is called by many different names all over the world, but nearly every traditional culture has the concept of some sort of life force. Some names by which it is referred to are: *prana, pneuma, ruach, spirit, od, azoth, sprowl, nwyfre,* and *orgone,* to name but a few. There are many esoteric techniques to concentrate and cultivate this force in the various spiritual systems worldwide. Earth magic includes the means of working with the life force that arises from the Earth herself, the landforms and the living plants, for the purpose of furthering growth, life, fertility, and healing, as well as limiting and constricting harmful forces, preventing them from causing further damage.

The heavenly qi, or celestial energy that flows from the stars and planets, interacts with the Earth qi according to the tides of the seasons, regulated by the Sun's angle and the length of day, which affect the quality and force of solar radiation that falls on the various parts of the Earth. The movement of the Moon, its phases and monthly path through the zodiac, as discussed previously, also change the quality of heavenly energy that the Earth receives. Added to this already complicated dance of the Sun and Moon are the whirlings of the planets as they race in their courses around the Sun, as well as the stars and other celestial objects of deep space. They make a constantly changing tapestry of aspects and interactions, all of which act on our Earthly mother and her energy field. As the alchemist

✦ *Michael the Archangel defeating Satan in the form of a dragon, illustrating
Revelation 12:7–10. Unknown medieval artist, late thirteenth century.*

Basilius Valentinus wrote, "The Earth is not a dead body, but is inhabited with a spirit that is its life and soul. All created things, minerals included, draw their strength from the earth spirit. This spirit is life, it is nourished by the stars, and it gives nourishment to all the living things it shelters in its womb."[2] It is to this spirit, the deep telluric current of the Earth, and to the elemental energies of the living biosphere, that our discussion will turn in this chapter.

THE POWER OF SACRED SITES

Many traditional cultures in the past have shown a deep sensitivity to these powers, constructing sacred sites on power spots of concentrated spiritual force. Many of the temples, earthworks, standing stones, megaliths, and pyramids of the ancient world are built on ley lines, or terrestrial networks or lines where the Earth energies are believed to be particularly concentrated.[3] These lines are similar to the *lung-mei*, or dragon lines, of the ancient feng shui system of China, which seeks to harmonize the building of temples, homes, and cities with the flows of land energy. These lines can be detected by studying the landforms and placement of sacred sites, which can be seen on a map to strangely line up, even on a continental scale. An example of this is the famous St. Michael's Line, which links sacred sites dedicated to the Archangel Michael and runs from the west of Ireland through Cornwall in the west of England, through France, Italy, Greece, Cyprus, and Israel.[4] The Archangel Michael, the archangel of the Sun, is often depicted piercing a serpent or dragon with a lance, and indeed the line passes through Delphi, the ancient sanctuary of the Sun god Apollo, and the site of his killing of the Python. As we will discuss later in the chapter, the serpent or dragon is often associated with the Earth spirit.

Earth mysteries researcher John Michell compared the building of temples and sacred sites with a sort of macrocosmic acupuncture, piercing them to channel their energies for beneficial purposes. Michael pierces the dragon, as the needle pierces the skin, to direct the energy elsewhere. As Michell writes: "The field of terrestrial magnetism, like the energy field of a plant or animal, exists only by association with a living body; and as traditional Chinese medicine treats the human body by regulating the currents of vital energy which flow through the skin, so the geomancer treats the body of the earth."[5]

To Michell, the adoption of the ways of civilization, including agriculture and mining, profoundly disrupted humankind's relationship with nature in general, and the Earth spirit in particular. According to him:

◆ *St. Michael's Line, linking sacred sites dedicated to the Archangel Michael.*

Whereas formerly, every part of the earth was inhabited and ruled directly by spirits, these are now placed in reservations that the world outside may be freed for the sacrilegious breaking of the earth for agriculture, building and mining. Even so, these activities are carried out in the knowledge that they are objectionable to the earth spirit and defy the gods. They must be attended with rituals designed to attract the gods' patronage. The farmer's year is regulated by the old deities, sun and moon, and its stages are sanctified by festivals and sacrifice; the sites of tombs, temples, and houses and all artificial features of the landscape are located in relation to the paths and centers of the earth spirit.[6]

So the cunning farmer is involved in the sacrilegious business of breaking the Earth for agriculture. This is not news to me. I have maintained since I've been involved in farming that tillage is profoundly unnatural, a necessary evil. It is, after all, according to the apocryphal Enochic tradition, one of the arts of civilization taught to humankind by the fallen angels, along with metallurgy, medicine, chemistry, cosmetics, and magic. So, having left the garden of Eden, cast out into the land of Nod, as it were, we are left to forge a new relationship with both the Earth and Sun and to honor the divinities of both. No longer nomads, instinctually following the paths of the Earth spirit wherever they lead, we must seek a balance with both the Earth spirit and humanity's Promethean tendencies. We are the exiled children of Eve and Cain as well, Cain being the first farmer. So we learn the sacred rites, offer up the first fruits of our land, and celebrate the turnings of the wheel of the year, the endless dance of the Sun, the Moon, the planets, and the stars with the Earth spirit. In this way, we restore the balance and repair the rupture.

Witches and cunning folk in Cornwall, according to Gemma Gary's *Traditional Witchcraft*, are adept at gathering the vital energies of the land and channeling them for use for magical and healing purposes. She writes that it is essential to develop a relationship with the unseen spiritual forces of the land in order to work effective magic, to "provide wisdom and divine people's fortunes."[7] It is this land-based magical ethos that really draws me to Gary's work, and following her advice has led me out into the forest on Full Moon nights to draw in the potent energy brought up from the Earth by the tidal pull of the Moon.

"For the Cornish witch," writes Gary, "one of the most potent and useful forces is known as the *Red Serpent* or *Sarf Ruth*. This is the spirit force or 'sprowl' that flows within the land, animates all living things, and empowers the spirit within all natural things." She continues: "Detecting and harnessing the serpentine flow is of great importance to the pellar [a traditional Cornish word for a cunning person], and they must know the ways to this and the places where this force will be best drawn forth." Intimate familiarity with the land as well as a natural talent for moving and drawing forth the summoned power are essential talents for the practitioner of this land-based magical tradition. The Earth magician, like the Cornish witch, must "walk out into the land to gather 'sprowl' and empower their craft, such journeys may be known as 'walking the serpent path,' a path of power and chthonic gnosis. The witches are very sensitive to the landscape in which they live and know well the places of power around them where the sprowl can best be drawn forth and saved for later use."[8]

It is, of course, no accident that the "serpent path" of Cornish magic is linked to the ley lines, also known as serpent paths, and the Earth spirit of John Michell, also often depicted as a serpent or dragon. Gary explicitly links the serpent of the land in her tradition with the more modern ley lines. As with Michell, the serpent force issuing forth from the ley lines is one of the main sources of ritual power and prophecy. As Gary writes: "Wisdom may also be gleaned from the serpent. At times, especially at the full of the moon when the serpent is most potent and generative, the wise are drawn down into the openings of the earth . . . for in these damp, dark, wombs of the land, a hypnotic force issues from the serpent and pools in abundance. This is known as the 'serpent's breath' or simple 'snake breath.' In such places haunted by the breath, the Wise drowse in this force to commune with the 'earth spirit' and receive visions, hear voices, and make magic."[9]

According to Michell, the building of a temple or the placing of a standing stone draws in and concentrates the earth spirit, allowing for easy communication with the gods, serving "to concentrate the telluric forces of the place on that one spot and to intensify the power of the earth spirit to generate prophecy." This is especially true when combined with auspicious alignments of celestial powers, such as solstices and equinoxes, as well as Full Moons. Michell describes this as a "technology" and as a "science of a more subtle, metaphysical order," the object of whose "attention was the animistic world of spirit," the Earth spirit, known the world over by many names.[10]

In contrast to the megalith and temple builders, witches and cunning folk do not always have standing stones and temples at their disposal, especially in North America. In addition, the tragic history of colonization unfortunately may make it inappropriate to perform magic at the historic sacred sites available to North American practitioners. I am not being prescriptive here; everyone has to follow their own guidance and intuition. Those who live in areas bereft of known ancient sacred sites will have to search their local environment for places with a wild and sacred feel to them, and caves and cliffs and springs issuing from the ground are good candidates. In Kentucky, where I live, the landforms are shaped by the underlying karst limestone, famous the world over for giving rise to eerie phenomena. The karst limestone is porous, and over vast stretches of geologic time, rainfall entering into the rock strata hollows it out, forming cave networks and sinkholes, which are natural power spots, seen as entrances into the underworld. The land I live and work on has several modest-sized sinkholes that seem to be mysterious underworld entrances, especially the farm I mentioned in the last chapter, the one I believe to be particularly haunted. It makes an ideal location for late-night expeditions on Full Moon nights to gather power from the Earth spirit. The cunning farmer or homesteader must make their own space sacred, wherever they may be, by enchanting and empowering the landscape of their home.

THE SACRED SERPENT

Next, let's examine the esoteric symbolism of the serpent as it pertains to the energies of the Earth, because it is a recurring motif. No animal is so misunderstood and disliked by humankind as the humble snake—not without reason, as many species are venomous, and at times a quick reaction is necessary to keep from getting bitten. The serpent is thus an ancient adversary. Paradoxically, it is also a symbol of the life force of the Earth itself. According to poet and mythographer Robert Graves, in the creation myth of the Neolithic inhabitants of Greece (the Pelasgians), Eurynome, the goddess of all things, dances the winds into creation. After dancing up the North Wind, she takes hold of it and twists it into the form of a serpent, Ophion, who becomes aroused by her dancing. She then mates with the cosmic serpent, becomes pregnant, and proceeds to assume the form of a dove. Brooding on the waters of the primordial sea, she lays the universal egg. Ophion coils seven times around the egg, until it hatches, and out of it come all things that exist. Graves writes, "Afterward Ophion vexed her by claiming to be the author of the Universe. Forthwith she bruised his head with her heel, kicked out his teeth, and banished him to the dark caves below the earth"[11]—where, presumably, he dwells to this day, his fertilizing power still giving life to all that lives.

Another, more familiar, mythic serpent is found in the Jewish creation narrative in the second chapter of the Book of Genesis, where the serpent is described as being "more subtle than any wild creature Yahweh Elohim had made." The serpent slithers up to Eve and tempts her to eat the fruit of the tree in the midst of the garden, the Tree of the Knowledge of Good and Evil, telling her, "You will not die. For the Elohim know that when you eat of it your eyes will be opened, and you will be like the Elohim, knowing good and evil."[12] She eats, offers the fruit to Adam, who likewise eats, and they realize they are naked for the first time, and they are ashamed. You know the rest: The primal couple are expelled from the garden, and the serpent is cursed for all time to eat dust and be the enemy of the children of Eve.

This mythic narrative is interesting because the serpent is punished by his creator—for being exactly how he is created to be, crafty and sly. In other words, he is a trickster figure. He is in the garden, which was created to be perfect, so he must have been part of the plan. Adam and Eve, likewise, didn't understand that disobedience was evil until after they ate the fruit, so they weren't truly capable of making a moral choice. It seems to me that this narrative, too,

◆ Eve Tempted, *by*
John Roddam Spencer
Stanhope, c. 1877.

is paradoxical: The serpent is the enemy of mankind but also a wise revealer and an essential part of the enfoldment of the whole creation. Furthermore, I suggest, to the contrary of most mainstream biblical interpretation, that this character is not the embodiment of evil he is made out to be. Yahweh Elohim, likewise, is in this story either strangely lacking in foresight or performing an elaborate ruse intended to initiate Adam and Eve into the next stage of human evolution. Without the serpent, the Garden of Eden is a beautiful dead end; there is no conflict and hence no growth.

In discussing the paradox inherent in this story in his work *Alchemical Studies*, Carl Jung mentions "the Creator's whimsical notion of enlivening his peaceful, innocent paradise with the presence of an obviously rather dangerous tree-snake 'accidentally' located on the very same tree as the forbidden apples."[13] Viewed from this perspective, it makes one wonder who in this narrative is actually the trickster—Yahweh Elohim for having set the serpent in the tree in the first place, or the serpent himself?

It is significant to remember that, while the curse of Eve was to bring forth children in pain, for Adam, the land itself was cursed:

> *Cursed is the ground because of you;*
> *In toil you shall eat all of the days of your life;*
> *Thorns and thistles it shall bring forth for you;*
> *and you shall eat the plants of the field.*
> *By the sweat of your face you shall eat bread*
> *Until you return to the ground*
> *For out of it you were taken.*
> *You are dust and to dust you shall return.*[14]

For Sigmund Freud, Jung's early mentor and predecessor in the theory of analytical psychology, the serpent is, of course, the phallic symbol par excellence, an obvious association, and one that strengthens the symbolic link to fertility and the passions. With this interpretation in mind, and recalling the story of the Fall in the Garden of Eden, the link between the symbol of the serpent and what St. Augustine called concupiscence, or lust, perhaps makes more sense.

In one Gnostic version of the story, contrasting sharply with the usual Augustinian interpretation, a specifically female spiritual principle, Sophia, inhabits the snake for a time and encourages Eve to eat the fruit of the tree that

✦ The Fall and Expulsion from the Garden of Eden, *by Michelangelo, 1509–1510.*

the envious archons, the wicked demonic rulers of the cosmos, had forbidden her to eat. Eve eats the fruit and encourages Adam to do the same, and the pair are expelled from Paradise, being ashamed of their spiritual nakedness and lack of gnosis. The archons "moreover . . . threw humankind into a great distraction and into a life of toil, so that they might be occupied by worldly affairs and might not have the opportunity of being devoted to the holy spirit."[15]

Echoing the biblical narrative, ancient Greek poet Hesiod tells us of yet another tree serpent, Ladon, that guards the golden apples in the garden of the Hesperides. And the Icelandic *Eddas* describe a monstrous serpent, Nidhogg, that gnaws at the root of the World Ash, Yggdrasil. In mythologies all over the world, dragons and serpents are associated with underworld lairs, caves, and trees. The serpent is also known, as we have seen, to be a benevolent symbol of

◆ *Votive relief of Zeus Meilichios, Athens, Greece, 350–300 BCE.*

the genius loci, the nature spirit of the home and farm. In ancient Greece, the benevolent serpent spirit was known as Agathodaimon, the good spirit, a god of good luck and prosperity, and the patron of the city of Alexandria in Egypt. In the Greek Magical Papyri, there are many hymns and spells addressed to the Agathodaimon.

Zeus, too, was known in serpent form, as Zeus Meilichios. Jane Harrison tells us that the epithet Meilichios means "Easy to be entreated, the Gentle, the Gracious One." The serpent form was associated with an earlier cult and was identified in later times with Zeus, not the familiar Olympian Zeus of the clouds and thunderbolts but rather a distinctly underworld, chthonic aspect of Zeus.[16]

Defeating the Serpent

There is a common theme in ancient mythology of earth serpents and dragons being defeated by solar heroes. As we discussed briefly earlier in the chapter, Apollo killed a monstrous earth-dwelling serpent, known as Python, who dwelled deep in the ground under Delphi:

> *A wild monster*
> *who practiced many evils,*
> *on the people of the earth,*
> *both on the people themselves*
> *and on their sheep with slender feet*
> *For she was a calamity of blood.*[17]

Euripides referred to the Python as a *drakon*, "a huge monster of the Earth" that "tended the chthonic oracle." The Sun god Apollo slew the monstrous snake and left her moldering corpse to rot at the sanctuary of Delphi. The ancient serpent cult center of Delphi became the holy site of the great sanctuary of Apollo—an old religion supplanted by a new one.[18]

Again and again in mythology, the divine hero defeats the dragon or serpent in battle. Zeus slew Typhon, Jason defeated the Hydra, Marduk conquered and dismembered Tiamat, and Yahweh defeated Rahab. What is the inner meaning of these tales? Is it, as some have suggested, a mythical memory of the defeat of a primordial cult of the Earth spirit by a conquering religion of Sun-god- or sky-god-worshipping invaders? That sounds like a plausible interpretation, especially in the light of John Michell's ideas about the association of monumental works

✦ Apollo Killing the Python, *by Hendrick Goltzius, c. 1589.*

and solar heroes who defeat Earth dragons. Mythologist Joseph Campbell saw in these myths the struggle of a new patriarchal order against an older cult of a primordial goddess. Indeed, most of the mythological dragons either happen to be female or the children of the Earth goddess herself. These myths of conquering gods and heroes of sky and sun stand, in the words of Jane Harrison, "first and foremost as a protest against worship of the Earth and the *daimones* of the fertility of the Earth."[19]

The mythology of the sons of light that come down to us in the form of tales of evil, deceitful serpents and dragon-slaying heroes, says Campbell, "is an effect of the conquest of a local matriarchal social order by invading patriarchal nomads, and their reshaping of the local lore of the productive earth to their own ends. It is an example, also, of the employment of a priestly device of mythological defamation, which has been in constant use (chiefly, but not solely, by Western theologians) ever since. It consists simply in terming the gods of other people demons, and enlarging one's own counterparts to hegemony over the universe."[20] It is the propaganda of conquerors writ large over the religious history of humanity. It is our task to restore the balance.

The collective psyche, the imaginal realm where these mythic dramas are played out, shapes our political, social, and environmental realities. Our world

✦ *Marduk slays Tiamat, the chaos monster. Drawing of a bas-relief originally from Nineveh, early first millennium BCE.*

order reflects the imbalance that is present in the hypertrophy of the solar ouranian consciousness and the denigration of the earthly and the feminine—in the mythic sphere as well as the worldly.

A first step is to reclaim our symbols from the conquerors. There have always been the heretics for whom the standard mythologies and their interpretations have rung hollow: the Gnostics, the alchemists, the magicians, and the witches. For them, the re-evaluation of mythic symbols and the subversion of the dominant mythology is the highest calling. First, we reclaim the mythology, and then we reclaim the ritual. For the cunning farmer, there is no life without the Earth and the Sun, and there is no enmity between the chthonic and the celestial, the transcendent and the immanent. Our task is to raise the Earth serpent and call down the solar and star powers with the aid of both the angelic and the *daimonic*, to practice both *anabasis* and *katabasis*, the ascent and the descent, respectively, of the cosmic axis. We stand on the Earth, between heaven and underworld. This

◆ The Bronze Serpent, *by Gustave Doré. Moses raises the bronze serpent Nehushtan in the wilderness in an illustration of Numbers 21:4–9 for* La Grande Bible de Tours *(1866). The Gospel of John 3:14 explicitly identifies the bronze serpent with the crucified Christ.*

is our place, in the midst of the four winds and the elements, working for the enchantment of all things.

In that spirit, let us first reclaim the serpent, following the great Carl Jung, for whom, "The snake is not just a nefarious, chthonic being; it is also . . . a symbol of wisdom, and hence, of light, goodness, and healing."[21] The serpent of Genesis is intimately connected with the symbol of the tree and is often depicted in art coiled around the Tree of Knowledge. It is "an illustration of the personified tree numen."[22] The serpent is identified with Mercurius, who we met in the last chapter. "Mercurius is hidden in the roots of the great oak-tree, i.e., in the earth, for it is in the earth that the Mercurial serpent dwells."[23] So the Red Serpent, the Genesis serpent, the Earth dragon, and the land spirit that quickens vegetative life and causes growth in nature are all forms of the shape-shifting trickster god Mercurius/Hermes. "As *serpens mercurialis*, the snake is not only related to the god of revelation, Hermes, but as a vegetation numen, calls forth the 'blessed greenness,' all the budding and blossoming plant life. Indeed, this serpent actually dwells in the interior of the earth and is the pneuma that lies hidden in the stone."[24] In the serpent we meet, yet again, our patron, Hermes.

THE SERPENT AS LIFE FORCE

As we have seen, the serpent is a multivalent symbol that is many things to many people. It is god, goddess, land daemon, and, in certain Gnostic myths, both Christ and Satan.[25] It dwells in the Earth, but there are also celestial dragons, dwelling in the stars. It is a destructive monster, a poisonous wild animal, a cunning tempter, a wise teacher, and a divine being. Mythology and mythological thinking defy simplistic explanations and interpretations. The fact of the matter seems to be that the more alienated a civilization is from nature, the more threatening it finds the serpent, as if the relationship with the serpent somehow personifies that alienation.

The next level of serpent symbolism is the serpent in our own subtle anatomy, the serpent goddess of our own life force. To be alienated from our own vital energy is a tragic consequence of our alienation from nature at large, a symptom in the microcosm of our estrangement from macrocosmic nature.

In addition to the serpent symbolism of the Earth spirit and its energy field, serpent symbolism is applied to the human subtle energy body as well. Many systems around the world postulate the existence of an energy body, spiritual body,

◆ *Mesopotamian caduceus in a drawing of a libation vase from* The Seal Cylinders of Western Asia *by William Hayes Ward (1910).*

subtle body, or astral body, which underlies and animates the physical body. This body mediates between the physical body, the soul, and the energy field of the cosmos itself. In the esoteric physiology of the yogic system of India, the serpent power, or *kundalini*, lies coiled at the base of the spine in the lowest of seven chakras, or energy centers, of the subtle body, which lie along the spine and in the head. The goal of certain esoteric yogic practices is to raise this serpent power up the subtle channel, or *shushumna*, from its resting seat at the base of the spine and up to the highest chakra, thereby provoking a transformation of consciousness.[26] Along the central channel are two other subtle channels, or *nadis*, known individually as *ida* and *pingala*, depicted in iconography as twin serpents whose bodies cross at each of the seven chakras, exactly corresponding to the caduceus, or the staff of Mercury, with its two snakes intertwined around a winged staff. This image of the caduceus is depicted in ancient art from India into Mesopotamia and Greece, and it is even believed by some that Plato and the Orphics were familiar with the idea of kundalini. Mercury and the Goddess just keep coming up in this study.[27]

Plato's writings mention a doctrine of three souls, or perhaps three levels of soul, localized like the chakras in three bodily centers:

1. Logos, or the intellect or reason, localized in the forehead.
2. Thymos, the spirit principle, the animal soul, localized in the heart.
3. Eros, the vegetative or appetitive soul, localized in the sacrum.[28]

These correspond to three of the seven centers of yogic subtle physiology and are also the three main centers of Taoist subtle physiology. The subtle body is shaped and organized by will and attention, therefore it assumes the form expected of it by the practitioner. Many systems of practice have varying numbers of energy centers. In the practice that follows, we will work with a three-center model.

The Neo-Druidic system recognizes three energy centers, which are referred to as cauldrons, deriving the doctrine from a medieval Irish text known as

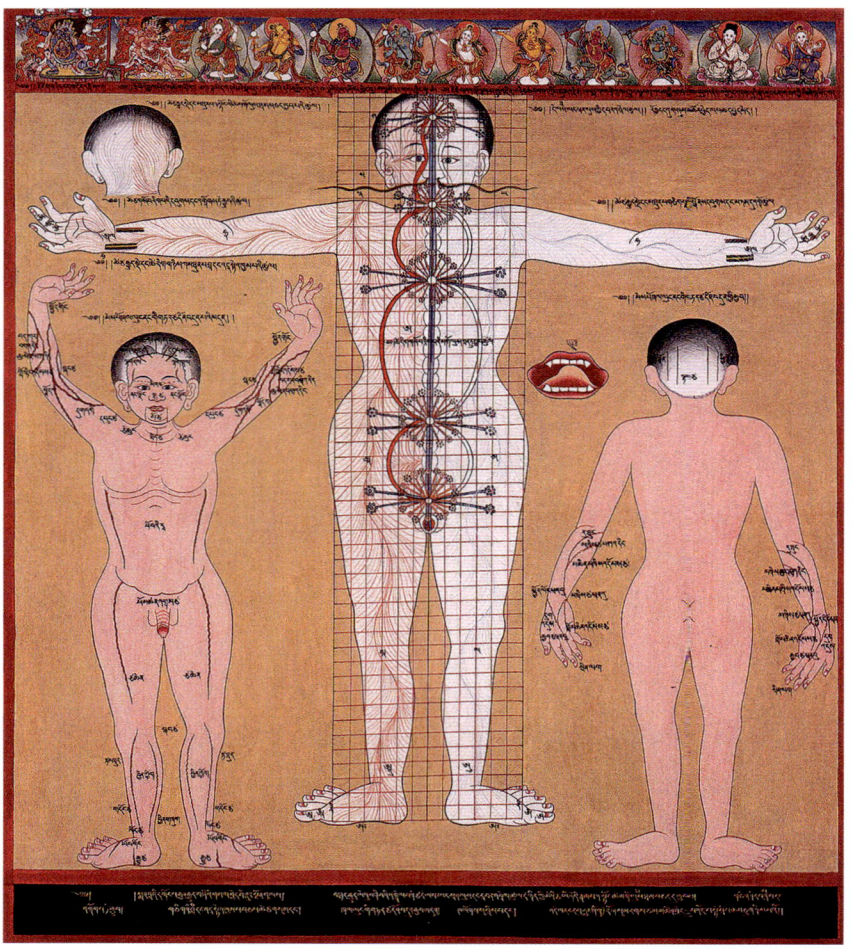

◆ *A Tibetan painting of the human subtle anatomy system showing the chakras and the nadis arranged like a caduceus. Image by dockedship.*

The Cauldron of Poesy.[29] The lowest, at the level of the sacrum, is known as the cauldron of the Earth, and it is the repository of the telluric current and seat of vitality. It corresponds to the lower *tang tien* of Taoism and to the eros of Plato. The second, the cauldron of the Sun, is the repository of the solar current and is localized in the center of the chest, the heart. It corresponds to the middle tang tien and to the thymos and is the seat of the emotions. The third, the cauldron of the Moon, is located in the head. According to modern Druid John Michael Greer, the three cauldrons are associated with the different divisions of *nywfre*, or universal life energy. The solar cauldron, the heart center, is associated with the solar current that flows from the Sun and varies according to the strength of the Sun at different times or seasons. The cauldron of the Earth is the receptacle of the telluric current, which flows from the land itself. It is the Earth spirit force we have been discussing, represented in a serpent or a dragon, and in the human body it represents the drives and passions. Says Greer: "Popular religion, here as so often, takes a partial view and calls one of these currents good and the other evil; thus the solar current is often equated with God and the telluric current with Satan. The reality is less simplistic. The telluric current corresponds to what may be called the animal self, the dimension of the self which has been built up during previous cycles of evolution and incarnation."[30] It is not the passions and desires themselves that are evil but rather their imbalance. The solar current, on the other hand, represents the mind, and the powers of thought, will, and memory. When the solar current of the heart center and the telluric current of the sacral center are in balance, they give rise to a third center, the cauldron of the Moon, which corresponds to the soul or higher self. As Greer puts it, "Where the telluric current works through the body and its passions, and the solar current works through the mind and its perceptions, the lunar current appears through the soul and its powers. When the lunar current awakens in an individual person, it awakens the inner senses and unfolds into enlightenment. When it awakens in the land, it brings fertility, healing, and plenty."[31] This last part is what I find especially interesting about these practices.

The focus on maintaining a balance between the chthonic or telluric energies and the solar or ouranian is especially important and is an appealing part of this system for me. Many magical traditions emphasize one or the other current, and while I am not one to judge the path of another, and I think we are called by fate to our particular path, exclusively emphasizing one current seems to me to be one-sided. The magician stands at the center of the cosmos, ascending or descending the world axis at will, calling on allies from all the realms as needed. The task of the magician, as with the shaman, is to maintain balance.

The first part of this practice, which I mentioned in the last chapter, is called the Sphere of Protection. It is a good daily practice for cleansing the subtle body of energetic contamination and situating oneself in the center of the spiritual cosmos, as well as invoking the beneficial powers of the four elements and directions, while banishing their potentially harmful and negative manifestations.

The four traditional elements are, according to ancient Greek philosophy, the four primal substances from which all things are made. First mentioned in the writings of Empedocles of Akragas, who called them by the names of the four deities who personified them for him, they are: Zeus, personifying fire; Hera, personifying earth; Aidoneus (Hades), personifying air; and Nestis, personifying water. They exist in various quantities as a mixture in all things, and even non-physical things partake of elemental qualities, with fire being likened to desire, earth to physicality itself, air to thoughts, and water to emotions. To Empedocles, who referred to them as "roots," the elements were brought together by a force he called Love and associated with the goddess Aphrodite, and they were dissolved and broken apart by a force he called Strife.[32] It's important to remember that the elements are archetypes and are not identical with the ordinary earthly substances that we call by the same names. They are combinations themselves of the primary qualities of hot and cold, moist and dry. As discussed previously in parts two and three, the elements are also traditionally associated with the four cardinal directions and the four winds, as well as the signs of the zodiac and the planets. It is their association with the directions that we are concerned about here. There are different systems that associate the directions with the elements, and some practitioners get rather passionate about which elements go with which direction. But that need not concern us at this point, because the Sphere of Protection uses the most traditional way of assigning elements to directions, that of the winds. In this system, the east is the domain of air; south, fire; west, water; and north, earth.

Interestingly, in 2005, a Romano-Gallic *turibulum*, or incense burner, was found in Chartres, France, in an excavation of some Roman-era ruins. The turibulum bears an inscription, a magical prayer to the divine powers of the four cardinal directions and the sacred names belonging to each of those powers. The incense burner and the prayers are believed to be associated with a genuine Druidic ritual.[33] If that is so, then it is compelling indeed that modern Neo-Druidic ritual replicates that same emphasis on the four directions.

The following is an adaptation, paraphrased and personalized, of the Sphere of Protection Ritual of the Ancient Order of Druids in North America, of which I am a member.

The Sphere of Protection
The Elemental Cross

Stand facing east (in daily practice, south in a larger working), with knees slightly bent. Visualize the Sun directly above your head. Bring your arms up from the sides in a wide arc, touching your palms above your head, and draw your hand downward so that your thumbs touch the point between your eyebrows. Say: "By the sky above me" while visualizing a golden beam descending from the Sun to the point between your eyebrows.

Bring your palms down to touch the point at your solar plexus, at about the level of your stomach, saying, "By the Earth beneath me," and visualize the ray of golden light descending from the point at your eyebrows, continuing through your body to your solar plexus, and continuing down to the center of the Earth.

Pivot your right hand out from the elbow, angling it down and outward from the shoulder, and turn your head to the right while visualizing a beam of light extending from the sphere at your solar plexus and extending out into infinite space at your right, saying, "By the Fire at my right hand."

Pivot your left arm in the same way, so that now both arms extend out from your body at an angle, while turning your head to the left. Visualize the light extending out from the sphere at your solar plexus and out into infinite space to your left, saying, "By the water at my left hand."

Cross your arms at your chest, right over left, and visualize rays of light shooting out into infinite space before you and behind you from your solar plexus. Say, "May the powers of nature bless and protect me / this place now and forever."

Lower both arms and intone the sacred Druidic word "Awen," drawing it out thus: "AAh-OOh-En."

Astute readers with some magical background will note some similarities to the "Kabbalistic Cross" from the Golden Dawn magical tradition. This is no accident. There is also a Celtic version of the Lesser Ritual of the Pentagram that starts out in a similar fashion. I feel that the general approach of the Sphere of Protection is designed to be more at home in a natural setting, whereas the Lesser Banishing Ritual of the Pentagram is more intended to purify a space as well as the operator for beginning a magical work. I use both at different times for different purposes, and now that I am familiar with the Sphere of Protection, I use it exclusively when I perform work out in the woods. Also, the practitioner is invited to substitute any elemental or deity names they may find more personally meaningful. I actually use the names of Hellenic gods and goddesses for the Elemental Cross and the Calling of the Elements. Many Druidic practitioners use names drawn from the Welsh or Irish pantheons. I prefer the Hellenic as I have a long history with them.

The Invocation and Banishing of the Elements
Calling the Air

Facing east, in the eastern quarter of the space, draw the elemental symbol of air with the first two fingers of the right hand (using what is known as the "sword hand mudra" in Taoism), visualizing it in bright yellow. The symbol of air is a yellow circle with a line extending from the top. Starting at the line on top, draw it clockwise, and then draw the line on top. Visualize the center of the circle filled with a paler yellow.

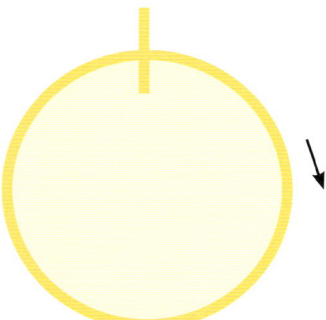

Say: "By the yellow gate of the rushing winds, by the hawk of May in the heights of morning, I call upon the air, its gods, its spirits, and its powers. May I receive the blessings of air this day, that I may attain illumination in this life."

Visualize the elemental power of air, billowing clouds, wind, and yellow light streaming into your solar plexus, infusing your being with aerial freshness and vigor.

Say: "I thank the air for its gifts."

Draw the sign of air again, this time counterclockwise from the line at the top, then draw the line upward.

Say: "With the help of the powers of air, I banish from within me and around me and from all my doings all harmful influences and unbalanced manifestations of air. I banish them far from me."

Visualize a fresh breeze driving away stagnant air like smoke, dust, or fog, leaving you feeling fresh and invigorated.

Calling the Fire

Facing south, in the southern quarter of the space, draw the elemental symbol of fire, which is a red triangle, pointed up. Draw it in bright red, clockwise from the top point, with the sword hand. Visualize the triangle filled with red in the center.

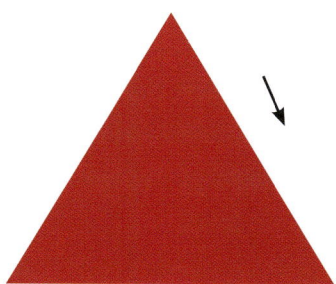

Say: "By the red gate of the burning fire, by the bright Sun at midday, by the white stag in the height of summer, I call upon the fire, its gods, its spirits, and its powers. May I receive the blessings of fire this day that I may attain illumination in this life."

Visualize the elemental power of fire—burning flames, glowing coals, and the bright Sun—drawing the red energy into your solar plexus, infusing your being with power and warmth.

Say: "I thank the fire for its gifts."

Draw the red triangle of fire again, starting at the top point, counterclockwise this time.

Say: "With the help of the powers of fire, I banish from within me and around me and from all my doings all harmful influences and unbalanced manifestations of fire. I banish them far from me."

Visualize a refiner's fire heating yourself to glowing and all impurities blowing away like ash, leaving you renewed and strengthened.

Calling the Water

Facing west, in the western quarter of the space, draw with your sword hand the elemental symbol of water—a triangle, pointed down—clockwise, starting from the point, in royal blue, the triangle being filled with a pale blue.

Say: "By the blue gate of the mighty waters, by the salmon of wisdom in the sacred pool, by the primal ocean of creation, I call water, its gods, its spirits, and

its powers. May I receive the blessings of water this day and always, that I may attain illumination in this life."

Visualize yourself standing before a vast ocean of blue water in a cool, steady, refreshing rain. Draw the elemental power of water into your solar plexus as a blue light, cooling, healing, refreshing.

Say: "I thank the water for its gifts."

Draw the inverted blue triangle again, starting from the downward point, counterclockwise this time.

Say: "With the help of the powers of water, I banish from within me and around me, and from all my doings, all harmful influences and unbalanced manifestations of water. I banish them far from me."

Visualize the water washing away all impurities, leaving you clean and fresh.

Calling the Earth

Turn to the north, and the northern quarter of the space. With your sword hands, draw the elemental symbol for earth, a green circle with a line extending down from the bottom. Draw the circle clockwise starting at the bottom, then draw the line. Visualize the circle filled with a lighter green.

Say: "By the green gate of the sacred stones, by the mighty oak tree, by the great bear in the starry heavens, I call the earth, its gods, its spirits, its powers.

May I receive the blessings of earth this day and always that I may attain illumination in this life."

Visualize the land in verdant green, mountains in the distance, a forest of tall trees surrounding you on all sides. Breathe in the elemental power of earth as a green light into your solar plexus. Imagine yourself growing in strength, solidity, and stability with each breath.

Say: "I thank the earth for its gifts."

Draw the elemental earth symbol, the green circle with the line at the bottom, this time counterclockwise from the bottom.

Say: "With the help of the powers of air, I banish from within me and around me and from all my doings all harmful influences and unbalanced manifestations of Earth. I banish them far from me."

Visualize all remaining tension and stress melting from your body and being absorbed by the stones and moss beneath you.

Calling in Spirit Below

Turn to face east, standing once again in the eastern quarter of the space. Trace the symbol of spirit below (it is like a Venus or female glyph) with your sword fingers, in vibrant orange, filled with a paler orange, clockwise starting at the bottom.

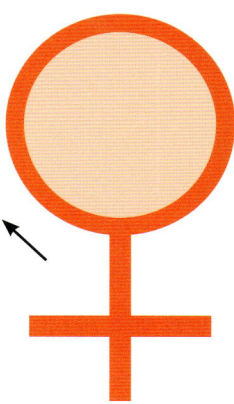

Say: "By the orange gate of the heart of this land, by the roots of the hills and by the bright heart of the Earth Mother, I invoke spirit below, its gods, its spirits, its powers. May the telluric current rise into me this day and bless me / this place with the power of the earth."

Breathe in and feel the telluric current ascend as a bright green light from the depths of the earth with each breath, filling your sacral center, the earth cauldron, with green earth energy.

Say: "I thank the spirit below for its gifts."

Calling in Spirit Above

Still facing east, draw the symbol of spirit above (an inverted Venus glyph) clockwise, in vibrant purple with the sword hand.

Say: "By the purple gate of the skies above, by the Sun in its glory, the father of light, I invoke spirit above, its gods, its spirits, its powers. May the solar current descend into me / this place, this day and always, blessing me / this place with the power of spirit above."

Visualize the heavens above, yourself standing at the center of the cosmos, the power of the Sun shining onto you. Breathe in and as you breathe, draw the golden light into your solar plexus, and, rising up, fill your heart center with the cauldron of the Sun with golden light.

Say: "I thank spirit above for its gifts."

Still facing east, become aware of the six powers you have invoked in the six directions.

Say: "By the six powers here invoked and here present, and by the great word 'Awen,' may a lunar current arise and bless and protect me / this place, this day and always. May it establish a sphere of protection around me / this place."

Breathe in and visualize a silver light rising with your breath from the cauldron of the Earth, through the Cauldron of the Sun, into the Cauldron of the Moon, in the head center.

Allow a drop of light to fall back into your solar plexus, expanding into a brilliant sphere of light that draws on energy from the solar and telluric current. The sphere grows and draws on the colors from all of the elements and directions, expanding to take in all of the area you want to extend its protective influence over. Hold this visualization for some time, imagining a wall of force that blesses, heals, and protects all within.

Closing Out the Ritual

End the ritual by saying: "I thank the powers for their blessing."[34]

As stated, deity names appropriate to the elements and directions may be added by practitioners to suit their taste, taken from the pantheon of their choice. Experimentation and practice are necessary for success with this, as with anything else. Building on this basic framework, other rituals can be performed to raise and work with the three currents—the lunar, solar, and telluric—for a variety of magical purposes.

One of the rituals described in John Michael Greer's work is the Three Cauldrons Working, which expands on the visualizations and breathing exercises in the Sphere of Protection, to draw the solar and telluric currents into their respective centers and blend them into the lunar current and raise them to the cauldron of the Moon. This working can be followed by the Awakening of the Dragons, in which currents of energy in the form of a red dragon and a white dragon, symbolizing the two currents, solar and telluric, are visualized as rising with the breath, intertwining through the three Druidic centers and then out of the head. This raises a great deal of energy for the final working, the Tree of Life working, in which the operator visualizes themselves as a great tree of light, the world tree. Next the worker circulates the light with the in-breath and out-breath, from the leaves and branches of the tree to the deepest roots at the center of the Earth and back. The final step is to form the Sphere of Protection itself, expanding the raised *nwyfre* out to the very borders of the land under one's care, blessing and protecting it. These rituals can be used for re-enchanting and blessing damaged and degraded lands, for bringing fertility to farms and gardens, and for protecting homes and homesteads from malign spiritual forces. They can also be used to open any ritual and can easily be followed by appropriate invocations and prayers to any deity.

I refer the interested reader to John Michael Greer's excellent works, which detail the practices mentioned above. They are listed in the bibliography at the end of the book.

The preceding rituals and other similar ones from nature-based spiritual traditions offer the practitioner a means of entering into contact with the animate powers of the land, the air, the water, the heavens, and the sky. The formats are flexible, and once the basics are learned, the structure can be adapted and improvised on for a number of purposes. It is important to memorize enough to be able to ad lib the rest of the ritual, making sure to call in any of the relevant powers, like the four elements and quarters, your favorite saints, deities, allies, and ancestors, or all of the above to your aid. Various breathing methods, energy-moving exercises, and meditations can be added as needed. When I began doing this ritual work outside I would read various ritual scripts from my phone, which is handy, but can be a bit distracting. Once I had memorized enough of a few basic techniques, I found that I could dispense with the phone and have a much more powerful and unmediated experience by simply using my memory and imagination.

13
THE MAGIC OF THE UNDERWORLD

Earth feet, loam feet, lifted in country mirth
Mirth of those long since under earth
Nourishing the corn. Keeping time,
Keeping the rhythm in their dancing
As in their living in the living seasons
The time of the seasons and the constellations
The time of milking and the time of harvest
The time of the coupling of man and woman
And that of beasts. Feet rising and falling.
Eating and drinking. Dung and death.

T. S. ELIOT, "EAST COKER"

In the autumn in the Northern Hemisphere, we descend to the threshold of the underworld, the dark and shadowy world of the dead, home of the ancestors, the chthonic spirits, and the roots of the world tree, behind and beyond and underlying our physical realm. This realm has been described in nearly every mythology in the world. It is not a physical place, though countless places in our physical world are seen to be portals of entry into its shadowy depths. In reality, it is a place outside of space and time.

We have discussed the dual aspect of the deities of agriculture, that they are simultaneously the gods and goddesses of the soil and the underworld beneath it, of fertility and death alike. All things planted in the dark earth must die and rise

241

✦ *Old Presbyterian church cemetery on Edisto Island, South Carolina.*

again to the light, must pass into the kingdom of Hades to decay and be renewed in the growth of the new season. "Except as a corn of wheat fall into the ground and die, it abideth alone: but if it die, it bringeth forth much fruit," say the scriptures. This is the ancient realization of the first planters, the mystical meaning of the cycle of planting and harvesting. The seed is sown and dies, losing its form and individuality, becoming a sprout, which endures the snows and cold of winter and begins to grow when the land thaws in the spring; the lengthening days of sunshine draw the stalks of wheat ever higher; the grain forms in the ears; the green fades, the sap withers; senescence, drying, the reapers come and cut it down. It is brought to the threshing floor and flailed mercilessly, the chaff winnowed and the crop cleaned, ready to give life and to die again as it is sown in the next crop. Persephone, Attis, Osiris, Dionysus, Christ, the slain and risen. John Barleycorn must die. The never-ending cycle of earthly life, the wheel of necessity. Birth following death, death following birth, as certain as the Sun and the tides. Just as the individual endures for a time, bears fruit, withers, dries up, and is cut down by the reaper in the endless cycle, the soul is taken down to the underworld. It drinks from the river of forgetfulness and is sent back, again and again for purification. Life endures as the spirit, but the forms are in constant flux. Being, becoming, and back to being. The relentless rolling of the wheel. We have been here many, many times, and will return many more.

This assurance of eternal life in the grand flux of existence was probably the secret of the mysteries of Eleusis, revealed after days of fasting and a dramatic nighttime ritual, which probably involved the consumption of entheogenic plants. Those of us who are modern *epoptae*, or *mystai*, who have had our eyes opened to the mysteries revealed by such substances, will understand that the actual mystical understanding of such revelations in their felt sense is a truth far beyond the ability of mere words to convey. We are assured, in no uncertain terms, of the continuum of the life of the spirit on both sides of the great divide.

What is our relationship, as the living, to this cycle and to those who have gone before? Where they are, whether it is possible to contact them, and, if it is, whether we can still help one another across the veil—these are a few of the questions we will discuss in this chapter and the next. We will start first with mythology and folklore, with a bit of metaphysics, and then move to the historical, magical, and religious traditions for practical advice on how to proceed to work with ancestors and the dead, for magic and healing. We will try to tease out the boundaries of these categories, as well as sketch out the geography of their

✦ *Ancient Greek bas-relief of Hades and Persephone, the king and queen of the underworld, with grain and flowers, emblems of ephemerality.*

realms. I will also include a few anecdotes of my own humble experiences of what I believe to be contact with these forces from the other side, as well as some practices I have experimented with for connecting with ancestors and other entities.

THE MYTHOLOGY OF THE UNDERWORLD

As cunning farmers, we are most concerned with underworld folklore as it relates to the land and our life on it. Many cultures have relevant traditions of spiritual beings residing in or under the land, including, as previously mentioned, the *sidhe* of the Celtic nations, more properly known as the *áes sídhe*, literally meaning "people of the mounds," the entities associated with the Neolithic burial mounds found in abundance in Western Europe. The actual relationship of the fey, as they are also known, to the human cultures who built the mounds and once buried their dead in them is unknown, but what is interesting is that it seems like the long dead can actually become land spirits, dwelling in or under the land as guardians or even trickster figures. As noted in chapter 2, North America also has its own "people of the mounds" who are long forgotten by their present-day descendants and who perhaps bear a similar relationship to them as do the modern inhabitants of the Celtic nations with the builders of their mounds. But are their spirits still there? Can they still be felt by the sensitive and honored by propitiations and offerings? It seems like we can make a distinction regarding land spirits, between the once-human and the never-human. The European fey seem to partake of both classes of entities in the folklore.

We will attempt to sift through the historical primary sources and academic literature to extract hints of relevant authentic practice that fits our life and situation here on the land. Anthropological and other academic literature gives us an outsider's view, an etic perspective. As practitioners, we are interested in gnosis, intimate knowledge, and an emic perspective. We seek the living reality beyond the physical that all humans have access to, one that can be accessed by techniques of consciousness expansion that usually fall into the realm of magic and mysticism. That is the reason for the seeming unity of spiritual traditions across history and place. Spiritual reality is one and ineffable and can only be approached through myth, symbol, ritual, and higher states of consciousness. Practitioners of technologies of the sacred everywhere and in all times have achieved contact with the fundamental realities of existence itself and filtered them through their particular cultural frameworks and unique individual perspectives. This explains both the unity and the diversity of approaches to the sacred. The trick for us as modern explorers of these territories is to sort out what is relevant and necessary for us as individual practitioners. We have a guide in the inner voice of the daimon that is with us as an advisor throughout our lives; we need only to learn how to listen.

An intuition came to me as I walked on the land one night that some people, when they leave this life, stay on the land as spirits, due to attachment or even love for the place where they lived. This is maybe a different class of ghosts from the ones who are attached to their places from anger or passion. This type seems more of a guardian. And the longer they stay, human lifetime after human lifetime, watching the living come and go, the less individual they become, becoming less and less human with each passing year and becoming more and more something like the fey—weird and ethereal, shades, more powerful with age, almost semidivine. Their existence, maybe, is something like the shades in Sheol or Hades or one of the conceptions of the afterlife from the ancient world—dark, infernal, cavernous places, realms shaped by the imaginings of those who went before and by the land above.

Our ancestors are with us always. Whether we knew them or not, they gave us the forms and qualities of our mortal existence and they continue to shape our thoughts and feelings from outside of space and time. Their experiences shape our consciousness and that of our descendents and our communities. They are not shadowy ghosts, shady forms and apparitions, or *eidola*, although sometimes they can be experienced that way. They interact with us intimately. They no longer have bodies, are no longer tied to space and time; we are their bodies, and we host their consciousness. Many healing traditions speak of energetic attachments, whereby a spiritual entity attaches itself to a person and feeds from their essence. I would venture to say that this is more common than we think, that we all have attachments to our ancestors, and if we begin to consciously feed them, consciously attempt to send them healing, to give cult, as the ancients would say, that we can begin to heal the wounds of the past. Many of us experienced pain and suffering at the hands of those now dead, and the wounds are too fresh, too recent, and healing and forgiveness is sometimes not even desired by the living. The trick here is to go back further, back into the past to a more generalized sense of ancestry.

My Own Family History

In my family, on both sides—as, I imagine, with most families—there is a great deal of ancestral trauma, which has become lodged in patterns of behavior. I can look back at the stark facts of the family tree, the untimely deaths of mothers, fathers, children, and spouses, the survivors of war, the ones who left their homelands fleeing famine and persecution to seek a better life in a new world where they had to start at the bottom. I have heard stories of lethal accidents in generations past, and it has been given to me to understand that there are consequences, on a spiritual level, that have haunted and divided my family and led to suffering and loss in generations still living.

The approach I have taken, as I have said, is to go past, back before memory to a mythical ancestor, the *maiores* (elders) or *tritopatores* (great-grandfathers) of the *gens*, or clan. For example, my surname is Elliott, which comes originally from the border region between Scotland and England. This was a place where clan warfare and raiding were so endemic that in the early seventeenth century, King James I of England (James IV of Scotland), presiding over a newly formed United Kingdom, forcibly deported many of the Elliots (the original spelling) and many members of the other troublemaking border clans, known as Border Reivers, to Northern Ireland and other locations, and forbade them to own arms and horses. Previously, in the sixteenth century, around 1524, these same troublemaking criminal ancestors were the subject of a very literal ancestral curse by the archbishop of Glasgow, Gavin Dunbar, which he instructed to be read in all the congregations of the diocese. The curse, which is over one thousand words long, calls down upon the Reivers and their descendants all manner of hellfire and damnation in hair-raising biblical language in the Scots vernacular.[1]

The origins of our surname and *gens*, however, date back to Anglo-Saxon times, to the Kingdom of Northumbria and its King Aelfwald. The word *Aelfwald* itself is an epithet of the Germanic god Freyr, the Vanir god of fertility, or "rain and sunshine and thus of the produce of the earth,"[2] and *Freyr* literally means "elf-ruler." The name refers to the sagas in which Freyr is given rulership over Alfheim, the land of the elves. As mythic ancestors go, Freyr is a pretty good one, especially for a cunning farmer, so that is who I have adopted as mine. I imagine this adoption of a mythical ancestor is possible with a little research for nearly everyone. Even for those who don't know their blood ancestry, you can, with the wonders of genetic testing, find an ancestral tribe to connect with. Honestly, as you can see, even though I have adopted Aelfwald and Freyr as my mythic ancestors, I am just as inspired by the classics as by the sagas, but the Greeks and Romans are more in the way of spiritual ancestors than blood.

✦ *Mourning scene from ancient Greek terra-cotta funerary plaque, c. 520–510 BCE.*

THE POWER OF THE DEAD

Many cultures had, and some still have, a similar conception of the state of the dead in the afterlife. To the ancient Romans, the dead in the underworld were worshipped as *Di Manes*, the divine dead, somehow losing their individuality after several generations but becoming a spiritual power that required offerings and cult and could reward their living descendants with benefits and punish those who neglected them with misfortune and illness. The ancient Greeks used the word *tritopatores*, or "great-grandfathers," which I introduced above, to refer to the distant blood ancestors of one's clan or tribe, several generations removed from the living and fading from individual memory but worshipped and given regular offerings to keep them happy and benevolent. Sarah Iles Johnston, in her book *Restless Dead*, states:

theoretically, we can distinguish between two reasons that the living might choose to make offerings to the dead. On the one hand, they might do so out of affection. Expressions of this might include both "funerary rites," directed to an individual at the time of his or her death and on certain days thereafter, and civic "festivals of the dead," during which all of the city's dead received offerings and were honored at the same time. On the other hand, the living might bring offerings to the dead not out of affection but rather in fear that they would cause harm if not appeased. Some such defensive rituals might be undertaken

✦ *Women bringing offerings to a tomb. From an ancient Greek* lekythos *(oil flask), c. 470–460 BCE (in the National Archaeological Museum in Madrid). Image by Marie-Lan Nguyen.*

spontareously, by individuals who thought that ghosts were haunting them. Others might be undertaken by all members of a city during appointed "days of the dead," when such ghosts were imagined to wander freely in the upper world.[3]

Ancient people (here I am referring specifically to classical Greek and Roman cultures, but this way of seeing was and still is widespread) did not see death as a state in which the dead were removed from the affairs of the living, but one in which the dead still had great influence over the lives of the living and continued to make their presence felt. If the ancestors were happy, things went well for the living, and if they were angry or displeased—if funeral rites were neglected, for instance, or offerings were not made—misfortune, perhaps illness or a poor harvest, could result. In Greece, during the classical period, offerings of libations were an integral part of the funeral ritual. Walter Burkert says of the practice of offering libations: "The offerings for the dead are pourings, *choai*: barley broth, milk, honey, frequently wine, and especially oil, as well as the blood of sacrificed animals; there are simple libations of water. . . . As the libations seep into the earth, so, it is believed, contact with the dead is established and prayers can reach them."[4] There was a supper for the dead, known as a *deipnon*, and meals for the living and regular visits to the grave, in which offerings were made. Johnston speculates "that survivors made additional offerings whenever they wanted the help of the dead person, or whenever they wanted him to participate, albeit distantly, in a family occasion such as a wedding."[5] It was believed the deceased lingered on at the place of burial for a time. As Burkert says, "The cult of the dead seems to presuppose that the deceased is present and active at the place of burial, in the grave beneath the earth. The dead drink the pourings and indeed the blood—they are invited to come to the banquet, to the satiations with blood; as the libations seep into the earth, so the dead will send good things up above."[6]

In Rome, the Manes, the divine ancestors, were believed to help the living in several important ways, according to C. W. King in his book *The Ancient Roman Afterlife*:

1. The *manes* have control over the duration of life and the conditions under which death occurs, so they are able both to prolong life through their protection and to initiate death at their discretion.

2. The *manes* are able to monitor the actions (including future actions) of living

✦ *A Roman gravestone from Chester, England (the Grosvenor Museum), with dedication Dis Manibus (or "to the Manes"). The inscription, translated, reads: "To the spirits of the departed and the spirit of Flavius Callimorphus, aged forty-two, and his son Serapion, aged three years, six months . . ." Image by Wolfgang Sauber.*

persons and can intervene throughout the world of the living to employ their power in relation to those persons.

3. The *manes* can secure for their newly deceased former worshippers a favorable location or situation in the underworld community of the dead.[7]

The first two reputed powers of the dead form the basis of the magical practice of working with spirits of the dead, often called necromancy, at least in the West. The idea that the dead, being spirits, have access to power and knowledge that the living do not, and that since they were once human, they are responsive and sympathetic to entreaties by the living for assistance, is an ancient one. It still resonates today in many traditions and parts of the world, including the very common Catholic practice of venerating the saints. The saints, after all, are the beloved dead in a baptized form. This is a reworking of the pre-Christian idea, found in sources like *De Mysteriis* of the Neoplatonic philosopher Iamblichus, that the souls of heroes become godlike due to their greatness in nature and deeds. They were thus given worship and devotion of their own.[8] And like the saints and the divine ancestors of the Romans, they could perform miracles of their own to help their devotees.

As historian Georg Luck says, "Heroes form a special class among the dead. Some heroes were the ghosts of the kings of old, who were considered powerful even after death, at least for a time, because they had been powerful in life." Luck says that there were many heroes' tombs all over the classical world: "Some of them became objects of a cult that seems to have continued from the end of the heroic age down to classical times, but after a while the worship ceased, and finally even the location of the tombs was forgotten, until they were rediscovered in their splendor." He adds that "the hero belongs to the community he protects, but his power does not really extend beyond those boundaries."[9] The image of the forgotten tombs of heroes of long ago brings to mind the burial mounds of Europe and America, whose presence dominates the landscapes to this day. Many authorities believe that the sídhe may be the Celtic world's answer to something like the Manes of the Romans, the divinized dead, heroes and nobility of a forgotten people, who still reign in the underworld to this day.[10] Hilda Ellis Davidson discussed similar practices of a cult of the dead among the Norse peoples in her first book, *The Road to Hel*.

Henry Cornelius Agrippa describes the "Animastical Order" in his Third Book: "Strong and Mighty Men; the magicians of the gentiles call heroes and demigods, or half gods and half men." The heroes of ancient times were worshipped as gods. As Agrippa says, "These heroes have no less power in disposing and ruling these inferior things than the gods and angels, and have both their offices and their dignities distributed to them: and therefore no otherwise than to the gods themselves were temples, images, altars, sacrifices, vows and other mysteries of religion dedicated. And their names invocated had magical virtues for the accomplishment of some miracles." He concludes the discussion of the Pagan cults of the heroic dead saying, "but these are the follies of the gentiles."[11]

Without a trace of irony, Agrippa then begins his discussion of the Christian cult of the heroic dead, the saints, saying:

> But as concerning our holy heroes we believe that they excel in divine power, and that the soul of the Messiah doth rule over them, that is Jesus Christ, who by divers of his saints, as it were by members fitted for this purpose, doth administer and distribute divers gifts of his grace in these inferior parts [in the world], and every one of the saints do enjoy a particular gift of working. Whence they being employed by us with divers prayers and supplications according to the manifold distribution of graces, every one doth most freely bestow their gifts, benefits, and graces on us much more readily, truly, and also more abundantly than the angelical powers by how much they are nigher to us, and more allied to our natures, as they who in times past were both men, and suffered human affections and infirmities; and their names, degrees, and offices are more known to us.[12]

I think Agrippa knew well the similarity of the Pagan hero traditions he was describing with the cult of the saints and chose to downplay the latter at the expense of the former to avoid incurring the wrath of the Inquisition, which he had spent most of his life fleeing.

The beloved dead, well cared for and remembered with proper rituals and offerings and recalled in stories and fond memories, are not the only kind. Ancient people feared the angry dead—the unburied, those who were victims of violence, those who died young and resented the living for their full and happy lives. Many of the rites and festivals for the dead in the classical world were apotropaic in nature; that is, they were performed for the purpose of appeasing, soothing, or expelling the angry dead, who were known in Greece as *alastores*, vengeful ghosts; *ataphoi*, those who hadn't been given a funeral; the *aoroi*, those who died young; and the *biaiothanatoi*, the victims of violence. In ancient Rome, these sorts of spirits were known as *larvae*. An untimely death represents a disruption in the fabric of life, an instance of chaos creeping into the orderly life of a culture, a source of impurity known to the ancient Greeks as *miasma*, which needed *katharsis* or ritual purification to appease the spirit of the deceased.[13]

It seems to me that this kind of miasma is similar to what is meant by modern energy healers when they refer to "spirit attachments," in which the spirit of a deceased person attaches itself to a living person, drawing on their reserves of psychic energy and causing psychological and physical illness. Reiki master William Lee Rand describes spirit attachments this way:

> There are many kinds of spirit attachments. Not all spirits have malicious intent. Many consist of confused spirits who did not cross over into the light after death and may not know where they are or that they have attached to you. We call these discarnate spirits and [Reiki techniques] can help them cross into the light and go home to their loved ones. Some spirits even think they are helping the person they attach to when they are not. In other situations a spirit may come to help a child or a person in crisis and remain present after they are no longer needed. Regardless of a spirit's intent, spirit attachments will almost always cause problems. They can cause physical weakness and illness, mental, emotional, and spiritual dysfunction, problems with relationships, weaken one's creative abilities, cause disorientation and make debilitating conditions one might have worse.[14]

In the shamanic traditions described by Mircea Eliade in his monumental classic work *Shamanism*, the situation is analogous. Among the Teleut people of Russia:

> Feelings toward the dead are ambivalent. On the one hand, they are revered, are invited to funerary banquets, in time they come to be regarded as tutelary spirits of the family; on the other hand, they are feared, and all kinds of precautions are taken to prevent their reappearance among the living. Actually, this ambivalence can be reduced to two opposite and successive behaviors: the recently dead are feared, the long dead are revered . . . as protectors. The fear of the dead is due to the fact that, at first, no dead person accepts his new mode of being; he cannot renounce "living" and he returns to his family. And it is this tendency that upsets the equilibrium of the society. Not yet having entered

◆ Tiresias Appears to Ulysses During the Sacrifice, *by Henry Fuseli, c. 1780–85.*

into the world of the dead, the newly deceased man tries to take his family and friends and even his flocks with him; he wants to continue his suddenly interrupted existence, that is, to "live" among his kin. So what is feared is far less any malice on the part of the dead man than his ignorance of his new condition, his refusal to forsake 'his world.'"[15]

The shaman is the mediator between the world of the dead and the living, and, taking the role of psychopomp, it is their task to assure that the souls of the deceased are led to the next world.

KATABASIS: DESCENT INTO THE UNDERWORLD

In the magical practice of ancient Greece, a class of sorcerers, known in Greek as *goetes* (singular *goes*), specialized in, to quote Plato (who denounced them in no uncertain terms), "winning over the gods by the supposed sorceries of prayer, sacrifice and incantation."[16] What the goetes most specialized in was the invocation of spirits of the dead for the type of magic seen in the Greek Magical Papyri, for various purposes, from the making of curse tablets, or *katadesmoi*, to cause harm or misfortune on an enemy; to the "erotic binding spells" intended to attract a lover, without regard for free will or consent; to aims as mundane as winning at gambling or chariot races or even getting one over on the competition in business. An example of this type of spell, from the Greek Magical Papyri, is PGM IV. 1390–1495: "Love spell of attraction with the help of heroes or gladiators or those who have died a violent death." It advises the practitioner, after making an offering of bread in the place where the deceased were slain, to pick up some polluted dirt from that place and throw it in the house of the desired woman.[17]

The practice of using dirt as a sympathetic link to the spirits of the dead is ancient and has a long history in many cultures, and it is echoed in the use of *gris-gris*, or graveyard dirt, in Hoodoo practice to this day. The soil remembers the dead and holds their influence. This idea has resonance for us as workers of the soil particularly. If the land we live on and work as farmers has been occupied for millennia, the odds are that someone's mortal remains are part of the soil that we till and on which we raise our food, and their influence is with us and is part of the spiritual power of our land. All dirt is thus graveyard dirt. The earth is our sympathetic link with our corporality and our mortality; we are born from it, are sustained by it, and return to it again. It is a law as certain as gravity.

Johnston links the role of the goes with that of the initiator, the specialist

◆ *Greek binding spell engraved on a lead sheet. The spell is intended to bind the tongues of several people named on the sheet. Note the use of lead for cursing. From Thessaloniki, c. fourth century BCE (in the Archaeological Museum of Thessaloniki). Image by Gts-tg.*

and instructor in the eschatological mysteries of the next life. As we will shortly discuss, it is intimately linked with the Orphic mysteries and the idea of katabasis, the spiritual descent to the underworld, for the purposes of soul retrieval and communication with the spirits for divination and magic. The office of the goes is similar in many ways to the office of the shaman, enough so that the goetes have been described as "Greek shamans."* According to Johnston, "The *goes* [was] also understood to have the ability to initiate souls into mystery religions, or, in other words, to ensure through his superior knowledge of the Underworld and its work-

---

*This is not the place to debate the merits of applying the culturally specific word *shaman* outside of its particular cultural and historical context in Central Asia. It is enough for us that the word has entered common English usage and is a useful term for a specialist in ritual and altered states of consciousness, soul travel, and spirit communication.

✦ *The Derveni* krater (krater being a type of ancient Greek vase), found in a tomb in Macedonia, dated to the fourth century BCE (in the Archaeological Museum of Thessaloniki). The papyrus scroll quoted on page 258 was also found at Derveni in the same tomb. Image by Michael Greenhalgh.

ings that the souls under his care would receive preferential treatment after death. The real *goes* of the classical period, in sum, might best be described as a man who could negotiate a variety of relationships between the living and the dead through virtue of his ability as a singer, a communicator,"[18] and I would add, most likely also as a visionary and adept in altered states of consciousness. Related to the term *goes* is the foreign term *magos*, a term which refers originally to the Persian priests and is the root of our English word *magic*, who Johnston says "were specialists in the care and use of the soul along the same lines" as the goes.[19]

The Derveni papyrus, which is the oldest "book" ever found in Europe, dates from the fourth to fifth centuries BCE and was unearthed at a construction site in Macedonia alongside a grave containing the remnants of a funeral pyre and a stone tomb containing a beautifully crafted gold urn. It is a commentary on an Orphic theogony, a poem describing the birth of the gods, which was intended to be burned with the deceased in the pyre but miraculously was only partially destroyed. It contains tantalizing hints at ritual practices by Orphic goetes and magoi, including this fragment:

> Prayers and sacrifices placate souls. An incantation by *magoi* can dislodge *daimones* that have become a hindrance. *Daimones* that are a hindrance are vengeful souls. For this reason the *magoi* perform the sacrifice, as if they are paying a blood price. Onto the offerings they make libations of water and milk, with both of which they make drink offerings. They sacrifice cakes that are many-humped, because the souls are countless. Initiates make first a sacrifice to [the] Eumenides in the same way as *magoi* do. Eumenides are souls. Hence a person who wants to sacrifice to [the] gods must first sacrifice a bird.[20]

✦ Orestes Pursued by the Furies, *by John Singer Sargent, 1921.*

◆ *Orpheus surrounded by animals, from a Roman mosaic, originally from Palermo.*

The Eumenides, a euphemistic term meaning "the kindly ones," are equated with the Erinyes, or Furies, and are infernal goddesses of vengeance who are dangerous when displeased by impiety and impurity but beneficent when properly propitiated. In Aeschylus's drama, Eumenides are "able to infect Athens' soil, wither its plants, and make its animals and humans barren, but also say that they are able, once they are made happy and become 'Semnai Theai,' to assure that Athenian maidens will marry and reproduce and the crops and animals will thrive."[21] The Furies are the avengers of the dead, the punishers of crimes against blood kin in Greek mythology and drama. They are thought by some to have originally been spirits of the dead transformed into underworld goddesses of punishment in this life and the next.

The Orphic mysteries were the spiritual and religious tradition supposedly founded by Orpheus, the mythical Thracian musician and shaman known from Greek mythology for his journey to the underworld to retrieve his deceased bride, Eurydice, on the condition that he not look back when he led her back to the realm of the living. He looked back at the last minute and lost her forever. He was killed and dismembered by a group of Thracian women while grieving in the wilderness. The earliest mention of Orpheus is from a sixth-century BCE inscription. The Orphic mystics practiced an ascetic religion, were vegetarian, and believed in reincarnation and avoided animal sacrifice and any kind of miasma. They had sacred books, some of which come down to us in fragmentary form, such as the *Rhapsodic Cosmogony* and the *Orphicorum Fragmentum*. The *Orphic Hymns* were composed in a later period, likely in the first few centuries of the Common Era, but may contain some very ancient and authentically Orphic elements. The Orphic movement was allied to both the Pythagorean tradition and the goetes, and there was likely considerable overlap between the various movements at different times.

THE MANY PLACES OF THE DEAD

The passage from life to death is the passage from a physical existence to a spiritual one. Location and duration are both features of embodied life in time and space. To pass from this life is to pass outside of time and space, to be a spirit, a thing of pure consciousness. The individuality of the living gradually fades away. This passing out of space-time reality explains the multi-locationality of the dead—how they can be in their grave, in the afterlife (however one conceives it), or even in the next incarnation, and accessible to the goes, medium, or devotee who prays to them or gives them offerings. It is as if we are all in a river, floating with the current, willy-nilly, and when one dies, one crawls up on the shore and can access the river at any point, upstream or downstream.

Mircea Eliade discusses the multilocationality of the dead in his essay "Mythologies of Death: An Introduction":

> There is a widespread belief that the departed ones haunt their familiar sur-roundings, although they are supposed to be concurrently present in their tombs and in the netherworld. Such paradoxical multi-location of the soul is explained in different ways according to the different religious systems. Either it is asserted that a segment of the soul remains near the dwelling or the tomb, while the "essential" soul goes to the realm of the dead; or it is held that the soul tarries for some time in proximity to the living before ultimately join-ing the community of the dead in the netherworld. Notwithstanding these and other similar explanations, there is a tacit understanding in most reli-gions that the dead are present concurrently in the tomb and in some spiritual realm.[22]

To the folk of what Eliade refers to as traditional cultures, "The almost universal conviction that the dead are present both on Earth and the spiritual world is highly significant. It reveals the secret hope that, in spite of all the evidence to the contrary, the dead are able to partake somehow in the world of the living."[23]

Georg Luck, in his classic study of the magic of antiquity, *Arcana Mundi*, states the problem of multilocationality like this: "How can the dead be in Hades and in their graves at the same time? The ancients apparently believe that only their shades (*eidola*) went down to Hades, while their bones or their ashes retained, magically, a particle of the extinguished life force, at least for a time. Hence the theme of the 'grateful dead,' as expressed, for instance, by the Hellenistic poet Leonidas of Tarentum, in a bucolic epitaph, 'There are ways, yes, there are ways, in which the dead, even though they are gone, can repay your favors.'"[24] To me this seems to connect the idea of multilocational-ity to the sympathetic link with the grave, and logically, it's not too far of a step to jump from this "particle of extinguished life force" Luck speaks of the practice of collecting graveyard dirt while leaving offerings for use in magical favors.

Eliade quotes the anthropologist W. Lloyd Warner, who describes the mystery this way: "The personality before birth is completely spiritual; it becomes completely profane or unspiritual in the earlier period of its life . . . gradually [becoming] more ritualized and sacred as the individual grows older and approaches death, and at death once more becomes completely spiritual and sacred."[25] For most modern people, death is not sacred but rather an abyss of nonbeing, regarded with fear and denied by our modern cult of life and youth. For ancient peoples, however, in the words of Eliade, "It is the experience of death that renders intelligible the notion of spirit and spiritual beings." He goes on to state:

> Everywhere in the traditional world death is, or was, considered a second birth, the beginning of a new, spiritual existence. This second birth, however, is not natural, like the first, biological birth; that is to say, it is not "given" but must be ritually created. In this sense, death is an initiation, an introduction into a new mode of being. And, as is well known, any initiation consists essentially of a symbolic death followed by a rebirth or resurrection. Besides, any passage from one mode of being to another implies necessarily a symbolic act of dying. One has to die to the previous condition in order to be reborn into a new, superior state.[26]

But what kind of condition, what type of existence is this new spiritual state? The newly deceased soul (*psyche* in Greek), says Walter Burkert:

> Leaves man at the moment of death and enters the house of *Ais*, also known as *Aides*, *Aidoneus*, and in Attic as *Hades*. *Psyche* means breath just as *psychein* is the verb to breathe; arrested breathing is the simplest outward sign of death. In the dead man, or dead animal, something has gone missing—something whose presence and power in the living creature is never given a second thought; only when there is a question of life and death is there any question of *psyche*. *Psyche* is not the soul as a bearer of sensations and thoughts, it is not the person, nor is it a kind of *Doppelgänger*. Yet from the moment it leaves the man it is also termed an *eidolon*, a phantom image, like the image reflected in a mirror which can be seen, though not always clearly, but cannot be grasped: the dream image and the ghostly image, the forms in which the dead man can still appear, are identified with the breath which has left the body. Thus the *psyche* of a dead man can on appropriate occasions be seen and at all events can be imagined.[27]

In Plato's dialogue *Phaedo*, Socrates, awaiting his execution, discusses death and the fate of the soul with his disciples in what is to me the most moving of the Platonic dialogues. Socrates tells his student Cebes that the soul of an initiate in the mysteries or one versed in philosophy departs the body and ascends immediately to the divine realm, but a different fate awaits the souls of the worldly and sensual. Says Socrates: "And we must suppose, dear fellow, that the corporeal is heavy, oppressive, earthly, and visible. So the soul which is tainted by its presence is weighed down and dragged back into the visible world, through fear, as they say, of Hades or of the invisible, and hovers about tombs and graveyards. The shadowy apparitions, which have been seen, are the ghosts of those souls which have not got clear away, but retain some portion of the visible, which is why they can be seen."[28]

The land of the dead is described in many, many mythologies, particularly in the ancient Mediterranean and Middle East. According to Franz Cumont, "This belief in the nether world is found among most peoples of the Mediterranean basin: the *Sheol* of the Hebrews differs little from the Homeric *Hades* and the Italic *Inferi* [italics in original]."[29] In particular, he says:

> The ancient Greek conception, going back to Orphism was . . . that of a subterranean kingdom divided into two contrasted parts—Tartarus, where the wicked, plunged in a dark slough or subjected to other pains, suffering the chastisement of their faults, on the one side; on the other, the Elysian Fields, those flowered, luminous meadows, gay with song and dance, in which the blessed pursued their favourite occupations, whether they were allowed to dwell there forever, or whether they awaited there the hour fixed for their rebirth on earth. This eschatology, which had become the common possession of the Hellenes, was certainly that of the mysteries of Greece and in particular the mysteries of Eleusis.[30]

Hades, then, had regions allotted for both punishment and reward.

For the Hebrews, the place of the dead was called Sheol, a place of darkness and forgetfulness, under the Earth, where, as the Book of Ecclesiastes states: "There is no work or thought or knowledge or wisdom."[31] You might as well, then, enjoy your life in the here and now, according to Koholeth, the preacher, author of Ecclesiastes. Job, in his affliction, said, "He who goes down to Sheol does not come up."[32] Bearing many similarities to the Land of Mot, the Mesopotamian land of the dead, the word *Sheol*, like Hades, is perhaps the name of an Assyrian god of the underworld, as well as the name for the place itself.

The ancient Greeks and Romans, at least the common folk among them, saw the afterlife as a literal underworld, existing in a vast hollow cavern beneath the earth. According to Franz Cumont:

> Among most peoples the idea of a primitive afterlife in the grave was enlarged into the common existence of the dead in the depths of the earth. The dead man does not stay confined in the narrow dwelling in which he rests; he goes down into vast caverns which extend beneath the crust of the soil we tread. These immense hollows are peopled by a multitude of shades who have left the tomb. Thus the tomb becomes the antechamber of the true dwelling of the spirits who have departed; its door is the gate of Hades itself. Through the tomb, the great company of the beings who have been plunged in the darkness of the infernal regions remains in communication with those who still sojourn in our upper world. The libations and offerings made by the survivors on the grave descend to this gloomy hypogeum and there feed and rejoice those for whom they are intended. Until the time of the Roman Empire, nay, to the end of antiquity, the common man believed in this wonder.[33]

And many modern people still do believe in a literal Hell, which may or may not be situated in the interior of the Earth.

THE GEOGRAPHY OF THE UNDERWORLD

Cumont was a skeptic and an academic, not a practitioner, and so he regarded all of this as antiquated nonsense rather than a mythic worldview underpinning spiritual exploration and practice. It is our interest here to form a picture of the landscape of the underworld for the sake of informing the practice of visionary katabasis, to have a clear picture of the imaginal world below for the sake of exploring it in vision and interacting with its denizens for the sake of gnosis and revelation. We descend into the depths of the underworld, the land of the dead, undertaking (pun intended) a shamanic journey to meet both our ancestors and ourselves, for in them we find ourselves. As Mircea Eliade put it:

> In all probability many features of "funerary geography," as well as some themes of the mythology of death, are the result of the ecstatic experiences of shamans. The lands that the shaman sees and the personages that he meets in his ecstatic

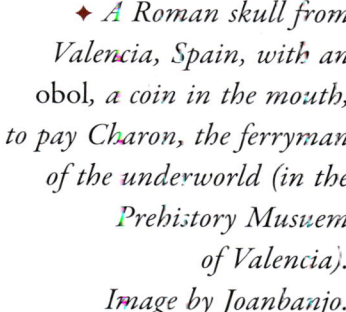

◆ *A Roman skull from Valencia, Spain, with an obol, a coin in the mouth, to pay Charon, the ferryman of the underworld (in the Prehistory Musuem of Valencia). Image by Joanbanjo.*

journeys to the beyond are minutely described by the shaman himself, during or after his trance. The unknown and terrifying world of death assumes form, is organized in accordance with particular patterns; finally it displays a structure and, in course of time, becomes familiar and acceptable. In turn, the supernatural inhabitants of the world of death become *visible*; they show a form, display a personality, even a biography. Little by little the world of the dead becomes knowable. In the last analysis, the accounts of the shamans' ecstatic journeys contribute to "spiritualizing" the world of the dead, at the same time that they enrich it with wondrous forms and figures.[34]

So, like the heavenly journeys we discussed in previous chapters, these underworld geographies also have their origins in the visionary states of specialists in altered states of consciousness, who take these voyages into mind space, making them available to the rest of the community. Eventually, it is forgotten that these were the visions of one particular seer, and the visions, written down, become the common property of the whole culture, like the Homeric epics, and finally religious dogma.

The underworld, like the world above, has a geography, complete with plains and rivers and different regions for the different qualities of souls. As Cumont puts it, "Although in our sources the infernal topography is somewhat confused, certain essential features . . . can be recognized in it." First, upon arriving from the upper world, the shades of the dead encounter a waiting area, says Cumont. This is "an intermediate region through which all of them pass but in which some are kept for a considerable time." The god Hermes, the *psychopompos*, the guide of the souls, leads the deceased to the underworld through one of its many entrances in the upper world. The people lived their mythology in their landscape, and many physical places were associated with portals to the nether world. Upon reaching the banks of the infernal river Styx, the deceased had to pay the fee—a coin known as an *obulus*, often placed on the eyes during the funeral rites—to Charon, the spectral boatman, to ferry them across, or else wander eternally on the bank, the fate of all who die without the proper rites. Cumont continues:

> They then cross the Styx, and a road which is also common to all of them leads them to the court which determines their lot. This judging of the dead is foreign to Homeric poetry: the idea of it was perhaps borrowed by Greece from Egypt, but from ancient Orphism onwards, it was an essential element of infernal eschatology. Infallible judges, from whom no fault is hid, divide into two companies the multitude of the souls appearing before them. The guilty are constrained to take the road to the left, which leads to dark Tartarus, crossing its surrounding river of fire, the Pyriphlegethon. There, those who have committed inexpiable crimes are condemned to eternal chastisement. But the road to the right leads the pious souls to the Elysian Fields where, among flowering meadows and wrapped in soft light, they obtain the reward of their virtues, whether, having attained to perfection, they are able to dwell for ever with the heroes, or whether, being less pure, they are obliged to return later to the earth in order to reincarnate themselves in new bodies after they have drunk the water of Lethe and lost the memory of their previous existence.[35]

The underworld as a place of punishment and reward is an idea that has had a long history, continuing particularly in Christian notions of the punishments of Hell and Purgatory. Dante's vision of Hell, depicted in the *Inferno*, was inspired by Virgil's *Aeneid*. And let us not forget the parallels with Christ, who harrowed Hell, like Orpheus, to release the righteous souls of the Old Testament, forever commemorated in the Creeds of the Church to which many Christians give assent. The story of the katabasis of Christ is depicted in the apocryphal Acts of Pilate and related texts.[36] These texts seem like an ancient version of a Marvel superhero film in which Jesus descends in light, bursts asunder the gates of Hell, and literally tramples death, defeating Hades, Satan, Beelzebub, and the rest, and leading Adam, David, Isaiah, and the other righteous of Israel up to Heaven. It is an awesome story and makes for exciting reading. Ultimately, in the Pagan version of the underworld, one avoided punishment by living a good pious life and avoiding the extremes of intemperate sensuality, while in the Christian version, one avoided Hell by professing one's faith in the saving power of Jesus, which points to the ancient Chrsitian debate about whether one is saved by faith or works. That Orthodox Christianity promised easy salvation and a happy afterlife to anyone—no matter how badly they had lived, provided that they converted and confessed—is a major reason for its success as a doctrine.

A Guidebook to the Afterlife

In his classic book, *The Center of the Cyclone*, pioneering psychonaut John C. Lilly made this statement regarding his experiences with LSD and sensory deprivation, "In the province of the mind, what is believed to be true is true or becomes true, within limits to be found experientially and experimentally. These limits are further beliefs to be transcended. In the province of the mind there are no limits."[37] If we may indulge in some provisional and temporary dualism for the sake of this discussion, without a body to tie it to time and space, the consciousness is "in" the province of the mind (I would say that we are always in the province of the mind, but that is a discussion for another time), and one's beliefs about the state one finds oneself in shape the experience of the dying process. This is why the different religions and cultures, as well as near-death experiencers, have such different accounts of the process.

The Orphic tradition, similar to the Tibetan Buddhist tradition of today, had its own answer to salvation in the afterlife: a guidebook.

In excavations of various burials in what was known as Magna Graecia (Greater Greece, now Sicily and Calabria, a stronghold of Pythagorean thought), gold tablets dating mostly to the third and fourth centuries BCE were found. They were rolled up and placed in the deceased's hands, on their chests, or even in their mouths. The tablets all bear similar inscriptions of instructions on how to proceed in the land of the dead. Here is the text of one, found in the Greek city of Hipponion in what is now Calabria:

✦ *One of the gold Orphic tablets, from Thessaly, dated to the fourth century BCE
(in the Getty Villa Museum in Pacific Palisades, California).
Image by Remi Mathis.*

*This is the work of Mnemosyne [meaning "memory"]. When he
    is on the point of Dying*
*Toward the well-built abode of Hades, on the right there is a
    Fountain*
and near it, erect, a white cypress tree.
There the souls, when they go down, refresh themselves.
Don't come anywhere near this fountain!
But further on you will find, from the lake of Mnemosyne,
water freshly flowing. On its banks there are guardians.
They will ask you, with sagacious discernment,
why you are investigating the darkness of gloomy Hades
Say: "I am a son of Earth and starry Heaven;
I am dry with thirst and dying. Give me, then, right away,
fresh water to drink from the lake of Mnemosyne."
And to be sure, they will consult with the subterranean queen,
*and they will give you water to drink from the lake of
    Mnemosyne,*
*So that, once you have drunk, you too will go along the sacred
    way*
by which the other mystai and bacchoi advance, glorious.[38]

These tablets were not just buried with the dead, but read, practiced, and memorized in mystery rituals, intended to prepare the initiates for the next world. Actually, the underworld journey of the soul makes an excellent pathworking meditation:

> Imagine yourself as a shade, descending with bright Hermes through a dark cave, and coming to the river Styx, paying dark, hooded Charon the *obulus*, and getting into his boat. Next, you meet the hideous three-headed dog Cerberus and offer him a cake (or three?) for safe passage beyond to the Asphodel Meadows where the numberless shades of the dead mill around muttering and whispering. Who do you see there? You are overcome with thirst but struggle not to drink from the waters of the river Lethe. You may meet a departed relative, a famous person from the past, Hades, Persephone—the possibilities are as endless as your imagination. The goal of the Orphic initiate was to get to the Elysian Fields, the goal of the blessed. Your goal in this exercise can be whatever you need it to be, or you can just let the experience unfold.

The major importance of all this discussion of the underworld for the cunning farmer, and the ongoing project of re-enchantment we are engaged in, is in how we put this into action in our own lives, our own personal practice, and our relationship with the sacred and the spirits of the land. So I will close this chapter with a quote from the masterful work of the late Jake Stratton-Kent (requiescat in pace), *Geosophia*, which is a must-read for anyone interested in these topics. Stratton-Kent introduces the idea of creating a mythic geography in our own lives. In his words:

> Elasticity of mythic geography, which undoubtedly served various needs and roles in ancient society, has profound implications for modern practice. It is only if we permit it that the secularized landscape of the modern world is emptied of myth and magic. After all, this is not the inevitable impinging of the supposedly real world on our fantasy life; on the contrary, an irreparable separation of the inner and outer worlds is both unreal and undesired.
>
> Mountains, burial mounds, crossroads, monuments, graves, trees, streams, and rivers were ancient locations of the numinous. They are no less full of Power today, if we but reclaim them. If communities and individuals have lost the sense of power attached to places—a very real loss—nevertheless the

magician's work requires them: this crossroads for offerings to Hecate or the spirits of the Underworld; this hollow tree to hide and isolate the image of a foe; this mountain, cave, or lake to court the favor of the otherworld. Included as well are more routine tasks such as disposal of ritual by-products and remnants, cutting of herbs and gathering water at auspicious places, or rods at suitable ruins. This extends even to suitable stores for obtaining mace, olive oil, and other sundries, not to mention the gathering of dirt or clay from banks, police stations, prisons. Employ mythic thinking to invest the mundane with the magical.

The magician looks about them and sees the magical potential in all things. Has this river no nymph, this mound no hero, this mountain no god? Perhaps under no name known today, but the magician is like a second Adam—replete with the Power of Naming. Many locations have magical uses or associations, awaiting our use of mythic language. If, say, a prehistoric burial mound is associated with no name known now, then ask your spirits which of them or their companions and allies dwells there. What matter if no one called the resident by this name before? Names change, but the ancient magic continues regardless. This extends to new places as much as old or rural ones—to any place with meaning for you. Reclaim the landscape, reinvest it with power and significance; be aware of the innate power and significance inherent in every place.[39]

In the next chapter, we will discuss the intersection of the spirits of the dead with the fairy traditions of northern and western Europe, as well as the funerary practices of the ancient Greeks and Romans. From this lore we can gather some hints for practices that are relevant and valuable for us today as modern people seeking to gain contact with the spiritual forces on the other side and under the land. We will look at the ways ancient people understood the world of the unquiet dead and the underworld to affect the lives of the still living, by being agents of disease and misfortune.

14
PRACTICES INVOLVING SPIRITS OF THE UNDERWORLD

With the theme of the previous chapter and this one very much on my mind, I went out for a walk one night, on the Full Moon, a few days before Halloween. It was mostly overcast, with swiftly moving low stratus clouds, but very bright, warm, and breezy. A deliciously spooky night. The cloud cover occasionally parted to allow the Moon to peek through, and on several occasions I caught a glimpse of Jupiter, which was a few degrees ahead of the Moon. I walked down the gravel road that runs through my farm, passing by a stand of very late sunflowers that were struggling to bloom far out of their season. A jimsonweed was blooming in the midst of the sun-flower plants, its lavender flowers suddenly illuminated by a beam of moonlight that shot through the parting clouds. I had a sudden inspiration to pick three flowers from the plant and add them to the offerings of beer and tobacco I was carrying to the abandoned farm where I was going to talk to the spirits. I carried them in my hand for a few minutes as I walked, until remembering that the alkaloids could be absorbed through the skin and that I had better place them in my pocket, just to be safe.

I went up to the abandoned outbuildings. Loose boards on the barn were rat-tling in the breeze. The milking parlor and the tobacco stripping room, where the farm family had spent much of their time tending their animals and preparing their crops for sale, were eerily quiet, darkness staring out of the shattered windows. I laid the jimsonweed flowers in the grass, poured out the beer, and crumbled the tobacco, and voiced my prayer to the spirits of the place and the souls of those who had lived there. As I often do, I said, "If you can hear me, knock." I immediately heard a thump from a clump of trees about fifty feet away. I stated my intentions to honor the spirits and to care for the land, as I asked for their favor and assistance in doing so.

I wandered slowly, tired from the week's farm work and the late hour, trying to decide how far I wanted to walk that late in the night. I stood in the middle of the clearing, listening to the sound of the wind in the trees, the distant barking of dogs and howling of coyotes, the uncanny squeaking of branches in the trees sounding like voices on the wind. I just stood and thought, leaning on my forked staff, picturing in my imagination the people who had lived there, imagining myself talking to them. I drifted into a reverie. I could hear the old couple who had lived their lives there and had run the farm tell me that they liked me, that I was a hard worker, that they were proud to see such good crops being grown on their land, and that they liked that I came to talk to them. I pictured the people who had lived there before in the nineteenth century, clearing the land and living in small frame houses and log cabins now gone, who lived and died on this land, the mothers, fathers, children, and elders, both enslaved and free, who passed whole generations here, whose bodies rotted into the land in now-forgotten family grave plots. I could see the settlers who squatted on the land claimed by the Shawnee, who were involved in the bloody guerilla war of genocide that was fought so violently in Kentucky in the last several decades of the eighteenth century.

And before the Shawnee, their ancestors, whose names have been forgotten by history, who had farmed the land, raising corn, beans, squash, and tobacco in the creek valleys, and lived in larger villages, and were wiped out by the epidemics brought from Europe and spread by Spanish explorers who traveled through the North American backcountry in the sixteenth century. This culture was entirely erased, and I could picture the epidemics killing so many so quickly, with the dead left unburied in the abandoned villages that had fallen silent and overgrown. In my mind, I went back to an earlier time, the time of the ones who built the mounds in the valleys, whom the archaeologists call the Hopewell and Adena people, the stargazing culture of astronomer shamans who made massive earthworks in Ohio and Kentucky to track the solstices and equinoxes and to predict eclipses, to perform unknown rituals to maintain the cosmic balance, whose farthest-western outposts included our land. I could clearly see the shaman priests, dressed in wolf skin, embodying the wolf, leading the ceremonies on this ridge in a clearing with an unobstructed view of the sky. They used the very substances I left them as offerings, the Datura and rustica tobacco, to fuel their visions. The leader, the one I call the "Wolf Shaman" in chapter 1, turned and looked at me with an intense and intimidating lupine stare.

Back in time I went, to mammoths and mastodons and bison, and the first humans on this land, at the end of the Ice Age. Back further still, the pictures became dimmer and out of focus, as I came back to myself standing in the clearing, leaning on my forked staff in the moonlight. An hour had passed as if it were a moment. I thanked the spirits and went home to bed. In the morning, I woke up and the light that came into my eyes had a strange and unreal quality, and I wondered what was wrong with my vision. I looked in the mirror and my left pupil was fully dilated, while the right was normal. I panicked a bit, and then I remembered the jimsonweed flowers that I held in my hand a little too long and the vision that I had while standing in the field. I must have rubbed a bit of the juice into my eye, enough to dilate the pupil and perhaps give the vision of the previous night some of its intensity and power, allowing me to see things previously hidden.

The underworld is memory, the memory of the land and the memory of the cultures that live on the land, the people whose lives form its stories. These stories exist out of time and space in a place of memory, accessible to the sensitive and intuitive as spirits, ghosts, fey folk. This underworld of memory is held in the land, in the rocks and soil, but also in the collective soul of humanity, as a dream that we all dream together, a space at once within us and all around us, which we can make contact with in order to form a relationship across time with those who have gone before and with those who will come after us. To form a true culture, one that honors its ancestors and makes wise decisions for generations to come,

✦ *Jimsonweed,* Datura stramonium. *This is a plant very much linked in global folklore to the underworld and the dead.*

we must make peace with our past, starting as individuals, with our ancestors and the place in which we find ourselves.

In previous chapters, we made a basic sketch of the folklore and mythology of nature spirits and the spirits of the dead and the underworld, mostly focused on classical mythology. In this chapter, I want to examine the parallels between the classical Roman concepts of *lemures* and Di Manes, as well as the Greek concepts of *lamiai*, Erinyes, and Keres—some of whom we met in the last chapter—and the fey traditions of Celtic Europe and the fairy tradition of the medieval and Renaissance periods that derive from them. There are many similarities between these traditions, although they are widely separated by geography and time. They seem to describe the actual human experience of contact with a preternatural reality at work in the landscape, both imaginal and physical, of people who live in intimate contact with the land, throughout all times and places. Throughout, we will emphasize a practical and rooted perspective, to try to extract a praxis from these folktales and myths that will enable us to reconnect with this hidden level of reality.

ORIGINS OF THE UNDERWORLD SPIRITS

In much of Europe, indeed all over the world, many of the practices and beliefs that the ancient Romans applied to Di Manes, the divine dead, could easily apply to the semidivine spiritual race known variously as the sídhe in Ireland, the Tylwyth Teg in Wales, the fey or fairies in England, the *vile* in eastern Europe, and many other names. There are many folk theories regarding the origins of these spirits, one being that they are fallen angels, cast out of heaven in the primal battle between the hosts of angels. In Ireland, it is thought that they are the former gods of the Celts, the Tuatha Dé Danann, diminished in power and stature by the arrival of the Christian faith but still powerful in their own right and ruling over the underworld. Some fairies are no doubt the elementals and nature spirits of the animistic conception of the world we have discussed in previous chapters. Another theory is that the fey are the spirits of the dead in the land, souls in purgatory, neither good enough for heaven nor bad enough for hell, or those who have chosen to remain as guardians of the land and people they love. There are some profound implications in this idea that if an individual soul is given a choice, deciding between heaven and purgatory or even the underworld is not always so simple, and many cultures have not seen the idea of ascent to celestial afterlife as something to be desired.

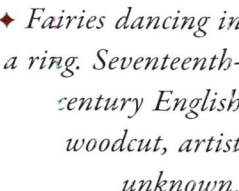

⬥ *Fairies dancing in a ring. Seventeenth-century English woodcut, artist unknown.*

It is very likely that such spirits are all of these things to some degree, as the imaginal realm does not allow for precise taxonomy. It is always shifting and must be interpreted through the highly subjective lens of each individual who comes into contact with it, as well as the way it is shaped by individuals' cultural expectations.

According to folklorist W. Y. Evans-Wentz in his classic study, *The Fairy-Faith in Celtic Countries*, people in many Celtic cultures made little distinction between the dead and the fairy folk. He quotes multiple informants in many locales who explicitly equate the souls of the departed with the fairy folk, such as this one, "The old people in County Armagh seriously believe that the fairies are the spirits of the dead: They say that if you have many friends deceased, you have many friendly fairies."[1] Another informant told Evans-Wentz, "The good people in this mountain are the people who have died and been taken." He concludes later in the work that "the animistic character of the Celtic Legend of the Dead is apparent; and the striking likenesses constantly appearing in our evidence between the ordinary apparitional fairies and the ghosts of the dead show that there is . . . no distinguishable difference between these two orders of beings, nor between the world of the dead and fairyland."[2]

In discussing the similarities between the Roman ancestral spirits we discussed in the last chapter and the Irish ancestral spirit, the banshee, he says:

The Roman Lares, so frequently compared to house-haunting fairies, are in reality quite like the Gaelic banshee; that originally they were nothing more than the unattached souls of the dead, akin to Manes; that time and custom made distinctions between them; that in the common language Lares and Manes had synonymous dwellings; and that, finally, the idea of death was little by little divorced from the worship of the Lares, so that they became guardians of the family and protectors of life. On all the tombs of their dead the Romans inscribed these names: *Manes, inferi, silentes,* the last of which, meaning the *silent ones*, is equivalent to the term "People of Peace" given to the fairy-folk of Scotland. Nor were the Roman Lares always thought of as inhabiting dwellings. Many were supposed to live in the fields, in the streets of cities, at cross-roads, quite like certain orders of fairies and demons; and in each place these ancestral spirits had their chapels and received offerings of fruit, flowers, and of foliage. If neglected they became spiteful, and were then known as Lemures.[3]

ILLNESS CAUSED BY SPIRITS

Let's move on to the subject of practice. Generally speaking, the relationship between the living and the beings of the imaginal underworld, however one conceives of them, is fraught with fear and mistrust. As mortal beings with a limited time span, we cling to our mortal attachments tenaciously; we love our mortal life, as it is all we know. And it is fragile and temporary. Misfortune and mischance threaten at every turn. Sickness and accidents alike were believed by our ancestors, and by many even now, to have their origins in spiritual causes, among these evil spirits bent on human destruction, including the angry dead. Hungarian folklorist and ethnographer Éva Pócs writes of the evil vile, the storm demons who bring bad weather and hail and are associated with the angry restless spirits of suicides and the unbaptized dead. Even the good vile punish the unwary trespasser who stumbles into their domain with sudden illness, pain, or paralysis.[4]

Typically, propitiation by means of sacrifice and the practice of apotropaic magic (practices to banish and expel evil forces) were two common means early magicians and shamans used to maintain the delicate balance between the realms of the living and the dead, the disturbance of which threatened to unleash the forces of chaos into the human realm. In the ancient Near East, as in much of the ancient world, according to Assyriologist Reginald Campbell Thompson, magicians "generally paid rites and made offerings to the dead, under the impression that they were feeding the ghosts, and that if they ceased to offer sustenance to the souls of their ancestors they might render themselves liable to affliction or

◆ The Wild Hunt of Odin (Åsgårdsreien), *by Peter Nicolai Arbo, 1872. A folk-motif common to most of Western Europe, the wild hunt was seen as a nocturnal procession of a host variously conceived as demonic or ghosts of the restless dead, riding on a storm wind, led by the Devil or a Pagan god or goddess—in this case Odin, or Óðinn. It was seen as a harbinger of death and misfortune.*

even possession by the hungry souls of the departed."[5] If propitiation by means of sacrifices and offerings failed to prevent sickness and misfortune, the only means open to the magician was to adjure and exorcise the entities responsible, banishing them from the realm of the living.

In order to protect themselves from fairy mischief, the peasantry all over Europe employed magical charms and substances, such as hanging nails or bits of iron over cradles or doors. In the Balkans, says Pócs, "Protective magical nostrums were used. . . . The most important—and universally best known—is garlic, as well as the various iron objects, certain plants, or parts of such plants (e.g., artemisia, elder, or linden tree, etc.) In Greece, especially, religious amulets were also used. One important protective object of the Rumanians was a horse's

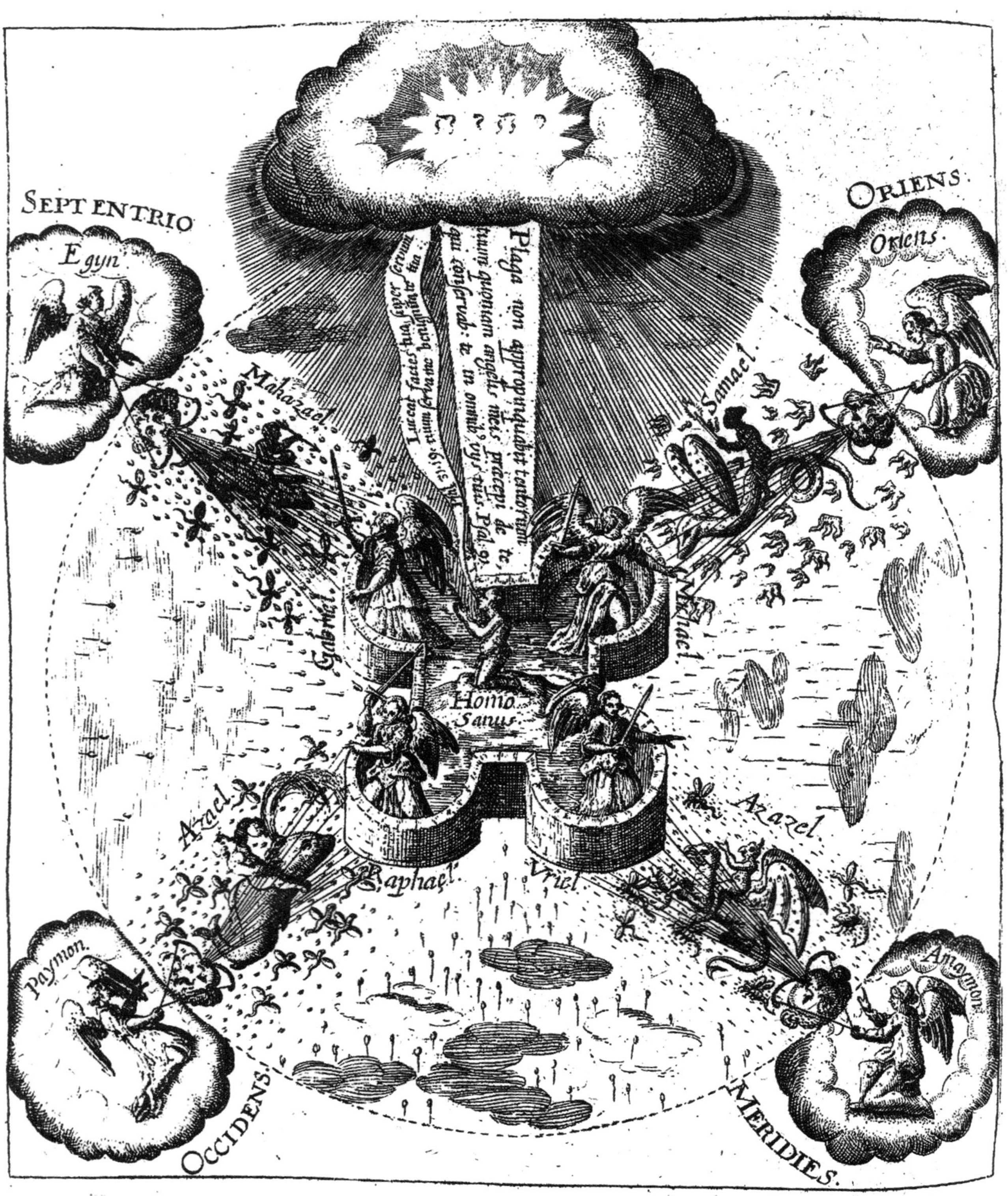

♦ *Demon kings of the air from the four cardinal quarters bringing disease and assaulting the castle of health. Illustration by Robert Fludd, 1631.*
Image by Fæ.

skull (which the fairies were especially afraid of), but a horse head was known in Southern-Slavic practice as well."[6]

As Lewis Spence put it in *British Fairy Origins*, the fairy mythos "has drawn upon and elaborated the very primitive notion that death is an unnatural thing, a calamity brought about by the malevolent magic of supernatural beings, spirits, ghosts, or wizards. Early man could not account for death; it seems to have puzzled him exceedingly. Illness, in his purblind view, was due to the interference of an ancestral spirit or to the great malignant host of the dead, who desired his presence in the underworld for the mere sake of his company."[7] I'm inclined to disagree with Spence on whether so-called "primitive" humanity was any more ignorant about the nature of death and its causes than modern people.

METHODS OF HEALING

What causes illness seems to be more of a breakdown in the vital defenses of the healthy organism in many cases than an incursion of malignant microbes. We should probably be studying the healthy more than we study the sick so that we can unlock the secrets of why some people don't succumb to illness. It seems likely to me that there are nonmaterial holistic factors at work in maintaining health, and that when they fail or break down, which can easily be due to the addition of other nonmaterial causative factors, illness results, and harmful microbes or inherent weaknesses in the organism get the upper hand. If we expand our notion of the possible, these nonmaterial causative factors acting on a psychic or spiritual level could easily be angry ancestors or malicious sorcery. Modern holistic medicine recognizes the existence of ancestral patterns of trauma passed down through family lines as well as the possibility of negative spirit attachments.[8]

In ancient Sumeria and Babylon, disease was believed to be caused by negative entities. As Lewis Bayles Paton wrote in his work *Spiritism and the Cult of the Dead in Antiquity*, "The Sumerians, or primitive Babylonians, believed that all disease was caused by the obsession of malignant spirits that entered into the bodies of men. Three classes of evil spirits were recognized, first, ghosts of those who had died unnatural deaths, or had remained unburied; second, vampires that were half-human and half-demon, third, fiends who were of the same nature as the gods."[9] An incantation inscribed on a Babylonian clay tablet translated by Reginald Campbell Thompson in his classic two-volume translation, *The Devils and Evil Spirits of Babylonia*, names the following sickness-causing entities:

◆ *The Mesopotamian wind demon Pazuzu. Bronze statue from the first millenium BCE (in the Louvre Musuem). Image by PHGCOM.*

Fever unto the man, against his head, hath drawn nigh,
Disease unto the man, against his life, hath drawn nigh,
An evil Spirit against his neck hath drawn nigh,
An evil Demon against his breast hath drawn nigh,
An evil Ghost against his belly hath drawn nigh,
An evil Devil against his hand hath drawn nigh,
An evil God against his foot hath drawn nigh,
These seven together hath seized upon him,
His body like a consuming fire they devour.[10]

The text continues, with many lacunae, to describe a ritual and an incantation involving the sacrifice of a kid goat and the burning of incense in order to expel the evil spirits and ghosts that are causing the sickness:

> Go, my son,
> Take a white kid of Tammuz,
> Lay it down facing the sick man and
> Take out its heart and
> Place it in the hand of that man;
> Perform the Incantation of Eridu,
> The kid whose heart thou hast taken out
> Is *li'i*-food [the translator notes that *li'i* is not understood] with which thou
>   shalt make an "atonement" for the man,
> Bring forth a censer and a torch,
> Scatter it in the street,
> Bind a bandage on that man,
> Perform the Incantation of Eridu,
> Invoke the great gods
> That the evil Spirit, the evil Demon, evil Ghost,
> Hag-demon, Ghoul,
> Fever, or heavy Sickness,
> Which is in the body of the man,
> May be removed and go forth from the house!
> May a kindly Spirit, a kindly Genius be present!
> O evil Spirit! O evil Demon! O evil Ghost!
> O Hag-demon, O Ghoul!
> O Sickness of the heart! O Heartache!
> O Headache! O Toothache!
> O Pestilence! O grievous Fever!
> By Heaven and Earth may ye be exorcised![11]

In medieval Western Europe, similarly, when propitiation failed to keep the capricious fairies from working their mischief, charms and exorcisms were used to treat the patients suffering from a fairy stroke or elf-shot. Elves were more or less the Germanic equivalent to the Celtic áes sídhe and were known in Old English as ælfe and in Old Icelandic, the language of the Norse Sagas, as ælfr. A spell from one of the medieval manuscripts known variously as *leechbooks*,

◆ *A page from* Bald's Leechbook, *a c. tenth-century Old English medical manuscript, showing a recipe for eye salve.*

which was composed in the late tenth or early eleventh century and is known to modern scholars by its first two words, *Wið færstice*, contains a number of ritual procedures to combat illnesses caused by the malicious activities of elves, which included a number of common ailments that, before the advent of modern diagnostic techniques, were regarded as inexplicable. According to medical historian Charles Joseph Singer, "A large amount of disease was attributed . . . to the action of supernatural beings, elves, *Æsir*, smiths, or witches whose shafts fired

◆ *Illustration from the Eadwine Psalter (originating in Canterbury), showing the psalmist besieged by disease demons, often cited as pictorial evidence of a folk belief in elf-arrows. Closer examination of the picture reveals that Christ is shooting the arrows, and that the psalmist is afflicted by God himself, a common theme in the psalms. From the twelfth century.*

at the sufferer produced his torments. Anglo-Saxon and even Middle English literature is replete with the notion of disease caused by the arrows of mischievous supernatural beings. This theory of disease we shall, for brevity, speak of as the *doctrine of the elf-shot*. The Anglo-Saxon tribes located these malicious elves everywhere, but especially in the wild uncultivated wastes where they loved to shoot . . . at the passer-by."[12]

Techniques used by the healers included the use of herbal remedies, but primarily these were used within a context of what we would now call ritual magic, which fused Christian elements with Pagan elements from traditional Anglo-Saxon practice. A typical example from *Bald's Leechbook*, as described in *Anglo-Saxon Magic* by Godfrid Storms, to treat "elf-sickness" or "elf-disease":

> Against elf-sickness. Take bishop's wort, fennel lupine, the lower part of enchanter's nightshade, and lichen from a hallowed crucifix and incense, take a handful of each. Tie all the herbs in a cloth and dip them three times into hallowed baptismal water.
>
> Have three Masses sung over them, one *Omnibus Sanctis*, the second *Contra Tribulationem*, the third *Pro Infirmis*.
>
> Then put live coals in a chafing dish and lay the herbs on them. Smoke the man with those herbs before 9 a.m. and at night.
>
> And sing the Litany and the Creed and the Lord's Prayer.
>
> And inscribe the sign of the cross on each limb.
>
> And take a small handful of herbs of the same sort, hallowed in the same way, and boil them in milk; drop a little hallowed water on them three times and let him take some of it before his food.
>
> He will soon be better.[13]

Interestingly, the previous charm contains instructions for expelling malicious spiritual entities by driving them away using the smoke of burning herbs, much like the practice of "smudging" in Native American religious contexts. The practice of fumigation with the smoke of burning herbs for exorcistic healing practices was known in Old English leechbooks as *rēcels*, where it was a major treatment for elf-disease, which was seen as various physical conditions where the health and wholeness of the body was disturbed by entities of a spiritual nature—namely, elves. In the words of modern Heathen witchcraft practitioner Cat Heath, "To clear a person with *rēcels* is to introduce something so holy that the unholy flees."[14] The practice of *rēcels*, and indeed the underlying philosophical framework of Heathen medicine in general, is tied to Germanic soul lore and the idea of *hæl* (in Old English) or *heill* (in Old Norse), which are cognate to our modern English words *health*, *holy*, *whole*, and *hale*, and have a meaning connected with all of these but also additionally signifying luck and spiritual health. Sickness was understood by the culture that produced the leechbooks and their charms as an outside force that attacked the hæl of a person, reducing it, or introducing an *unhæl* (unhealthy, unholy, unlucky) element into their person. The outside forces could be spiritual entities like elves, or hags, or human-induced magic such as curses. The *leech*, the Anglo-Saxon healer, was thus engaged in a spiritual battle against the unhæl forces causing the sickness.[15] In her excellent book *Elves, Witches & Gods*, Cat Heath gives a full ritual recipe for using rēcels, as well as a great deal of other useful information pertaining to the topic of this chapter.

Another charm from *Bald's Leechbook*, a recipe from plantain and red nettle salve for rheumatism or any sudden pain, runs thus:

> Boil feverfew and the red nettle that grows through [the wall of] a house and plantain in butter.
>
> Loud they were, lo loud, when they rode over the mound,
> they were fierce when they rode over the land.
> Shield yourself now that you may survive their ill-will.
> Out little spear, if you are in here!
> I stood under the linden-wood, under a light shield,
> where the mighty women betrayed their power,
> and screaming they sent for their spears.
> I will send them back another,
> a flying arrow from in front against them.
> Out little spear, if you are in here!
> A smith was sitting, forging a little knife,
> Out little spear, if you are in here!
> Six smiths were sitting, making war-spears.
> Out spear, not in, spear!
> If there is a particle of iron in here,
> The work of hags, it shall melt!
> Whether you have been shot in the skin, or shot in the flesh,
> Or shot in the blood,
> Or shot in a limb, may your life never be endangered.
> If it be the shot of Aesir, or the shot of elves,

> or the shot of hags, I will help you now.
>
> This as your remedy for the shot of Aesir, this for the shot of elves,
>
> this for the shot of hags, I will help you.
>
> Fly to the mountain head.
>
> Be whole. May the Lord help you.
>
> Then take the knife and dip it into the liquid.[16]

The similarities between the example from Mesopotamia above and the second charm are striking, including the coincidence of the orders of disease-causing entities. In this case, elves must correspond to the Babylonia; demons, or ghosts, and the Aesir (or Æsir) to the gods; and the hags, well, to the hags. Folklorist Ronald Hutton, in his recent book *Queens of the Wild*, goes about his usual work of deconstructing the nineteenth-century academic myths of "Pagan survival" in contemporary Britain. He says of this charm: "'With Faerstice' [Wið færstice], the charm 'For a Sudden Stitch,' [at] one point calls upon 'the Lord' (in Anglo-Saxon sources, conventionally meaning the Christian God) for help. It ascribes the pains which it is designed to cure, however, to a range of non-Christian entities: 'mighty women' who ride across the land and hurl magical spears; the 'Aesir' (pagan deities); elves; and a hag. These verses could be taken to prove the existence of a dual system of religion, the old continuing alongside the new."[17] Hutton mentions the condemnations by churchmen, such as Aelfric of Eynsham, who, around the year 1000 CE, "condemned healing magic as 'heathen worship.' He furthermore accused those who practised it (and whom he termed *wiccan*, i.e. witches) of teaching people to make offerings to stones and trees to achieve their needs, in 'heathen' fashion. Some ten centuries later, the kind of folk magicians whom Aelfric denounced were still around, under the name of wise or cunning folk."[18]

The transformation of the Anglo-Saxon elves into the medieval fey was a process that took several centuries and reflects influences from classical Rome being reintroduced into British culture from France. According to historian Francis Young, "The terms 'elf,' 'elves' and 'elvendom' continued to be used in medieval English until the borrowing of the French words *fée* and *féerie* into English in the fourteenth century (as 'fay' and 'faerie')—and indeed 'elf' continued to be used . . . even after the widespread adoption of the French term. By the early modern period 'faerie' (properly the realm of the fays, or the state of fay enchantment) had become conflated with 'fay' to produce the word 'fairy.' The word *fée* itself is the French rendering of the Latin *fata*, signifying 'fate' or a goddess who controlled fate."[19]

❦ *Alberich the Dwarf defeated by Dietrich, an illustration of the*
Nibelungenlied *(a German epic poem). Illustration by Johannes Gehrts,*
late nineteenth century.

ENLISTING SPIRITS FOR AID

To go from expelling the spiritual creatures responsible for causing illness and misfortune by ritual exorcism and banishing or gaining their favor with propitiations and sacrifices to binding them with conjurations in order to enlist their support for various magical purposes is not such a large step. Like the uses made by Greek magicians of the ghosts of the restless dead, the magicians, cunning folk, and witches of the Renaissance and early modern period consorted with the fairies for sorcerous aims, either as familiar spirits, or as spirits summoned by the methods of conjury inherited from the magical practice of the Middle Ages, the exorcistic binding by oaths and magic circles. Many Renaissance era and early modern period grimoires list procedures for invoking various fairy beings. One of the more well-known manuscripts, *V.b.26* from the Folger Shakespeare Library, and recently published by Daniel Harms, James Clark, and Joseph Peterson as the *Book of Oberon*, gives ritual invocations for several fairy beings, notably Oberon.

Oberon is the king of the fairies famous for his appearance in *A Midsummer Night's Dream*. As a denizen of the imaginal realm and a literary character, he has had a long history in European literature under several different transformations, beginning his career in Germanic mythology as Alberich the Dwarf.[20] The Folger manuscript also includes a ritual to summon Mycob, the queen of the fairies, as well. Oberon's primary offices are to assist the magician in the finding of treasure and confer invisibility, as well as instruct the magician in natural magic.

There are a range of approaches among practitioners between those who seek to form a relationship with their familiar spirits and those who would command them in the name of an Almighty God. As witch and author Lee Morgan says in his excellent recent book on fairy magic, *Sounds of Infinity*:

> The ceremonial magician could deal with demons but walk away unscathed and unsullied due to the use of magic circles and prayers to Jehovah and his angels to dominate and subdue them on the magician's behalf. The powers of the old world had been drawn into a framework that by its inherent nature relied on faith in God, circle barriers and enforced compliance. As a philosophy it could

✦ The Quarrel of Oberon and Titania, *by Joseph Noel Paton, 1849.*

be accepted within church-dominated culture because it drew the wild powers of the heathen world into a demon/angel binary and then subjugated all via the power of God's Word. As a form of meta sorcery it made perfect sense.[21]

The circle and the prayers serve to protect the operators from the inherent dangers of calling on the amoral powers of fairies, giving them a layer of protection. The witch, according to Morgan, needs no such protection; her transgressive, liminal nature gives her relationship with the fairies the power of intimacy.

The sixteenth-century English Protestant Reginald Scot, author of the anti-witchcraft (and anti-Catholic) work *The Discoverie of Witchcraft*, which describes period practices of ritual magic in great detail and became a major sourcebook for later generations of cunning folk, gives a ritual for summoning the fairy Sibylia. This involves first enlisting the help of the shade of a recent suicide or victim of hanging, who is summoned in order to act as an intermediary to fetch the fairy spirit herself. Sibylia herself is called "to appear in that circle before me visible, in the forme and shape of a beautifull woman in a bright and vesture white, adorned and garnished most faire, and to appear to me quicklie without deceit and tarrieng and that thou faile not to fulfill my will and desire effectuallie. For I will choose thee to be my blessed vigine and will have common copulation with thee. Therefore make haste and speed to come unto me, and to appear as I have said before: to whom be honour and glorie, forever and ever. Amen."[22] The spirit is bound with oaths by the power of God and contained within a magic circle, a sex slave for the lonely magician. Sexual relationships between humans and preterhuman entities have always been part of mythology and folklore, and indeed fairy folklore is rich in these kinds of encounters. That is nothing new.

It is the exploitative nature of the relationship that one objects to. It seems impious and disrespectful. It brings to mind the famous quote attributed to Francis Bacon, younger contemporary of Reginald Scot, "I am come in very truth," he declared, "leading to you nature with all her children to bind her to your service and make her your slave."[23] The fairy as nature spirit dominated and subjugated by the conjuror is very akin to how the Baconian scientist would "conquer and subdue her [nature] to shake her to her foundations"[24] in order to compel her for his own purposes. This fairy magic of the learned magician described by Scot is a far cry from the way of the peasant cunning folk, who entered into their relationship with the fey on better terms, with respect and caution for an ancient and formidable power.

To speak plainly, this case illustrates to me the very modern attitude toward nature that was developing at the time of *The Discoverie of Witchcraft*. If fairies symbolize the animist notion that nature is alive with ensouled invisible beings, then surely the act of a male magician exploiting them for sex, however fantastic it may seem, is rich in symbolic implications for what would happen over the next few centuries of the human relationship with animate nature. The fairies' role, either as nature spirits or as deep ancestors, is as mediators between humanity and the natural world, and cultivating reverence and respect toward them is necessary for the cunning farmer.

Agrippa gives us a bit of a gentler method for interacting with nature spirits, reserving the harsher techniques known to the Renaissance ritualist for those nature spirits that become unruly. He describes the method as follows (in James Freake's seventeenth-century English translation, with eccentric spelling preserved):

There is another kinde of Spirits, which we have spoken of in our Third Book of Occult Philosophy, not so hurtful, and neerest unto men; so also, that they are effected with humane passions, and do joy in the conversation of men, and freely do inhabit with them: and others do dwell in the Woods and Desarts [deserts]: & others delight in the company of divers domestique Animals and wilde Beasts; and othersome do inhabit about Fountains and Meadows. Whosoever therefore would call up these kinde of Spirits, in the place where they abide, it ought to be done with odoriferous perfumes, and with sweet sounds and instruments of Musick, specially composed for the business, with using of Songs, Inchantments and pleasant Verses, with praises and promises.

But those which are obstinate to yield to these things, are to be compelled with Threatnings, Comminations, Cursings, Delusions, Contumelies, and especially by threatning them to expel them from those places where they are conversant.

Further, if need be, thou maist betake thee to use Exorcisms; but the chiefest thing that ought to be observed, is, constancy of minde, and boldness, free, and alienated from fear.

Lastly, when you would invocate these kinde of Spirits, you ought to prepare a Table in the place of invocation, covered with clean linen; whereupon you shall set new bread, and running water or milk in new earthen vessals, and new knives. And you shall make a fire, whereupon a perfume shall be made. But let the Invocant go unto the head of the Table, and round about it let there be seats placed for the Spirits, as you please; and the Spirits being called, you shall invite them to drink and eat. But if perchance you shall fear an evil Spirit, then draw a Circle about it, and let that part of the Table at which the Invocant sits, be within the Circle, and the rest of the Table without the Circle.[25]

In her fascinating work, *Cunning Folk and Familiar Spirits*, scholar Emma Wilby relates the folk beliefs in the power of the fairies, who "possessed a semi-material or 'astral' form, putting them somewhere on the spectrum between human flesh and pure spirit."[26] This semimaterial form made the fairies capable of great powers of magic:

> Fairies could, for example, live long beyond the allotted human span; they could become visible or invisible at will; they could shapeshift between animal and human form and they could fly, sometimes travelling vast distances at great speed. They could also divine the future, heal the sick and possess knowledge of objects and events in far off places.[27]

Importantly, for our purposes, much of the magical work carried on by cunning folk with the fairies was the ordinary sort, intended to propitiate these capricious supernatural beings for the sake of good fortune in agricultural and domestic activities. Although less organized than the cult of the lares and lemures in ancient Rome, regular offerings to the fairies served many of the same purposes. So much of the life of a peasant farmer is entirely up to the whims of fortune that everyday magic of this sort is of great consolation. As Wilby says:

> Although the majority of ordinary people did not encounter the fairies directly, many still considered their relationship with these spirits to be of the greatest importance. Fairies were believed to be able to use their supernatural powers to influence almost any aspect of the natural world, including the lives of humans, and as a consequence people were very anxious to curry fairy favour. At night, housewives across the length and breadth of Britain would leave out bowls of water or milk and plates of bread or cake on the kitchen floor for the fairies. By day, out in the fields and the animal sheds, their husbands would tie cords and bury bones and mutter charms in an effort to please these capricious spirits of the land. In return for this solicitude, the fairies might help the housewife to a fine batch of butter or a strong brewing of ale, or may even leave fairy silver in her boots; while they might treat the farmer to a fine August, a good lambing season, or help him turn up a crock of fairy gold with the plough. While human attempts to please the fairies were motivated by the desire for good fortune, however, they were just as strongly motivated by fear. These same housewives strove to keep their kitchens clean and be generous to their neighbours in the belief that the fairies abhorred slovenliness and meanness, and their husbands avoided ploughing in a

spot of land or felling a tree if it was believed to be sacred to the fairies, for fear of upsetting them. Any of the mishaps which occurred around the homestead could be attributed to fairy displeasure.[28]

Anyone who works a farm for long quickly learns how much fortune plays a hand in agricultural endeavors. Sometimes a crop suddenly seems to gather momentum and vigor, roots taking hold in soft loam and leaves reaching up to the heavens, spreading out in vigorous green as if the spirits who govern the growth of green things are smiling on them. Conversely, just as easily, how a crop can mysteriously wither, victim of mysterious wilts and blights, despite the interventions of modern chemistry. If the vital energies of a place are out of balance, the crops will not thrive spiritually, machinery will break mysteriously, and the animals and people will get sick. A farm, to begin with, is an artificial order imposed on the landscape, as the ancient people knew. The natural tendency is for the imposed structure to break down unless effort, spiritual as well as physical, is put into maintaining the balance. The fairies, in this way of looking at things, are the natural spiritual forces that maintain both the order and the disorder. They punish laziness and slovenliness, and they reward attentiveness and offerings with their favor and good fortune. I am an unabashed adherent of the fairy faith. I know that this goes beyond the ordinary flow states achieved by a well-ordered operation (which my farm definitely isn't). It is by the favor and cooperation of the fey, or conversely their disfavor, that the fortunes of the farmer rise and fall.

Can You See the Fairies?

I am not one who is extraordinarily gifted with clairvoyance. I do not see the fairies as fully formed humanoid beings, but in part of the reverie described at the beginning of the chapter there was a moment when, with eyes closed, I could "see" the serpent-like veins of earth energy flowing under the land, being carried up the roots of the trees, into their trunks and out through their branches as a flow of scintillating light. I was seeing the internal flow of vital qi through the plants and in the land. Many clairvoyants have described the work of the fairies, known to them as devas, in a similar fashion. When I see the leaves of plants animated, trembling, on windless days, I like to think of the play of the sylphs, or air elementals. At night, I see sudden flashes of blue or red light, which could be a trick of my aging eyes, but maybe they are something else, and as they are entirely subjective phenomena and no one has ever been around when I have seen them, no one will ever know. "Faith," as St. Paul said, "is the substance of things hoped for, the evidence of things not seen,"[29] and this applies as much to the fairy faith as to any other. There is evidence that can't be seen or measured with the physical senses. Let's remember that many of the intangibles that make life meaningful—love, poetry, music—can't be measured but have effects on the organism that can't be accounted for except in the meaning conveyed by sound, marks on a page, or a lover's glance. We know we are not alone in a dark wood on a moonlit night, when we speak the words of invocation into the wind and call the spirits to witness, and know we have been heard.

The cunning folk described in Wilby's book enlisted the direct cooperation of fairies as personal familiar spirits. The use by magicians of familiar spirits has an ancient provenance in magical practice, deriving from the most ancient levels of shamanic practice common to most cultures on the Earth. Familiar spirits are imaginal beings who appear to the shaman in dreams or visions or are experienced as a helping or teaching voice that teaches the magician the techniques and mysteries of magic and the invisible world. They are experienced in some cultures as angelic or divine beings and are clearly related to the phenomenon of the guardian angel in Western magic, or the daimon or *paredros* in Hellenistic magic. Some shamans have animal familiar spirits, others have dead relatives for helpers, and according to Wilby, cunning folk have fairy familiars, whereas witches have demonic spirits. In my opinion, this division into witches with demonic familiars who practice malefic magic and cunning folk with fairy familiars who practice benefic magic sounds too simplistic to be accurate. In reality, I imagine, there have always been "two-handed" practitioners who were willing to work both kinds of magic when necessary. Indeed, as Lee Morgan says, responding to this very statement by Wilby, "The entities named demons are often darker types of faery creatures."[30]

The practice of magic and the focus on the spiritual realm and occult forces, properly done—with reverence and humility—opens one up to messages from "higher" or "inner" realms. The discernment of these messages is of paramount importance, as both delusions and intuitions arise from the unconscious, and discerning between the two isn't always easy and requires a certain amount of wisdom. I am comfortable stating that my experience of phenomena of this type, though limited, is a bit like the daimon of Socrates, an internal voice that warns me and advises me, that wakes me up from a night of wild dreams with a new insight on a problem I have been thinking about, or even offers a spiritual insight. I believe that true vocation comes from deep within, from your personal daimon, your higher self, and the sooner discerned the better.

If you are a witch or a practitioner of magic, it's possible your daimon is your familiar spirit, and an intimate friend and advisor, maybe an angel or a fairy. If you're a devout Christian, your daimon can appear as Christ or an angel, and actually mine has dressed up as Jesus on occasion, and convincingly, I might add. I call him "Teacher," and he speaks with love and compassion and uses New Testament language. Emma Wilby writes of cunning folk with Christ-like or angelic familiar spirits, so there is a precedent for this in the tradition. But the Teacher can take other forms as well. Carl Jung famously had a spirit teacher named Philemon. My own take on this is that the spirit world is amorphous and shifting, and the spirits can appear however you expect them to appear, according to your individual and cultural expectations. This is why Tibetan Buddhists don't have visions of the Virgin Mary in a blue mantle holding the baby Jesus. As John Lilly said, "In the province of the mind the limits of belief are the limits of reality." And when are we not in the province of the mind?

This malleability of the spirit world also accounts for the shifting nature of fairy mythology over the long centuries of time; they adapt to the human culture that they interact with. Beginning as chthonic deities and divinized ancestors or fallen angels and morphing into the capricious elves of the Anglo-Saxons and Scandinavians, who could be both the dark and evil *duergar* as well as the bright and beautiful ælfar, shifting and evolving still, they became the fairies of the Renaissance and early modern periods, the familiar spirits of the cunning folk and the spirit girlfriends of lonely magicians. Transforming further, they became the winged children of the Victorians and Disney movies as well as the nature spirits of Rudolf Steiner and the Theosophists. And now, they are becoming the personifications of our desire for reconnection and re-enchantment. And they are more real than ever. But as with so much of the spiritual realm, we can't see them as they truly are, but only through the veils of our concepts and categories that obscure so much of the mystery.

CONNECTING WITH THE UNIVERSAL FAIRIES

We have so far focused on the diversity, the differences, and the shifting nature throughout time of these creatures whom we call the fairies, but there is a striking unity all over the world in the tales of diminutive and mischievous spiritual folk who bring either good fortune or ill, depending on whether they have been properly treated or not. There are tales of them from every continent and culture, especially where people live in close contact with the natural world. What changes from culture to culture and across the broad expanse of time are the particulars—the dress, the language, the appearance, the names. The numerous examples of fairy phenomena worldwide are the subject of Joshua Cutchin's recent work *Ecology of Souls*, in which he links the "fae phenomena" not only to the dead but to shamanism and UFO phenomena.[31] The proliferation of eyewitness reports as well as nearly universal folklore worldwide suggests an underlying reality that people all throughout time and in all places have borne witness to. What this underlying reality is must remain a mystery, a mystery in the heart of nature and our relationship with it. It is subtle, but when one starts to probe its depths, it becomes undeniable.

Doing so requires a shift in belief, as I have said elsewhere. One must divest oneself of the wrong kind of skepticism, the kind that tells you fairies and the like are nonsense, impossible. Open to the possibility. Court them. Go out on the land, far out, at night, when no one is around, and bring offerings like milk, butter, liquor, tobacco, and sweets. Call them quietly. Most likely, you will hear something, but you will tell yourself it is only an animal—and maybe it is. But maybe it isn't. Sit quietly by a place where there is an opening in the Earth, a cave, a rock, or a spring. Imagine yourself entering the spring or sinkhole in astral form, descending deep into the earth, down a moss-covered tunnel where bioluminescent fungi provide the only light. Deeper still, until the tunnel opens into a large cavern, so large that the roof is as distant as the sky, and the space is illuminated by an eerie lavender light. There is a table spread and dancing people, beautiful and strange, both somewhat monstrous and strangely attractive, uncannily beautiful, possessed of an erotic magnetism that feels almost threatening. You are ravenously hungry, and the food smells tantalizing, but you dare not eat. You are thirsty, and the ruby wines and amber liquors are enticing, but you dare not drink.

One of the dancers notices you and takes you by the hand and brings you to a pair of gilded thrones at the far end of the chamber. Who sits on the throne? What do they say? What do you ask them?

There is ample room for experiments of this type in guided visionary work. Your imagination is the limit.

PART FOUR

PLANT MAGIC AND SACRED AGRICULTURE

15
ENTHEOGENIC PLANTS AND THE PLANT PATH

In the next few chapters, we will examine the role of plants in magic with a view toward seeing them as spiritual entities with their own particular place in the cosmos, with a unique role in our practice as magicians, healers, and gardeners. Cunning farmers are plant people. A green thumb is as necessary to this path as a knowledge of the arcane secrets found in ancient books of sorcery and occult science. The plant world is first and foremost an intermediary between the human realm and the rest of nature. We use plants for food, clothing, medicine, and for the beauty of their presence. But they can also open our eyes to the world of the spirit, revealing the living, breathing, energetic heart of the cosmos and helping us to find our place in it. So to begin our discussion on plant magic, and to give it some personal context, I would like to discuss the role that entheogenic plants have played in my life and journey, as well as some background on their history, mythology, and lore.

The plant world is for many of us an entry point into the mysteries. At least it was for me. I didn't know I had a green thumb until my early twenties. As a young child, I was fascinated with insects; I particularly loved butterflies, beetles, and grasshoppers. I would keep praying mantises and feed them grasshoppers, watching them eat them like bananas, grasping them in their claws and eating them head first. This did not make me popular in elementary school, as I would spend recess catching bugs in a butterfly net instead of playing football like "normal" kids. My nickname was "nature boy," which was not intended to be flattering. I loved then, as I do now, wandering in wild places, which near my childhood home in suburban Philadelphia were to be found in the marginal

✦ *The Tree of Life from the German alchemical manuscript* Gemma Sapientiae et Prudentiae *(c. 1735). Image by Wellcome Images, London.*

places and vacant lots, near abandoned railroad tracks and ruined factories, or on the sprawling grounds of a colonial-era mansion. As a teenager, I never lost my love of nature, though I was more sensitive to the fact that carrying a butterfly net did nothing for my social life. On summer evenings, I would ride my bicycle out to a historic and picturesque church dating from the eighteenth century and spend the hours of the gloaming in the cemetery reading poetry among the fireflies. The English Romantics like Blake, Shelley, and Keats, whose radical views and love of nature really resonated with me, were my favorite companions on these trips. In short, I've always been a bit odd.

EARLY EXPERIENCES WITH ENTHEOGENS

But my first brush with the magic of the vegetable kingdom came from the realization that some of these little creatures could alter consciousness. I had tried marijuana a few times as a young teenager, but I always got paranoid and it made me feel so guilty, because I was basically a good boy who had internalized the drug war propaganda of the 1980s. By the time I was nineteen, however, I was reading Robert Anton Wilson, and I wanted to consciously experiment with consciousness alteration via cannabis and LSD. Around that time, in 1993, a friend of mine who was into magic wanted to visit Montauk on Long Island, because Aleister Crowley had apparently had a vision there. So we drove up there. I had obtained some weed and smoked a joint all by myself, and being naive to marijuana, I was strongly affected. It was perhaps the fourth time I had ever partaken and the only time I made an informed decision to do so, and it had been several years since the last time. We were on the shore with low dirt cliffs on one side and a beach made of the most beautiful pebbles on the other. With the doors of my perception cleansed, each pebble glowed like a jewel, the ocean glittered like burnished gold, the Sun and sky were glorious, and there were the most gorgeous goddess-like women walking around naked (it was a clothing-optional beach, apparently.) And that's all I remember. The experience made a profound impression on me of the power of a plant to transform consciousness.

I went on to become a serious student of all things psychedelic. I read everything Timothy Leary and Terence McKenna had ever written, while devouring Mircea Eliade, Joseph Campbell, Jung, Crowley, Alan Watts, Aldous Huxley—anything pertaining to the mysteries I could get my hands on. I wanted to start my own revival of the mystery cults of the ancient world. I took LSD occasion-

✦ Haoma (Peganum harmala), *also known as Syrian rue, growing wild in Iran. Image by Ninara.*

ally, but I preferred psilocybin mushrooms, as the experience seemed to involve more of a contact with a benevolent presence, and there was almost always a deeper spiritual meaning to the trip. I read everything I could about ethnobotany and shamanism, reading the works of pioneering ethnobotanist Richard Evans Schultes, plant chemist Jonathan Ott, and the psychedelic poet and master of plant lore, Dale Pendell, whose trilogy of works, beginning with *Pharmako/ Poeia*, practically became my Bible. A battered copy of the first edition, which I still have and treasure, rode around in my hippie bag for years.

I learned about ayahuasca from those books in 1993. Ayahuasca, though it comes in many regional variations, is a mixture composed of the woody vine, *Banisteriopsis caapi*, a source of the alkaloids harmine and harmaline—which are inhibitors of an enzyme in the human body known as monoamine oxidase, or MAO*—and a wild member of the coffee family, *Psychotria viridis*, which contains the potent entheogenic compound DMT, which is not active orally, being deactivated by MAO in the body. The resulting brew is strongly entheogenic and orally active, owing to the MAO inhibition of the vine component of the brew combined with the DMT. To scientists, it is a mystery how these

---

*You've likely heard of the class of antidepressant drugs that inhibits monoamine oxidase, known as MAOIs.

"primitive" people discovered MAO inhibition, perhaps thousands of years before Western science (but maybe not, as we will see). In the early nineties you couldn't find an ayahuasca retreat in the United States like you can now, and travel to South America was not in my budget, so I set out to do some research.

In his 1994 book, *Ayahuasca Analogues, Pangaean Entheogens*, Ott proposed that it would be possible to produce an herbal compound with similar chemistry using different plants that were legal and could be found in other parts of the world, as the alkaloids harmine and harmaline and DMT are widely distributed in the plant world. One potential source of the MAO-inhibiting portion of the experimental brew was the Eurasian plant known as Syrian rue, *Peganum harmala*, which was known in the ancient world as a legendary magical plant and was even suggested by a leading ethnobotanist to have been the source of the Vedic entheogenic potion *soma*, the Zoroastrian version of which was called *haoma*. I managed to obtain some from a catalog specializing in "poisonous nonconsumables," but I had done enough research to know it was relatively safe, provided you didn't mix it with certain drugs. The harmaline-containing Syrian rue, being a potent inhibitor of an important brain enzyme, could be dangerous. I was still lacking the DMT portion of my potion, but I ordered some seeds of a grass, *Phalaris aquatica*, reported to contain high levels of the substance. It was going to take a while to get a tiny packet of grass seeds to yield any quantity of plant material. In the meantime, I had heard that the rue could potentiate the psilocybin contained in magic mushrooms, *Psilocybe cubensis*, which is very close, chemically speaking, to DMT. So I put the theory to the test. It turns out that it was correct.

I made an awful-tasting brew of carefully measured doses of the rue and the shrooms, and on an evening which, according to my journal, synchronistically happened to be January 6, the Feast of the Epiphany. According to Brian Muraresku's excellent work, *The Immortality Key*, the date of the Epiphany was taken over from a major Dionysian festival in antiquity where entheogenic substances were almost certainly consumed.[1] My girlfriend at the time and I had no idea of the significance of the date, which was chosen by chance. We took the mixture together in our room and had our own epiphany. The events of the night are pretty blurry, but I remember that we held each other and cried tears of sadness, empathy, and joy for all the beauty and tragedy of the whole cosmos. Every word was laden with multiple layers of meaning and feeling. At one point, she mentioned, tearfully, a dead chipmunk that she had seen, which for her epitomized all of the needless suffering in the world, while I echoed the sentiments of Julian of Norwich—of whom I was not at that point aware—and told her that everything was going to be OK. "All manner of things shall be well," said Julian, and we knew that they would be, that they already are, even now. In the midst of the violence and hatred in the world, there is infinite love and peace at the center of all things.

I heard the word *synousia* whispered in my head as if to describe what was happening. The word literally means a merging of essences and often refers to sex, but since we were engaged in a more spiritual union at that particular time, in this case it referred to the state of communion with whatever benevolent god, daimon, or angel was visiting us that night. A book I was reading at the time, *The Mystery-Religions*, by Samuel Angus, has this to say: "Mystics of all ages have seen therein the most adequate symbol of the ineffably intimate union of the soul with God. Such *synousia* had a double underlying idea: first, an erotic-anthropomorphic, in which the *synousia* has the character of an offering or sacrifice (of purity); secondly, the magical, whereby the worshippers participated in the god's mana and secured life and salvation."[2] Words cannot describe the ineffable experience of that night, but one thing is sure: My life was changed forever.

◆ *Typical specimen of the hallucinogenic mushroom,* Psilocybe cubensis, *growing wild in Mexico. Image by Alan Rockefeller.*

It remains one of the most powerful spiritual experiences of my life. We set out to recreate the ancient mysteries, and in that moment, we came very close.

The experience convinced me of the power of plants and their magic and made me, if possible, a more dedicated student of what I saw to be the plant path. I prowled the asphalt jungles of West Philadelphia, where I lived, in search of ethnobotanical specimens, stealing ornamental tobacco and poppy pods from the gardens of my neighbors, not to ingest, but just to spend time with them. Inspired by the chapter in pioneering ethnobotanist Richard Evans Schultes and LSD chemist Albert Hofmann's book *Plants of the Gods*,[3] a work detailing the halluci-nogenic plants of the world, I dug up jimsonweed, *Datura stramonium*, and took it home and planted it in my backyard, just to bask in its solanaceous ambience. That was the first plant I ever tended. As a farmer whose farm is plagued by the little spiny-podded, hallucinogenic buggers, it seems insane to ever cultivate one. To this day, the various *Datura* species—especially *D. meteloides*, made famous by painter Georgia O'Keefe—and their tropical cousins from South America, the *Brugmansias*, are still some of my favorite plants, with their intoxicatingly fragrant night-blooming flowers and their generally sorcerous vibe. I have always made sure to have several in my garden, babying the *Brugmansias* so they bloom profusely.

Living the Plant Life

From that point on, my trajectory was leading me away from civilization and modernity. My housemates and I, a polyamorous bunch of anarchist misfits and artists, lived a bohemian lifestyle in a large rented Victorian duplex house in West Philadelphia. In August of 1995, we piled into my housemate's very underpow-ered VW Rabbit pickup "truck" to traverse the Appalachian Mountains, very slowly due to our trusty chariot's lack of horsepower. We were en route to a little spot, not really a town, on the Cumberland River in southern Kentucky, very close to the Tennessee line. Our hosts were, like ourselves, hippie artists and mis-fits, though a generation older. They lived in an ancient "holler," a creek bottom valley, which emptied out into the Cumberland River. The place is an absolute paradise, and it spoiled me for city life forever. They lived a life of freedom in iso-lation in handmade hippie houses, and they grew gardens and did odd jobs to get by. We spent a week there, all of us. At one point, we ate the sacred mushrooms on the bank of the river, on a rocky little spit of land that jutted out into the water, skinny-dipping, free and happy, playing like wild children in the mud and the warm water, under the August Sun, and feeling the layers of civilization peeling off of us like shedded skin. My journal entry about the experience sums up my enthusiasm: "I saw the forests of Eden in those Kentucky woods, the flowers and trees and tiger swallowtail butterflies all aglow in preternatural radiance, and I was very happy. I realized that this experience was holy, that for a little while, I had banished history and profane time and returned to a timeless innocent existence. It felt so natural and seemed more real than my everyday existence in the city." I really was convinced for a time, and a bit obsessed with the idea, that we could actually undo the Fall and get back to some sort of Edenic existence, on the land, in communion with it. If you have read my writings up to this point, you can probably tell that I'm still living that myth to some extent, almost thirty years later.

I moved to Kentucky for good within a year of the experience of the golden age on the riverbank, in June of 1996. I was quickly confronted with the realization that undoing the Fall is impossible. My youthful conceit of returning to a life in the primordial garden shattered, as I quickly learned that hard work is necessary to get by in the country, and that (according to biblical mythology) raising plants for a living is exactly the fallen life, the curse of Adam. My partner and I cleared land, felled and burned trees to put in a garden, built a cabin, apprenticed with local vegetable growers to learn the ropes, and learned to plant our own crops. All that work didn't bring us any income, and so we hired ourselves out as farm hands to tobacco farmers to get by. The demanding life of a farmer left precious little time for experiments with entheogenic plants, and that course of study was postponed. But I learned how to work with plants, make them thrive, and how to make them happy. I studied horticulture and herbalism, learned how to grow many different kinds of plants for food and medicine, how to improve the soil, and, eventually, how to make a living from growing plants as a farmer.

My youthful self was enamored by the idea that there was a *prisca sapientia*, a primal wisdom that ancient people had, which has been largely lost but that has come down to us in fragmentary fashion for those who know where to look. One aspect of that myth that I found (and still do find) completely compelling was that there were entheogenic potions used by ancient people—like the Vedic Hindu soma and the Zoroastrian haoma, and the mysterious *kykeon* drank by initiates at Eleusis—and that some of the medieval European alchemists and witches preserved this occult plant lore in secret, and that some of their elixirs, potions, and ointments could conceivably be a European survival of these more ancient entheogens. This is an idea that has gained some traction lately, with several notable books on the subject having been published, including *The Immortality Key* (mentioned above), *The Witches' Ointment* by Thomas Hatsis, and *Angels in Vermilion* by P. D. Newman.

ENTHEOGENS IN TRADITIONAL RITUALS

Before mycologist R. Gordon Wasson, the modern "discoverer" of magic mushrooms, brought the sacred fungi of Mexico to the American and Western imagination in a 1957 article in *Life* magazine,[4] such mushrooms were largely unknown outside of their native context. Wasson published an account of an expedition—funded by the CIA—to Oaxaca, Mexico, where he and others sat in on a sacred mushroom healing ceremony, called a *velada*, led by celebrated *curandera* (healer) Maria Sabina. Recounting all the details of that expedition is beyond the scope of our tale here, but I refer the interested reader to the article itself, available on the website of the Multidisciplinary Association for Psychedelic Study.

Here is a sample of Maria Sabina's lyrical poetry from the excellent work by the late Dale Pendell, *Pharmako/Gnosis: Plant Teachers and the Poison Path*:

> *I am the Morning Star woman, says*
> *I am the Cross Star woman, says*
> *I am the Moon woman, says*
> *I go up to heaven, says*
> *I am the woman of the great expanse of the water, says*

✦ *Mazatec shaman Maria Sabina. Photo by Don Juan Peralta.*

I am the woman of the expanse of the divine sea, says
There I am asking you for your principal herb, your sacred
  herb, says
Your clean herb, your well prepared herb, says
I am going there, says.
On my knees for hands, on my knees for feet, says
On my knees of tortillas, on my knees of water, says[5]

When the shaman repeats the word "says" in the Mazatec language, "*tzo*," (you can hear her repeating the word after every line in the recording), she is quoting the mushroom, which speaks through her, giving her words. The poet Henry Munn, who was present at veladas with Maria, writes in his essay "The Mushrooms of Language":

The Mazatecs say that the mushrooms speak. If you ask a shaman where his imagery comes from, he is likely to reply: I didn't say it, the mushrooms did. . . . He who eats these mushrooms, if he is a man of language, becomes endowed with an inspired capacity to speak. The shamans who eat them, their function is to speak, they are the speakers who chant and sing the truth, they are the oral poets of their people, the doctors of the word, they who tell what is wrong and how to remedy it, the seers and oracles, the ones possessed by the voice. "It is not I who speak," said Heraclitus, "it is the logos." Language is an ecstatic activity of signification.

Munn continues, "For the shaman, it is as if existence were uttering itself through him."[6]

In my humble experience, as a partaker from another culture, it is more as though they were speaking to me than through me, as if a teacher of great wisdom were patiently explaining mystical truths of great personal significance, though in a very humble manner that wasn't in any way grandiose and messianic. It gave me a mission to care for my family and friends, for my land, rather than to be a savior of the world.

For the Mazatec people, Munn says the mushroom experience is

inseparably associated with the cure of illness. The idea of malady should be understood to mean not only physical illness, but mental troubles and ethical problems. It is when something is wrong that the mushrooms are eaten. If there

◆ *Mid–first century BCE frescoes from the Villa of the Mysteries in Pompeii, Italy.*
Image by shakko.

is nothing the matter with you there is no reason to eat them. Until recent times, the mushrooms were the only medicine the Indians had recourse to in times of sickness. . . . According to the Indians, syphilis, cancer, and epilepsy have been alleviated by their use; tumors cured. They have empirically been found by the Indians to be particularly effective for the treatment of stomach disorders and irritations of the skin.[7]

I can attest to this latter property of the mushrooms, because in the spring of 1997, I was suffering from a terrible case of a persistent intestinal parasite after drinking some dodgy water. I had lost a lot of weight and couldn't eat. I had been sick for a month. One evening, some friends and I ingested the mushroom / Syrian rue combo. It turned out to be the last time I tried the mixture. After an intense and not very enjoyable trip, in which I was convinced I was dying, I woke up fully cured.

Wasson compared the velada he attended in 1957 explicitly to the classical mysteries of Ancient Greece, writing:

> If our classical scholars were given the opportunity to attend the rite at Eleusis, to talk with the priestess, what would they not exchange for that chance? They would approach the precincts, enter the hallowed chamber, with the reverence born of the texts venerated by scholars for millennia. How propitious would their frame of mind be, if they were invited to partake of the potion! Well, those rites take place now, unbeknownst to the classical scholars, in scattered dwellings, humble, thatched, without windows, far from the beaten track, high in the mountains of Mexico, in the stillness of the night, broken only by the distant barking of a dog or the braying of an ass. Or, since we are in the rainy season, perhaps the Mystery is accompanied by torrential rains and punctuated by terrifying thunderbolts. There, indeed, as you lie there bemushroomed, listening to the music and seeing the visions, you know a soul shattering experience, recalling as you do the belief of some primitive peoples that mushrooms, the sacred mushrooms, are divinely engendered by Jupiter Fulminans, the God of the Lightening-bolt, in the Soft Mother Earth.[8]

The Eleusinian mysteries of the grain goddess Demeter and her divine daughter Persephone were one of the most enduring expressions of the sacred in the ancient world, possibly celebrated since Minoan times at a site near Athens, Greece, until they were ended in 396 CE after the invasion of Greece by the

◆ The Exaltation of the Flower, *a bas-relief from the 460s BCE from Farsala, Greece (in the Louvre Musuem). Two women, thought to be the goddesses Demeter and Persephone, holding flowers suspiciously reminiscent of the* Psilocybe *mushrooms. Photo by Françoise Foliot.*

Visigothic king Alaric. Many notable Greeks were initiates. The rites were kept so secret that to this day no one knows exactly what the mysteries consisted of. The poet Pindar said of initiates into the mysteries, "Blessed is he who hath seen these things before he goeth beneath the earth; for he understandeth the end of mortal life, and the beginning of a new life, given by god."[9] Whatever was seen and done, the initiates conquered their fear of death.

The reader will recall our discussion of Demeter and Persephone as chthonic goddesses, and of the connection of agriculture with the mysteries of death and rebirth. As philologist Walter Otto said of the theme of the mysteries:

> Generation and fertility, and particularly the growth of grain, are indissolubly bound up with death. Without death, there can be no procreation. The inevitability of death is not a destiny decreed by some hostile power. In birth itself, in the very act of procreation, death is at work. It is at the base of all new life. . . . This then is the core myth of Persephone, to which the Eleusinian mysteries attach. Man receives the fertility which is indispensable to him from the hands of death. He must appeal to the Queen of the Dead. And this he can do; for here in Eleusis her divine mother mourned for her, here she returned to her mother, and here the goddesses created agriculture.[10]

Central to our inquiry here is the 1978 book Wasson coauthored with classicist Carl A. P. Ruck and Albert Hofmann, whom we have met briefly, titled *The Road to Eleusis: Unveiling the Secret of the Mysteries.* The authors propose

✦ *Barley grain infected with the hallucinogenic fungus ergot,* Claviceps purpurea.

that one of the secrets of the Eleusinian Mysteries in ancient Greece was the consumption of an entheogenic compound. The mysteries were reported to result in a very powerful and moving experience of spiritual transformation following a dramatic ritual and ingestion of a drink, known as the kykeon, or mixture, the ingredients of which are listed in the Homeric hymn to Demeter as barley, water, and *glechon*, identified as the herb pennyroyal, *Mentha pulegium*. In the opinion of the authors, the barley was ergotized, which means infected with the parasitic fungus ergot, *Claviceps purpurea*. This fungus is common in all grain crops and is the source of a number of psychoactive alkaloids chemically related to Albert Hofmann's "problem child" LSD, which is prepared by modifying one of these compounds chemically.[11] The authors determined that this was both chemically and culturally possible, and the resulting thesis caused shock waves in the intellectual world, where it was considered scandalous that the ancient Greeks, inventors of reason itself, could have indulged in anything as irrational as using psychedelic drugs.[12]

Poet Robert Graves, author of the creative masterpiece *The White Goddess* and friend of the Wassons, advanced the theory that Demeter's kykeon contained hallucinogenic mushrooms. He claimed that the ingredients for the potion, as listed in the Homeric hymn to Demeter, spelled out in an acrostic a Greek word for mushrooms. One argument in favor of this theory is that there were no poisonings reported in the entire long history of the Eleusinian Mysteries, which would be unlikely if a dangerous substance like ergot was the active ingredient.[13]

Whatever the agent imbibed, the initiates saw something of incredible transformative power. Otto quotes Aristides, who said, "Of all the divine things which exist among men, it is both the most terrible and the most luminous." The initiates saw something that impressed upon them, in the words of Otto, "the immense life-giving power of the earth mother," possibly an ear of grain, at the climax of a dramatic ritual performance under the influence of the entheogenic kykeon.[14] Lucius Apuleius said of his own initiation into the mysteries, "I approached the very gates of death and set one foot on Persephone's threshold, yet was permitted to return, rapt through all the elements. At midnight I saw the sun shining as if it were noon; I entered the presence of the gods of the underworld and the gods of the upper world, stood near and worshiped them."[15]

The reports from initiates in the mysteries sound very much like some of the modern claims for the benefits of the psychedelic experience, including that it reduces anxiety about death, particularly in the terminally ill. A clinical trial at Johns Hopkins Medicine reported that among patients with a life-threatening cancer diagnosis, a significant reduction of anxiety about death was reported following a treatment with a high dose of psilocybin and supportive psychotherapy. Sixty-seven percent of patients reported it as being one of the top five meaningful experiences in their lives. In our secular society, the potential of these medicines to reduce anxiety and improve lives is finally getting the attention it deserves, but only removed from its sacred context.

My particular youthful interest in the alchemical connection to entheogens was also triggered by my chance finding of a copy of Carl Jung's last book, *Mysterium Coniunctionis,* providentially, in a secondhand book store. The text is a treatise on the symbolism of opposites and their unification in alchemy and in the human quest for wholeness. While perusing it one day, I came across a reference to a certain mythical plant, the *arbor philosophica,* the tree of the philosophers. Jung quotes Galen, who says of this plant, "There is a certain herb or plant, named *Lunatica* or *Berissa,* whose roots are metallic earth, whose stem is red, veined with black and whose flowers are like those of the marjoram; there are thirty of them corresponding to the age of the moon in its waxing and waning. Their color is yellow. Another name for *Lunatica* is *Lunaria,* whose yellow flowers (alchemist Gerhard) Dorn mentions attributing to them miraculous powers."[16] In the footnotes, Jung quotes Galen and various Greek herbals which identify *Lunatica* or *Berissa* with *Peganum sylvestre,* an older name for

◆ *Alchemical Rebis from the 1617 alchemical treatise* Theoria Philosophiae Hermeticae *by Heinrich Nolle.*

P. harmala, the harmaline-containing plant with which I had been experimenting. I was intrigued when I recognized the name of my plant ally in the footnotes and quickly theorized that it could have been part of a primordial entheogenic potion, a sort of ancient Middle Eastern version of ayahuasca, when combined with a DMT-containing admixture plant.

Several medieval manuscripts containing versions of a Latin Hermetic text give additional magical properties of the mysterious lunar plant identified as *P. harmala*. The manuscript "Trinity College MS 0.2.48" contains a version of the text quoted above, complete with prayers to be said while harvesting it, including *nomina* of the kind seen in other texts of medieval magic, such as *The Sworn Book of Honorius*. Harvesting the plant without the proper prayers and ritual procedures will reportedly cause it to disappear. The juice of the herb is said to be able to affect the transmutation of various metals, including making gold. It is an elixir of life said to reverse the aging process, and gold made with it can be used to make a magical ring.[17]

When exaggerated legendary claims are made for a humble desert weed, to me that hints at a history of ritual use. The ability to transmute lead into gold can be seen as a metaphor for spiritual transformation. In many alchemical texts, the Moon is to be united with the Sun to complete the work. As Mircea Eliade wrote in his study of alchemy, *The Forge and the Crucible*, "It is interesting to note that the *coniunctio* and the ensuing death is sometimes expressed in terms of the *hieros gamos*: the two principles—the Sun and Moon, the King and Queen, unite in the mercury bath and die (this is the *nigredo*): their 'soul' abandons them to return later and give birth to the *filius philosophorum*, the androgynous being (Rebis) which promises the imminent attainment of the philosopher's stone."[18] This description of the alchemical process could easily apply to the psychedelic experience or the mystical experience in general, as well as the process of creating the medicine itself.

In the Hermetic manuscript mentioned above, one is instructed to boil Mercury in the juice of *Berissa* in order to produce a red stone. Then the instructions say to put some of the red stone on an unknown substance (the word is unfortunately lost by a lacuna in the text) in order to produce the Sun. If *P. harmala* was the plant of the Moon, then was there a DMT-containing plant or tree identified with the Sun? If so, my bets would be on one of the species of acacia known to the people of the ancient Near East, possibly the Egyptian *Acacia nilotica*, which is known to contain DMT and is heavily invested with sacred symbolism in Egyptian mythology. One recently published book, *Angels*

◆ *Alchemical Rebis from the 1706 alchemical treatise* Compendium alchymist[ae] novum sive Pandora . . . *by Johann Michael Faust.*

in Vermilion by P. D. Newman, proposes that European alchemists knew how to extract the DMT from acacia bark and that Renaissance British magicians John Dee and Edward Kelley used it in their scrying rituals. Sounds plausible to me. Adding to this theory, my own personal myth about a lost Western version of ayahuasca is maybe not too much of a stretch. We will likely never know for sure.

Did Early Christians Use Psychedelics?

It is likely that entheogenic and hallucinogenic plants were used by cultures all over the world, and that many of the practices associated with these substances were never recorded at all, either due to having been a closely guarded secret or due to having been practiced by preliterate people. Aside from the Eleusinian Mysteries, and soma (and its Persian relative haoma), another contender for ancient entheogen use exists in the West. There are serious scholars who believe that some early Christians originally used some sort of psychedelic compound in their communion ritual. Brian Muraresku makes the case that entheogenic wine, spiked with henbane, opium poppies, and other substances, was central to the cult of Dionysus in the Greco-Roman world, and that further, Greek converts to Christianity, themselves attracted by the Dionysian symbolism and language of the new religion, could have celebrated the love-feasts and communion rituals in much the same fashion.[19] St. Paul, in his letter to the Church in Corinth, reproves the Corinthians for improperly celebrating the Eucharist, accusing them of drunkenness (1 Corinthians 11:21), and goes on to warn them, in 1 Corinthians 11:27, "Whoever eats the bread or drinks the cup of the Lord in an unworthy manner will be answerable." He then tells them he is aware of some who ate and drank the sacrament in an improper manner and that those who do "drink judgment against themselves. For this reason many of you are weak and some have died [literally fallen asleep]."[20] This chapter of First Corinthians is crucial for the Christian religion; it contains the "words of institution" said in every celebration of communion. But if Muraresku is right, St. Paul is not warning the Corinthian congregation to examine their consciences before taking communion, which is the standard interpretation of these passages, but he is warning them against the dangerous practice of drinking drugged wine at their rituals, wine known to contain powerful and potentially toxic plant substances.[21]

All this lore and speculation make the case that these consciousness-expanding substances were extremely important for our ancestors. With them ancient peoples confronted the mysteries of the next world and performed rites and rituals to unite themselves with the gods and transcend the ordinary limitations of profane reality. If we take our cues from entheogen-using cultures and use these sacred plant teachers in the context of seasonal rituals, they can help us come to terms with our alienation from ourselves and deepen our connection to the natural world. As part of the program of re-enchantment and rewilding of the imagination, they definitely have value. They certainly had a hand in setting me on the path. The powerful experiences that I have had, especially when I was young, clarified my values in favor of a life lived on the land and among natural things, tending plants and living with the seasons, pursuing deeper knowledge away from the commercial values of the modern world. They helped me to become the person I was meant to be.

The ability of psychedelic plant medicines to shift a person's values and priorities in favor of a more mystical worldview and allow them to see through cultural programming makes them threatening to entrenched power structures. The entheogenic experience can give one a more open sense of self, expanding the narrow confines of the ego and allowing for the realization of mutuality, the sense that we are parts of a greater whole, a living and ensouled cosmos. These medicines can open one up to imaginal spaces unfathomable to the uninitiated, peopled with fantastic creatures of shifting light. They have brought me face to face with both angels and demons, as well as inner guides with surpassing wisdom. The plant teachers show us that the divine is within us, accessible and intimate, as well as transcendent and otherworldly. They make possible the Gnostic realization of direct experience of the divine, without mediation by any institution.

But they are not without risk. The temporary dissolution of the ego produced by these substances is often experienced with fear and panic, especially to those without experience. The fear accompanying these experiences can be so unpleasant that the unprepared may end up in the emergency room or worse. There are some very dark programs lurking within the unconscious that may make themselves known without warning. The only way out is through, to hold to your mantra, pray to your gods, and trust. You also have to make sure you are in a safe place around people who care for you and where you will not be disturbed for some time. And it is precisely the medicine's propensity to shift values and priorities that gives me another reason to warn the uninitiated: The

◆ *Alchemical Tree of Knowledge from the German alchemical work* Gemma Sapientiae et Prudentiae *(c. 1735). Image by Wellcome Images, London.*

medicine will change you, so if you don't want to be changed, then it is best to leave it alone. If you have a sublime experience of contact with spiritual reality (from whatever source, this counts for religion too), that will become very important to you, perhaps to the exclusion of your old values (think of converts of all kinds). That is why taking the time to integrate these experiences is so necessary. The divine is addictive, even though the medicine is not. A taste of the divine will leave you wanting more. So count this as my official disclaimer: I am in no way encouraging anyone to embark on this path.

In my youth, I came to have an almost messianic faith in the salvific power of entheogens, similar to the claims made by the alchemists for their elixirs. I began to associate the two, believing that this must have been the universal medicine they sought, the elixir of immortality. Maybe the immortality wasn't literal, physical immortality, it was spiritual liberation, salvation. I hear similar claims made for potential benefits of these substances today. It does give me hope that they may be of some help in overcoming not only our personal blocks and traumas but the immense challenges we face as a species. The mature me, however, thinks that the real world isn't that simple and that our collective karma will need to be faced. And that Gnosis always gets banned. The moment the psychedelic renaissance starts to make any headway in healing our collective psychopathology, the archons may be waiting in the wings to give us a reality check. I hope I am proved wrong, but this is a fallen world, after all.

Not only can they be psychologically and spiritually risky, they are also illegal in many states and countries. With all of the progress made in legalizing entheogenic plants in the United States and around the world, the very traditional and conservative state where I live is still one of the holdouts and probably will remain so for the time being. Medical marijuana is now legal here in Kentucky, and my community is home to large crops of the variety of cannabis raised for its medicinal CBD content, so there is hope. However, I have friends who have been persecuted legally for the sacramental use of sacred plants, which is a tragic miscarriage of justice. For myself, I will have to hold off personal research into entheogens until the legal climate improves.

16
BANEFUL PLANTS
IN MAGICAL PRACTICE

Plants are spiritual beings. Like us, they are encapsulated in a physical form. By cultivating and interacting with plants, even just being in their presence, we come into contact with them as spirits. We are dependent on them for life, and we exist in a relationship of mutuality with them. But with some plants, the relationship is deeper, darker. By ingesting the more potent among them, we can become possessed by them, at least for a time. But some plants leave their partaker indelibly marked, imprinted for life with a new perspective, a new habit, or even a new hunger. Some of our green friends are wholesome, some are angelic, and some are fiendish, summoning daimonic entities into our very souls. With some of these plant allies, a pact is made in blood and the cost can be dear to the mortal. It is the work of a lifetime to extricate oneself from the more severe of these pacts. The most dangerous of plants can be worked with as powers and allies without the risk of ingesting them.

The task of the cunning farmer is to communicate with the genius of the plant, to draw it forth and tame it, to make of it an ally and not a master, to forge a working relationship with it. To sit down in the presence of the plant, mind clear and open, and to be receptive to the subtle voice of what the northern European Heathen tradition calls the "plant wights," the animate intelligences that enliven the vegetable world. In this way, we can work with these plants safely, the way an animal trainer can work with dangerous animals or a goetic magician can work with dangerous spirits.

It is perhaps in these extreme and powerful plants—the difficult and the dangerous ones—that we see the genius that animates them most clearly. A far

✦ Datura, *poppies, and tobacco growing in the author's garden.*

cry from the humble potato or utilitarian carrot—sources of sustenance—the plants we are discussing here are *pharmaka*, a word whose meaning encompasses both medicine and poison. There is nothing subtle about the modus operandi of tobacco, which makes its devotees at first queasy and nauseous before enslaving them to a nearly unbreakable addiction that nevertheless offers them courage, emotional security, and, according to some cultures, magical defenses against malefic spiritual forces.

SATURNIAN BOTANY

In our last chapter, we discussed the role of entheogenic plants in the practice of the cunning farmer. In this chapter, we are going to discuss the darker entheogens of the nightshade family, the *Solanaceae*. This botanical family includes edible favorites such as peppers, tomatoes, and eggplants, which have a large presence on our farm in the summer, as well as tobacco, with which I began my agricultural career. Here we will be focused on the hallucinogenic nightshade plants with a history of use in shamanism, witchcraft, and magic worldwide: mainly the genuses *Datura*, the thorn apples and jimsonweeds; *Brugmansia*, their South American cousins, also known as angel's trumpets; *Hyoscyamus*, the henbanes; *Mandragora*, the mandrakes; *Atropa*, also known as belladonna or deadly nightshade; and tobacco, the genus *Nicotiana*, which was an important magical plant and entheogen in the Americas with a long history of use among Indigenous people.

I will put a disclaimer at the beginning of this chapter so there is no misunderstanding: I am merely reporting the ethnobotany, folklore, and something of the cultivation of these fascinating plants but am in no way encouraging their use. There are strict warnings associated with these plants for a reason. The user who experiments with ingesting these tricky plants risks personally experiencing the nasty hangovers, missing time, insanity, blindness, loss of mental functioning, and untimely death. That said, I have always found myself attracted to their presence. I keep specimens of some of these in my gardens, with the exception of belladonna, *Atropa belladonna*, which I have not grown in years because it has tempting-looking berries that are deadly if consumed, and is therefore not safe to be grown near small children.

TOBACCO

Let us begin with the member of this class that I have had the most personal involvement with: tobacco.

There are various species of the genus *Nicotiana*, named for the sixteenth-century French diplomat and scholar Jean Nicot, who first introduced tobacco to France. The cultivated species *N. tabacum* and *N. rustica* are the ones most used in magical and ceremonial work. Both species originated in the Americas, with *rustica* being cultivated by the Indigenous peoples of North America at the time of European contact and *tabacum* having been introduced from the West Indies

◆ *The author harvesting tobacco using the traditional method.*

by English settlers to Jamestown as a commercial crop for the European market. *Rustica* tobacco is commonly known as Indian tobacco or wild tobacco and is much smaller than *tabacum*.

The principal alkaloid in both is nicotine, an extremely poisonous substance lethal in minute amounts. One or two drops of pure nicotine, 60 to 120 milligrams, placed on the skin, would be a lethal amount, and the average cigar contains enough nicotine to kill two people if it were to be extracted and injected.[1] Farmworkers who work with the plant are frequently poisoned, usually only suffering nausea and vertigo and recovering after several hours. I can personally attest to this, having been poisoned by tobacco on several occasions, most notably the first week I ever worked harvesting the crop. The tobacco plant is covered with a sticky gum that accumulates on the hands and clothing of the harvester. After working hard in the Sun all day, I poured a bucket of cold water over my head, causing the nicotine-containing gum in my clothing to release its alkaloids into my system. The strange thing was that the intoxication, which consisted of vertigo and nausea, did not occur until after I had showered and changed my clothing.

Nicotine as administered in tobacco products is one of the most addictive substances known to humankind. According to the World Health Organization, tobacco addiction is one of "the biggest public health threats in the world, killing over 8 million people a year" and contributes to poverty and malnutrition by diverting family resources toward maintenance of the tobacco habit. In 2020, 22.3 percent of the global population used tobacco.[2] It is not my intention to turn this chapter into a public health announcement, though. To me, this is remarkable evidence of the power of this plant to enmesh its users in a habit that's very hard to break, similar to a pact made with a malevolent entity. As a former smoker who damaged my lungs permanently through years of tobacco use, I keep tobacco outside the magic circle, as it were, using it as an essential substance in spirit offerings and incense but never ingesting it.

When I began my agricultural career, my first job as a farmhand was harvesting tobacco, which was initially interesting to me due to my fascination with sacred plants. The crop is harvested by hand to this day. Mechanization is not practical, as the valuable leaves are damaged by careless harvesting. The plants are chopped at the base with a hatchet-like tobacco knife and then impaled on a wooden stake, six plants to the stake. After wilting in the sun for several days, the entire crop is loaded onto wagons and laboriously hung on racks in barns. There are many different varieties of tobacco for different products. At my first farming job, we grew the Burley type of tobacco, which is mostly used for cigarettes. I live in a state where tobacco production is a huge part of the culture and has been since colonial times. Indeed, tobacco has deep roots in the eastern United States, having been grown in the region since John Rolfe obtained Spanish tobacco seeds and planted them in Virginia in 1609.[3] Tobacco of the *rustica* variety was grown by the Indigenous people long before that. Although smoking tobacco in the modern world is a destructive vice that has claimed many lives and made handsome profits for dishonest corporations and individuals over the years, it has nevertheless allowed many small farmers in my region a steady and dependable income from a relatively small acreage for generations. In the last twenty years, that has all changed since a government-administered program that limited production to keep prices high was elimi-

✦ *Harvesting tobacco in Kentucky, c. 1940. Photograph by Marion Post Wolcott.*

nated due to public outcry over the corrupt tobacco industry, which responded by purchasing cheaper imported tobacco from overseas.

After working for other farmers for several years, in the year 2000, I raised my own crop of organic Burley tobacco, about 12,000 plants, grown on 1.3 acres in 43 rows that were each over 350 feet long. Organic tobacco farmers remove the shoots that branch out at every node where a leaf is attached as well as the flowers at the top of the plant to concentrate the energy of the plant into growing large leaves. It is very labor intensive, and gives one a good dose of nicotine, which is easily absorbed through the hands. I remember one beautiful summer morning (the work has to be done in the morning because the plants are more full of water then and the tops and branches snap more easily), walking down the rows of tall bright green plants, with large, lush, pungently scented leaves, snapping the clusters of flowers off before most had had a chance to open and dropping them on the ground, getting dizzier and dizzier from the nicotine as I went.

✦ *"Ancient tobacco,"* Nicotiana rustica, *growing in the author's garden.*

Other varieties, such as cigar tobacco, have very different methods of cultivation and are harvested as the leaves mature, starting at the bottom of the plant. The plants of the cigar varieties are very aromatic, scenting the air at quite a distance with the slightly musky scent of a cigar shop, as well as being ornamental, with large tropical leaves and attractive pink flowers. They have a place in every garden of magical plants.

A handsome plant, and very ornamental, the *rustica* type is smaller and has greenish yellow flowers, which it produces in abundance. It reportedly contains a higher percentage of nicotine. As Johannes Wilbert wrote about *Nicotiana rustica* in his informative work *Tobacco and Shamanism in South America*:

> As the hardier and richer in narcotic properties of the two cultivated species, it spread far beyond the tropical and subtropical belts to the very limits of New World agriculture between Chile and Canada. In fact, in its dispersal *N. rustica* rivaled even maize, along with such cultigens as cotton and the *Lagenaria* gourd. It has been the *petún* of Brazil, the *piciétl* of Mexico, and the *Indian tobacco* of eastern North America. Considering its wide range from the tropics to high latitudes, it is likely that *N. rustica* represents the older of the two principal tobacco cultigens, since *N. tabacum*, in pre-Columbian times, did not extend beyond the tropical climates.[4]

It is the tobacco of choice of many Indigenous American cultures and has been used by shamans to attain visions. Like its larger cousin, *N. tabacum*, it originated in tropical South America and was spread northward by trade among the tribes and nations.

Tobacco in Magical Practice

To understand the place of tobacco in magical practice, we must principally concentrate on the lore of the Indigenous peoples of the Americas, as the plant was a latecomer to the Old World, not arriving until the sixteenth century. It never attained a major role in the magical traditions there, being no more than an herb of martial or saturnine virtue, useful when such was needed, but not especially important as were the herbs with a longer history of magical use. Entheogenic plant specialist Jonathan Ott says, "Tobacco was the shamanic inebriant par excellence throughout the Americas."[5] It was smoked, chewed, snuffed, and ingested in infusions—taken in whatever manner appealed to the imagination of the partaker. Tobacco was thought by the shamans of some

South American tribes to strengthen their defenses against harmful magic, and they ingested it nearly constantly to feed their helping spirits. This is strong tobacco, ingested in a way that maximizes nicotine absorption and often pushes the user right to the edge of unconsciousness and trance, propelling them into an altered state of consciousness wherein they can communicate with the spirit world. This is a far cry from the addictive commercial product of modern tobacco companies.

Among the ancient inhabitants of the land where I live, tobacco was a crucial part of their culture and lifestyle, seen by the sheer number of pipes deposited as grave goods and offerings in Hopewell and Adena culture sites. One mound in Chillicothe, Ohio, contained over eighty beautifully carved animal effigy pipes that were ritually destroyed before being entombed.[6] I imagine, though physical evidence would have decayed long ago, that the pipes were buried with a copious supply of *rustica* tobacco in order to provision the dead for the afterlife.

✦ *Owl-shaped effigy pipe from the Hopewell culture c. estimated 200–500 CE. Found in Ohio.*

Anthropologist Charles Hudson relates the use of tobacco among the Cherokee for what he calls conjury. The traditions of the native peoples of the southeastern United States are of particular interest to students of American folk magic because of the influence they had on Appalachian magic, root work, Hoodoo, and conjury in general. Hudson, who was an authority on the Indigenous cultures of the southern United States, preferred the use of the word *conjury* to *sorcery* because of the particular resonance of the former term in the South. He defines conjury as "the use of oral formulas and ritual acts to affect spiritual beings, including those that governed the weather, game animals, cultivated plants, and most especially the health and well-being of people."[7] Witchcraft, by contrast, in Hudson's terminology, following that of anthropology of his time, "was the intrinsically evil action of mystical means to cause human suffering and death."[8] It is important to remember that Western anthropology is engaged in the practice of, among other things, fitting the terms of foreign cultures into Western categories and that the idea of a "good witch" is relatively recent.

The Cherokee conjurors used "ancient tobacco," *N. rustica*, "for serious acts of conjury." Hudson adds that tobacco was one of the most important herbs to the tribe and that it had many uses. The Cherokee "smoked it to suppress hunger, used it as a medicine, and smoked it as a kind of spiritual facilitator before councils of war and peace and before performing rituals and ceremonies." According to Hudson, and as stated by other authorities, "the Southeastern Indians sometimes experienced mind-altering effects from *N. rustica* far more than is experienced from our commercial *N. tabacum*."[9]

The tobacco was grown on small patches of land that had been prepared for cultivation by burning wood from trees that had been struck by lightning. The patches were secret. Nobody but those growing the tobacco could see them. To the Cherokee, Hudson says, "Ancient tobacco has no intrinsic spiritual properties; it was only an herb. It gained its power by virtue of the ritual act of 'remaking' which infused it with thought and power." This is indeed interesting, similar to the act of consecrating a talisman; the "remaking" is a ritual act of enchantment whereby the intention of the conjuror was imprinted onto the plant: "It was this act of remaking which transferred the thought onto the herb, making it into a medium whereby one person could affect another." The shaman—or as Hudson prefers, the priest—remade the tobacco

at dawn on the bank of a spring or a creek. A conjuror would face the sun rising in the east, hold up the tobacco in his left hand and recite a formula while kneading the tobacco in a counterclockwise direction with four fingers of the right hand. The conjuror often blew his breath or rubbed his spittle on the tobacco. If the tobacco was to be used for an antisocial purpose, it was sometimes remade at dusk or midnight, and it was rubbed in a clockwise direction. Remade tobacco could be used in four ways: it could be smoked near the person who was the target of the conjury, so that the smoke would actually touch him; it could be blown in the direction in which he was likely to be located; it could be smoked so that the smoke would pervade and affect everyone in a general area; and bits of tobacco leaf could be left where the person to be affected would come into contact with them."[10]

Reading Hudson's accounts of various practices of conjury among the Cherokee, it is easy to see how the people of European and African descent who lived alongside the Cherokee and other Indigenous nations exchanged a variety of ideas and practices with their native neighbors before their tragic forced removal in the infamous "Trail of Tears" in 1838 and 1839. The folk magic of the American South is a blend of African, Indigenous, and European traditions, in varying proportions, depending on the particular tradition. Tobacco was used by some of these practitioners as an ingredient for spellwork to help win court cases and in love magic. The use of tobacco in magic for success in court is attested to in several sources. It has been used for banishing negative energy and as a spirit offering.[11]

In my practice, tobacco is used as a spirit offering, either burned as incense or simply laid on the earth as an offering to earth spirits. As Dale Pendell said in *Pharmako/Poeia*, "Tobacco is the food of the gods."[12] And Daniel Schulke, a noted authority on the praxis of the poison path, states that spirits flock to it "in great joy."[13] At least where I live, tobacco is a highly effective offering to the land spirits, with an ancient history in this land, and the gods of this place have given me many signs that it is appreciated. Whenever I go to the woods to gather herbs or simply to pray and meditate, I always bring with me a pocketful of tobacco from my own harvest, often combining it with other offerings like cornmeal, beer, milk, or fruit.

Growing Tobacco

Tobacco, adapted to many climates, is an easy plant to grow. It is now grown all over the world, on every continent except Antarctica. Its cultural requirements aren't that different from those of tomatoes, a close relative. The tiny seeds must be started indoors early in most temperate climates and then set out when the weather has warmed in the spring. The plants are heavy feeders, requiring large additions of manure or compost for promoting vigorous growth and large leaves. The seedlings must be kept weed free until large enough to shade the ground. The traditional method for weeding is to keep the soil between the plants regularly stirred with a hoe (and larger farm implements) to keep it aerated, a process known to farmers in my region as "cultivating." Harvesting is done in several ways. The simplest method is called "priming" and involves removing the leaves that begin to wither or "ripen" at the base of the plant, in the morning, when the humidity is high, so they don't crumble. The harvester ties them in bundles, or hands, and stores them in a shed until completely dry. Alternatively, the whole plant may be cut with a machete when most of the plant is beginning to turn yellow, which indicates maturity, and then hung up in a barn or shed for curing.

I have grown many varieties and species of tobacco over the years, my favorites being a variety of *N. rustica* called "Hopi tobacco," a relatively large-leaved variety for a *rustica*; and a variety of *N. tabacum* called "West Indies," a tall, fragrant, cigar-type tobacco. I save my own seed and would be happy to provide packets to interested readers for a nominal fee.

JIMSONWEED (*DATURA* SPECIES)

These fascinating weeds have a long history of use as medicines, entheogens, and poisons. They grow worldwide and are represented by several species, which we will consider separately. They are responsible for multiple poisonings worldwide due to deliberate ingestion by people seeking an entheogenic experience—often teenagers—which turns into a trip to the emergency room, the psychiatric ward, and in some cases, the cemetery. People have also been poisoned by *Datura* contamination of food products, as *D. stramonium* is a very common agricultural weed and gets consumed accidentally from confusion with edible plants. However, the plant is used medicinally in many parts of the world to treat diarrhea, asthma, flu, and the common cold.[14]

The principal alkaloid in the plant is scopolamine, with atropine present in lesser amounts. Scopolamine and atropine are used in modern medicine in a

◆ Datura inoxia, *also known as moonflower, in bloom, Nuevo Laredo, Mexico. Image by Miguel Angel Omaña Rojas.*

variety of ways, including the dilation of pupils for ophthalmic exams, prevention of motion sickness, and prevention of nausea and vomiting in a variety of illnesses.

With a plant this dangerous (or one might say powerful), with a history of use and misuse, present worldwide as a variety of species, it should come as no surprise that it is used in magic and ritual all over the world. We will start with a few examples, with the first being close to home among the Indigenous cultures of eastern North America.

The Algonquian people, living in what is now Virginia, used *Datura stramonium*, called in their language *wysoccan*, as part of the *huskanaw* ceremony in which the boys of the tribe were initiated into manhood. The boys danced with the men for three days, said goodbye to their mothers, were given the *Datura* potion, and were left in the forest for an extended period of time, forgetting their lives as children and returning home as men.[15] According to Schultes and Hofmann's *Plants of the Gods*, the Yokut people of California gave *Datura* (most likely *D. inoxia*) to adolescents as part of a coming-of-age ceremony as well "to ensure a good and long life." The Tübatulabal tribe, also of California, use *Datura* in initiation ceremonies to acquire a spirit animal or a ghost helper, which is revealed in the visions triggered by ingestion of a decoction prepared from the roots of the plant.[16] It is interesting that many of the ethnobotanical reports of *Datura* use in North America focus on its use in initiation rituals for adolescents and that on the same continent in modern times most poisonings occur among teenagers and young people. Unlike its use by native people with millennia of experience, modern experimenters are meddling with an ancient power unprepared, one that nevertheless has a potent mystique and an undeniable pull. In these reports, we can see the agency of the genius of the *Datura* plant at work, calling to the young and uninitiated, promising unearthly visions and delivering madness.

Datura is used by the Navajo for divination, finding lost objects, and diagnosis of disease, as well as for love magic. According to Schultes and Hofmann, when used to reveal the cause of sickness, a healing chant may be given. *Datura* is called the "laughing medicine" or the "mind medicine." One informant told anthropologist Clyde Kluckhohn that only those who know medicine may eat *Datura* without going crazy, whereas others get crazy easily.[17]

A well-known literary source of *Datura* lore is in the work of Carlos Castaneda, especially his first book, *The Teachings of Don Juan: A Yaqui Way of Knowledge*. The book, which is presented as an ethnographic study of Castaneda's apprenticeship with a shaman of the Yaqui tribe of northern Mexico and was Castaneda's PhD dissertation, is widely reported to be fraudulent. Whatever the truth of the source of the content in the book, it has nonetheless contributed to the lore and fascination with *Datura* in the popular imagination, and the book is partly responsible for introducing the hippie generation to the plant as an entheogen. In the book, the shaman, Don Juan, instructs Castaneda on the preparation and use of "the devil's weed," which is what he calls *Datura*:

> The devil's weed has four heads: the root, the stem and leaves, the flowers, and the seeds. Each one of them is different, and whoever becomes her ally must learn about them in that order. The most important head is in the roots. The

power of the devil's weed is conquered through the roots. The stem and leaves are the head that cures maladies; properly used, this head is a gift to mankind. The third head is in the flowers, and it is used to turn people crazy, or to make them obedient, or to kill them. The man whose ally is the weed never intakes the flowers, nor does he intake the stems and leaves, for that matter, except in cases of his own illness; but the roots and the seeds are always intaken; especially the seeds; they are the fourth head of the devil's weed and the most powerful of the four.

My benefactor used to say the seeds are the "sober head"—the only part that could fortify the heart of man. The devil's weed is hard with her protégés, he used to say, because she aims to kill them fast, a thing she ordinarily accomplishes before they can arrive at the secrets of the "sober head." There are, however, tales about men who have unraveled the secrets of the "sober head." What a challenge for a man of knowledge.[18]

Don Juan then teaches Carlos to prepare a flying ointment from *Datura* seeds and roots from plants he has grown himself, using the fat of a wild boar as a base. Carlos experiences the joy of flying and awakens naked in the desert. Afterward, he interrogates Don Juan to ascertain whether or not he flew bodily or simply in his mind. As he does so often in these dialogues, Don Juan castigates Carlos for his lack of vision and pedestrian Western worldview, saying, "There you go again with your questions about what would happen if . . . It is useless to talk that way. If your friend, or anybody else, takes the second portion of the weed all he can do is fly. Now, if he had simply watched you, he might have seen you flying, or he might not. That depends on the man."[19] I tend to think of Castaneda's work as philosophical dialogues in which the author uses the characters to give voice to philosophical ideas of great profundity, not dissimilar to Plato, whose Socrates and other characters probably never said most of the things attributed to them. They are more along the lines of pseudepigrapha than outright forgery, but in the modern world there is no distinction.

The use of *Datura* as flying ointment echoes the folklore we will see when we look into the other herbs in this class—mandrake, henbane, and belladonna—which have similar chemistry and are often included as ingredients in the classic flying ointments of the European witches. The use of the term *devil's weed* also links Castaneda's account to European witchcraft. Though I am no expert in North American Indigenous use of *Datura*, most examples of North American use I have read involve the ingestion of various portions of the plant, rather than use of an ointment. Castaneda has been criticized for many things, among them fake ethnobotany, and in particular the use of *Datura* flying ointment by the Yaqui, borrowing the whole story from European historical reports. Even if Castaneda's Don Juan and his devil's weed ointment is a fabrication, it is good writing and still a highly influential myth contributing to the cultural perception of *Datura* and entheogenic plants in general. There is no reason to fabricate fantastic tales about *Datura*, because the truth is stranger than fiction, as we shall see.

Returning to *Datura* use in Western magic, we find the *Picatrix* has several hair-raising recipes, including one "for appearing in the form of any animal you wish," by taking the fat cooked out of the head of an animal combined with *Datura* seeds and making an ointment with it, burning it in a lamp, and then anointing the face of someone with it.[20] This would reportedly cause that person to appear as the animal whose head was used. This will not be the last time we will come across an example of these solanaceous plant pharmaka being associated with shape-shifting and theriomorphic transformation.

Another bizarre recipe was "for ruining the senses and thoughts." This involved a human head, spleen, heart, and liver, and the heads of various animals, calcined, reduced to ash, and mixed with *Datura* and henbane seeds, to be placed in the drink of the victim.[21] A modern person would think that all of this was bizarre and unnecessary and that all of the body parts and animal heads would only serve to dilute the active constituents, the *Datura* and henbane seeds. To the author of the *Picatrix*, however, the occult virtues of the various human and animal ingredients added potency to the mix. Indeed, the *Picatrix* states, "The human body possesses many marvels pertaining to magical operations." Maslama al-Qurtubi, the author of the *Picatrix*, can be seen as a proto-scientist (maybe a proto–mad scientist, an Andalusian Dr. Frankenstein), dissecting cadavers in an effort to find out what makes them tick.

It is important to remember that "natural magic" of the type represented in the *Picatrix* is the precursor to modern science, and that the idea that there are "occult virtues" or hidden properties of physical objects, like plants and animal parts in this case, is the forerunner to the science of pharmacology and much of modern medicine. Of course, occult virtue is much more than primitive science and bastard pharmacology; it is the sacred Hermetic science of the hidden sympathies of nature and speaks to a different order of truth than materialist science. This is a topic which we will explore in greater detail in the next chapter on plant magic.

The great Renaissance scientist, playwright, natural magician, and polymath

Giambattista della Porta, author of *Natural Magick*—a fascinating compendium of recipes and experiments in distillation, medicine, pharmacology, perfumery, cosmetology, the making of invisible inks, burning lenses, magnets, and more—has several experiments pertinent to our topic, including this one:

> With *Stramonium*, or *Solanum Manicum*:
> The seeds of which, being dried and macerated in Wine, in the space of a night, and a Drachm of it drank in a Glass of Wine, (but rightly given, lest it hurt the man) after a few hours will make one mad, and present strange visions, both pleasant and horrible; and of all other sorts: as the power of the potion, so doth the madness also cease, after some sleep, without any harm, as we said, if it were rightly administered. We may also infect any kind of meat [food] with it, by strowing thereon: three fingers full of the Root reduced into a powder, it causeth a pleasant kind of madness for a day; but the poisonous quality is allayed by sleep.[22]

In my practice, as I have said, *Datura* is best interacted with by enjoying the eerie beauty and presence of the plant. The species native to the southwestern United States and Mexico in particular—*D. inoxia*, *D. wrightii*, and *D. meteloides*, which are very similar in appearance (honestly, I'm not sure which one I grow here on the farm)—are beautiful and highly ornamental, having large white flowers that open at night and emit an intoxicating perfume. The Asian species, *D. metel*, is available in stunning double-flowered purple or yellow forms. Seat oneself next to a thriving plant with dozens of blooms open. As the Moon rises, they exhale their ethereal scent. Bask in its dark beauty with a clear and open mind. You can enter into communication with the genius residing in the *Datura* plant. Sit. Inhale. Listen. Something will come.

Growing Daturas

The *Daturas* are easily cultivated and are grown much like peppers, requiring warmth and a long season. The seeds are started indoors, with bottom heat if possible. When they germinate, they should be moved to four-inch pots until they are ready to be planted. The plants require good drainage and full Sun, high levels of fertility, and regular watering during dry spells. The sprawling, vine-like *inoxia-meteloides* group need plenty of room to grow and will self-sow if allowed to produce seed, which they do in great abundance. As a farmer, I do not recommend planting jimsonweed, or *D. stramonium*. Although it is an attractive plant with an interesting history, it is a noxious agricultural weed and will probably show up on its own in any garden eventually, especially if you import cattle manure.

A warning when working with all these plants: The alkaloids are easily absorbed through the skin. Special care should be taken to avoid getting the juice of the plant in the eyes, as I found out recently and described in an earlier chapter. Wash your hands after handling, and don't rub your eyes unless you want your pupils to be dilated for several days!

ANGEL'S TRUMPET (*BRUGMANSIA* SPECIES)

Related to *Datura* are the "tree daturas" now classed in the genus *Brugmansia*, which are native to tropical South America and are grown all over the world as ornamentals. These fascinating plants share similar chemistry and traditional use with *Datura* and are used by Indigenous shamans in the Amazon and upland regions of Columbia and Peru. As with *Datura*, the major alkaloid

✦ Brugmansia suaveolens *from the author's garden.*

is scopolamine, representing 0.3 to 0.55 percent of the dry weight.[23] According to Schultes and Hofmann, there is a consistent association of this plant with graveyards, evil spirits, and the dead, perhaps owing to the violent nature of its intoxication.[24]

Because of its use as an ornamental in subtropical climates, there has been a resurgence of use of *Brugmansia* among the uninitiated as well as the initiated in warm states of the United States, such as Florida—so much so that after a record number of 112 reported poisonings in the state in 1994, the municipal government of Maitland, Florida, banned new plantings of *Brugmansia* in 1995.[25] I recently listened to a podcast in which two Florida-based modern witchcraft practitioners discussed the making of a *Brugmansia* flying ointment. It sounded like the doses were low in this particular preparation, which was intended to enhance psychic ability for spirit contact, one of the traditional uses of these plants.

Growing Angel's Trumpet

The wild ancestor being extinct, the *Brugmansia* species now exist only as cultivated plants for their beauty and intoxicating properties. Many varieties no longer produce seed and are propagated solely by cuttings. I grow several kinds, principally (I think) *B. sauveolens*. They are gorgeous plants, fast growing and stately, and they produce occasional flushes of large, yellow, trumpet-like flowers, which hang downward from the plant and, as with *Datura*, open at night to attract the moths that pollinate them. *Brugmansia* are harder to grow than *Daturas*. They are tender to frosts and freezes. Potted specimens need to be overwintered indoors, and plants in the ground need to be cut back and mulched heavily in the winter. They are heavy feeders and require regular watering and fertilization with both manure and liquid fertilizer for best growth. They will let you know that they are not happy by wilting and showing yellowing of the lower leaves. They are easily propagated by stem cuttings that should be rooted in water before planting in pots.

In my garden, the *Brugmansias* are planted in the front of our house near the door and serve as vegetal guardians, protecting our home. The large, lush plants shade our bedroom windows as well as my study. On cool evenings in the late summer or fall, sometimes hundreds of flowers open at once, filling the night air with their intoxicating perfume. We leave the windows open to allow the scent to enter the house. As with *Datura*, merely being in the presence of these plants brings a sensitive person into the presence of the genius of the plants, which I picture as distinctly feminine, alluring, seductive, mysterious, and definitely dangerous, but unlike the *Datura*, with a vigor and power all her own.

MAGICAL MANDRAKE (*MANDRAGORA OFFICINARUM*)

With the mandrake, we begin our discussion of the hexing herbs proper, those solanaceous plants famously used by European witches to make the famous flying ointments used to propel the consciousness out of the body on fantastic journeys to the Witches' Sabbath. The mandrake is a low-growing herbaceous perennial, belonging, as with the other plants we have discussed so far, to the family *Solanaceae*. It is native to Europe, Asia, and the Middle East. It produces blue to purple bell-shaped flowers followed by yellowish sweet-smelling berries, which are the famous aphrodisiac "love apples." The root contains principally the tropane alkaloids hyoscyamine and scopolamine.[26]

An ancient plant whose use has been documented for millennia, the mandrake

✦ *A lovely specimen of* Mandragora autumnalis, *the autumn mandrake, blooming in Sicily. Image by tato grasso.*

is mentioned in sources ranging from ancient Egypt and Mesopotamia, and it is even mentioned in the biblical Book of Genesis in connection with ancient Hebrew fertility magic:

> In the days of wheat harvest Reuben went and found mandrakes in the field, and brought them to his mother Leah. Then Rachel said to Leah, "Give me, I pray, some of your son's mandrakes." But she said to her, "Is it a small matter that you have taken away my husband? Would you take away my son's mandrakes, also?" Rachel said, "Then he may lie with you tonight for your son's mandrakes." When Jacob came in from the field that evening, Leah went out to meet him and said, "You must come in to me; for I have hired you with my son's mandrakes." So he lay with her that night and God hearkened unto Leah and she conceived and bore Jacob a fifth son.[27]

✦ *Women, perhaps Rachel and Leah, gathering mandrakes. Painting by Robert Bateman, 1870. Image by Wellcome Images, London.*

The precise role of the mandrakes (called in the Hebrew language *dudaim*, meaning "love apples") in this transaction is unclear. What is clear is that the biblical women considered them extremely valuable and that they were involved in previously childless Rachel's sudden acquisition of extraordinary fertility. The author of this part of Genesis included mandrakes in this part of the narrative for a reason, and we must assume that the intended audience would have understood the reference better than we can nearly three millennia later.

Indeed, mandrakes have been included in love and fertility magic since ancient times. As the Greek herbalist Dioscorides wrote, "Mandragoras has a root that seems to be a maker of love medicines."[28] According to the *Dictionary of Plant Lore* by D. C. Watts, the mandrake is sacred to Aphrodite, the goddess of love, and must be gathered on Friday, the day of the goddess. Theophrastus recommends the root be scraped and steeped in vinegar for "sleeplessness, and for love potions."[29] Egyptian and Arabic colloquial names for the plant further reinforce its reputation as an aphrodisiac with fertility-bringing properties,

✦ *Male and female mandrakes. Illustration likely from the* Hortus Sanitatis, *a medieval herbal and medical text published by Jacob Meydenback in Mainz, Germany, in 1491.*

namely "phallus of the field" and "devil's testicles."[30] In the *Dictionary of Plant Lore*, Watts reports that in the early twentieth century, Jewish people in America imported mandrake roots from the Near East as fertility talismans.

The mandrake has been esteemed as an anesthetic as well as an aphrodisiac. Dioscorides reports that it is given to those "who cannot sleep, or are seriously injured, and whom they wish to anesthetize to cut or cauterize." The potion is made by adding three pounds of the root into thirteen gallons of sweet wine, three cupfuls being given to the patient as anesthetic.[31] The mandrake was the primary surgical anesthetic until the discovery of ether by Valerius Cordus, a pupil of the celebrated herbalist and alchemist Paracelsus, in 1540. Ether was not perfected as a reliable surgical anesthetic until three hundred years later. A fifth-century work on medicine quoted by Watts states, "If anyone is to have a member amputated, cauterized or sawed, let him drink an ounce and a half in wine; he will sleep until the member is taken off, without either pain or sensation."[32]

It is perhaps for its hallucinogenic and intoxicating properties that it is best known. The plant is rich in mythology. Its roots, which often branch in a shape that resembles a human body with arms, legs, and a body, reportedly emits a deadly scream when unearthed, requiring an odd ritual wherein a dog is employed to pull out the root and dies from hearing the scream. As we have discussed previously, a plant that is surrounded with mystique, ritual, and folklore is most likely one with widespread use as a ceremonial intoxicant.

Mandrake, perhaps because of its association with witches and magic, as well as the humanoid form of its roots, has other purely magical uses. Especially esteemed were the roots from plants that grew under gallows, which were said to spring from the urine or semen voided by hanged men,[33] and which were dug up in the dead of night on Midsummer Eve, "when the year is on the turn and many plants are invested with mystic, but evanescent virtues."[34] The mandrake is harvested by tying the tail of a black dog to the plant by a string and holding out a piece of bread at a safe distance. The plant then "gives an awful yell" and the "poor dog drops dead in the process,"[35] reports folklorist James George Frazer in *Folklore in the Old Testament*.

The gallows mandrakes were especially valuable as familiar spirits and would act as oracles and bringers of good fortune to the bearer. After harvesting, the roots were washed in red wine and wrapped in red and white silk, then cared for by bathing them every Friday (remember, this is the day of Aphrodite) and giving them a new white shirt on the New Moon. The mandrake, says

Frazer, "will answer any questions you like to put to it concerning all future and secret matters. Henceforth you will have no enemies, you can never be poor, and if you had no children before you will have a quiver full of them afterwards. Would you be rich? All you need is to lay a piece of money beside the mandrake overnight; next morning you will find the coin doubled. But if you were to keep the Little Gallows Man long in your service, you must not overwork him, otherwise he will grow stale and might die."[36] The mandrake must be handed down from father to youngest son. In parts of Germany, the word also means a helping spirit in the form of an elf or goblin. The legend of the mandrake, in French, *mandragora*, is conflated with the hand of glory, *main-de-gloire*, the legendary hand of a condemned criminal, said to grant invisibility and unlock any door.

In German folklore the mandrake is known as *alraun*, which is cognate with our word *rune* and means "all wise one," and has the connotation of "witch" or "wizard." Jacob Grimm, in his monumental study *Teutonic Mythology*, says, "I have identified . . . the personified plant with that of wise women in our remotest antiquity." He adds that the alraun is pictured as a goddess of the crossroads.[37] Developing this line of thinking, Grimm adds, "The wise woman of the ancient Germans is called *Aliruna*, because she is *alja-runa*, and speaking secret words not understood by the common folk, has skill in both writing and magic; hers is the Gothic *runa*, hers the [Anglo-Saxon] *rûncræft*. . . . And this holy name of the heathen priestesses could easily be transferred to the holy herb which perhaps pertained to their ritual."[38] Do we see here a folk memory of an entheogenic cult and its "Heathen priestesses" (to repeat Grimm's evocative phrase) in the word *alraun*? This idea is not so farfetched, especially considering the role that mandrake and the other "hexing herbs" of similar pharmacology played in the subsequent history of European witchcraft, which could well be itself a successor to the Heathen entheogenic cults of northern Europe.

Mandrake was frequently cited as being one of the principal ingredients of the celebrated and infamous witches' ointments of the European witch cults. It was combined with a rogues' gallery of toxic and hallucinogenic plants, supposedly combined with baby fat or some other such shocking (and likely false, but after reading the recipes in the *Picatrix*, one never knows) ingredients. Andrés Laguna, the sixteenth-century physician of Pope Julius III, gives this account of his experience with one such ointment, which came into his possession during the trial of some accused witches:

[Its] odor was so heavy and offensive that it showed that it was composed of herbs cold and soporiferous in the ultimate degree [referring to the classification of herbs according to their primary Aristotelian qualities, which we discussed briefly in connection with astrology], which are hemlock, nightshade, henbane, and mandrake. Of which unguent, by way of a constable who was my friend, I managed to obtain a good canister-full, which I later . . . used to anoint from head to toe the wife of the hangman, who because of suspicions about her husband was totally unable to sleep, and tossed and turned and almost half mad [from lack of sleep and worry; for this reason, Laguna's help was sought]. . . . On being anointed, she slept such a profound sleep, with her eyes open like a rabbit . . . that I could not imagine how to awaken her . . . I so hurried her that at the end of thirty six hours she regained her senses and memory: although the first words she spoke were: "Why do you wake me at such an inopportune time? I was surrounded with all of the pleasures and delights of the world." And casting her eyes on her husband (who was there all stinking of hanged men), she said to him smiling: "Knavish one, know that I have made you a cuckold, and with a lover younger and better than you," and she said many other and very strange things. . . . From all this we can conjecture that all of the wretched witches do is phantasm caused by the very cold potions and unguents: which are of such a nature as to corrupt the memory and the imagination, that the wretched ones imagine and even very firmly believe, that they have done in a waking state all that which they have dreamt while sleeping.[39]

We have in the mandrake a plant whose use is very ancient and which is everywhere associated with magic, witchcraft, death, and the more liminal aspects of herbalism, those that work the boundary between poison and remedy. The mandrake in small amounts is intoxicating, in larger amounts stupefying, and in still larger amounts fatal. As Daniel Schulke says, "Progressive doses of poison are ever larger gates into the spirit realm, the last of which is physical death."[40] This applies to mandrake as much as it applies to all of the drugs of this class. But it must be remembered that all drugs are poisons. Learning how to use these plants is essential, necessary knowledge. We may yet come back to a time when modern surgical anesthetics are scarce or unavailable and those with knowledge of the cultivation and use of these misunderstood plants may be in demand once more. I will come back to this topic at the end of the chapter.

Having a mandrake as a pet or a specimen plant, a sort of familiar spirit in houseplant form, is an easy and safe thing to do. When your pet is two or three years old, it can be harvested and treated as an alraun as described above. Taking mandrake internally is not something I'm comfortable giving advice on, but there are several books out on the subject currently, including Coby Michael's helpful and informative *The Poison Path Herbal*.

Growing Mandrake

In terms of horticulture, the mandrake is easily grown, but is not reliably hardy in cold winter climates. I start about thirty seeds in the spring, at about the time I start my pepper plants. I buy dried seed and use bottom heat like we do for tomatoes and peppers. About half the seed germinates within three weeks. Other people report that germination can take months. When the little seedlings are big enough to handle, I move them to four-inch pots. When they outgrow their pots, I move them into larger pots. They need plenty of room to thrive and make big strong roots. The plants go dormant in the heat of summer but resprout again in the fall. Don't worry, they're not dead! They will come back. I once tried to set a bunch of plants out in my herb garden and they instantly went dormant and I couldn't find them. The weeds took over, and I assumed they were dead. When I tilled up that part of the garden the following spring, I found, to my dismay, little bits of chopped-up mandrakes. If you plant them outside, make sure you know where they are and give them deep rich soil to grow in, perhaps partially shaded. I would recommend keeping most of your plants in pots. I like to keep a pet mandrake plant in a pot in the home office where I do my writing; it adds a touch of magical atmosphere to the room.

HENBANE (*HYOSCYAMUS NIGER*)

Another of the so-called "hexing plants" used in the witches' oils and unguents, henbane likewise has a long history of use as medicine, hallucinogen, and intoxicant. It is a low-growing annual, biennial, or perennial herb, "A coarse, clammy, ill-smelling wayside weed,"[41] according to horticulturist L. H. Bailey. It has attractive yellow flowers with purple veins, followed by curious funnel-shaped, spiky seed pods. It was an official medicine for years, used as a sedative and antispasmodic. The principal alkaloids are the tropane alkaloids hyoscyamine, atropine, and scopolamine. *Hyoscyamus* is derived from the Greek words meaning "hog bean."

The plant was an ancient adulterant to beer and wine, added to increase the

✦ *Henbane,*
Hyoscyamus niger,
growing in the author's
garden.

intoxicating properties. According to Brian Muraresku, henbane seeds were found in combination with sprouted barley seeds, essential for the production of beer, during an excavation of a Celtic settlement in Germany.[42] Up until this discovery, henbane use in beer among ancient peoples was theorized, but no evidence had ever been found. The use of henbane to adulterate beer was widespread in Germany, where the plant is known as *bilsenkraut* and reportedly was a major ingredient in pilsner beer. According to Dale Pendell, the famous *Reinheitsgebot*—the celebrated German purity law of 1516—was intended particularly to prevent the adulteration of beer with henbane. Henbane, associated with Heathen practices, witchcraft, and licentiousness, was not welcome in pre-Reformation Germany, coming as it did one year before Martin Luther's Ninety-Five Theses were nailed to the door of the cathedral in Wittenberg on October 31, 1517. The Reinheitsgebot, although not the first food safety law or the first regulation to ensure the purity of beer, is considered by many to be the first drug law in European history.[43]

Dioscorides lists many medical uses for this important herb:

Ten grains of the seeds (taken in a drink with the seed of poppy, honey and water) do the same things, and are also good for coughs, mucus, fluid discharges of the eyes and their other disorders, and for women's excessive discharges [menstrual flow] and other discharges of blood. Pounded into small pieces with wine and applied, it is good for gout, inflated genitals, and breasts swollen in childbirth. It is effective mixed with other poultices made to stop pain. The leaves (made into little balls) are good to use in all medications—mixed with polenta or else applied by themselves. The fresh leaves (smeared on) are the most soothing of pain for all difficulties. A decoction of three or four (taken as a drink with wine) cures fevers called *epialae* [sudden]. Boiled like vegetables and a *tryblium* [plateful] eaten, they cause a mean disturbance of the senses. . . . The root (boiled with vinegar) is a mouth rinse for toothache.[44]

Dioscorides wrote his *Materia Medica* in the first century BCE, and this useful plant has remained an important medicine for the intervening two thousand years. Tincture and extract of *Hyoscyamus* were listed as official medicines in my 1936 copy of the *United States Pharmacopeia*.

The same useful seeds of the henbane plant were also believed to have been the intoxicant inhaled by the Oracle of Apollo at Delphi, to induce her prophetic trance state.[45] The smoke of the seeds was inhaled by twentieth-century experimenter Gustav Schenk, who describes the experience like this:

My teeth were clenched, and a dizzy rage took possession of me. I know that I trembled with horror; but I also know that I was permeated with a peculiar sense of well-being connected with the crazy sensation that my feet were growing lighter, expanding and breaking loose from my body. (This sensation of gradual body dissolution is typical of henbane poisoning.) Each part of my body seemed to be going off on its own. My head was growing independently larger, and I was seized with the fear that I was falling apart. At the same time I experienced an intoxicating feeling of flying.

The frightening certainty that my end was near through the dissolution of my body was counterbalanced by an animal joy in flight. I soared where my hallucinations—the clouds, the lowering sky, herds of beasts, falling leaves which were quite unlike ordinary leaves, billowing streamers of steam and rivers of molten metal—were swirling along.[46]

It is not surprising, after reading Schenk's account of visionary flight, that henbane was one of the major constituents of flying ointments.

Growing Henbane

The henbane plant is moderately easy to grow. The tiny seeds should be started indoors early and the plants potted into a cell flat or small pots and set out when the weather warms and all danger of frost is past. The soil should be well drained. Raised beds would be appropriate, as this plant hates wet feet and excess moisture and will promptly wilt and die if unhappy. It should be harvested when the tops are still green and the seeds are beginning to ripen. As with tobacco, the whole plant is covered with sticky gum, which is probably loaded with alkaloids, so wash your hands after handling to prevent accidental ingestion.

BELLADONNA (*ATROPA BELLADONNA*)

The last of the Solanaceous "hexing herbs" we will cover in this chapter is perhaps the most alluring and mysterious of them all and the one with the most sinister reputation: *Atropa belladonna*, deadly nightshade, once known as *dwale*. An upright perennial herb with fleshy stems growing to a height of three feet, it has nodding purple flowers followed by juicy-looking purple berries (which are reported to be very sweet tasting). The plant has long been a favorite among poisoners and makers of flying ointments and witches' brews. The genus's name, *Atropa*, given for the Fate who cuts the thread of life, and the specific name *belladonna*, Italian for "beautiful lady," reflect the historical use of the plant to dilate the pupils of women to enhance the beauty of their gaze and to make them more alluring. The gaze was believed to have the power of "fascination," or an occult power to influence the onlooker, similar to the evil eye but with different intent.[47] The names of belladonna associate it with a dark feminine archetype, linking the plant to the more nocturnal and chthonic goddesses who preside over witchcraft and the poisoner's art.

The chemistry of belladonna is similar to the other plants in the class, as are the effects. The principal alkaloids are again the tropane alkaloids, hyoscyamine and scopolamine, with atropine also present. The uses, too, were similar

◆ *An attractive specimen of belladonna,* Atropa belladonna, *growing wild in Poland. Image by peganum and Mielon.*

to the uses of mandrake and henbane. The plant was employed in the flying ointments, with mandrake and henbane, and applied topically to the skin and mucous membranes to induce the feeling of flying and the experience of spirit flight, as we have discussed. There was, however, one additional recorded use of these and similar ointments: theriomorphic transformation, or shape-shifting.

Shape-shifting, Entheogenic Plants, and European Shamanism

Many accounts from historical sources and witchcraft trial records regarding lycanthropy and other human-to-animal transformations relate the use of ointments with tropane-alkaloid-containing plants as major constituents. Lucius Apuleius, the Roman writer who was once put on trial for sorcery, wrote of an ointment that transformed a witch into an owl and another that turned his protagonist into an ass in his comic novel, *The Golden Ass*. Giambattista della Porta, in his *Natural Magick*, records a recipe for a potion intended to make the victim think he has transformed into a beast. Witchcraft trial records from France and Germany in particular detail the use of ointments containing one or more of the plants we have been discussing in the context of shape-shifting and lycanthropy, with one investigator declaring "all shape-shifting is mere hallucination."[48] Such "hallucinations" sometimes led people to commit the most violent crimes, if the historical sources can be believed.

In Germany in 1590, one Peter Stumpp was executed for the horrific crime of killing thirteen children and two women, the latter by ripping the unborn children from their wombs. Stumpp reportedly committed his crimes in the shape

✦ *An engraving of the Beast of Gévaudan, a suspected werewolf who ravaged the French province of the same name between 1764 and 1767.*

of a wolf, effecting his transformation by wearing a belt made of wolf skin and an ointment made from wolf's fat that contained entheogenic herbs. Stumpp's lurid exploits are by no means the only example of the theme of the magic wolf-skin belt in northern Europe, but it is the most violent of these reports. There are many other instances of wolf transformation in the literature of *trolldom*, the folk magic tradition of Scandinavia, many of which involve belts along with charms, which bring about the transformation.[49]

There may be more to this lycanthropy than meets the eye, however. The magical transformation into animal form is a characteristic of shamanism all over the world. Harner lists several reports from South America of shamans using hallucinogenic plants for theriomorphic transformation. The South American shamans often take the form of jaguars, but in North America and Europe the shamans take the forms of wolves and bears and other animals. In folklore involving the use of plant intoxicants to effect animal transformation, we can see a linking of motifs common to European witchcraft with global shamanic practice. Indeed, scholars such as Emma Wilby and Carlo Ginzburg have noted the similarity between shamanism in Siberia and the Americas and the witches and cunning folk. And, let us not forget my neighbor, the so-called Adena Wolf Shaman, we discussed in chapter 1, who was buried with a piece of a wolf jaw protruding from between his missing teeth and whose body was wrapped in some kind of animal hide, thought to have been a wolf hide.[50]

The shaman imitates the animal spirits as part of his ecstasy, writes Mircea Eliade, and this imitation can pass into possession by the animal spirit, or perhaps more accurately the *"taking possession of his helping spirits by a shaman.* It is the shaman who *turns himself* into an animal, just as he achieves a similar result by putting on an animal mask. Or, again, we might speak of a *new identity* for the shaman who becomes an animal spirit, and 'speaks,' sings or flies, like the animal and birds."[51] Often, the shaman would don costumes made of animal skins to more effectively accomplish the transformation. The shaman and the witch both lived in a magical, animist universe in which animal transformation was possible and expected. It was not, in the view of the members of these groups, "mere-hallucination" to the practitioners; the transformation was real. As Eliade points out:

> We must not forget that relations between the shaman (and indeed "primitive man" in general) are spiritual in nature and of a mystical intensity that a modern, desacralized mentality finds it difficult to imagine. For primitive man

donning an animal skin was becoming that animal, feeling himself transformed into an animal. We have seen that even today, shamans believe they can transform into animals. Little would be gained by recording the fact that shamans dressed up in animal skins. The important thing is what they felt when they dressed up like animals. We have reason to believe that this magical transformation resulted in a "going out of the self" that very often found expression in ecstatic experience. Imitating the gait of an animal or putting on the skin of an animal was acquiring a superhuman mode of being. There was no question of regressing into pure "animal life"; the animal with which the shaman identified himself was already charged with a mythology, it was, in fact, a mythical animal, the Ancestor or the Demiurge. By becoming this mythical animal, man became something greater and stronger than himself.[52]

In the medieval Scandinavian context we find the phenomena of the úlfhéðnar, the "wolfskin wearers,"[53] shape-shifting warriors who would fight in wolf form to gain power in battle, much like the *berserkir*, who would assume the shape and powers of a bear in battle, becoming nearly invincible. Directed outward at the enemies of the tribe, this transformation served the purposes of society, unlike the much later Peter Stumpp. One wonders what methods the *úlfhéðnar* and berserkir warriors used to effect their transformation, to call into themselves the power of the animal spirits, and if they also involved the use of entheogenic plants.

We also find, alongside the warrior magic practices of animal transformation, practices of soul journeying in animal form, which were part of *seiðr*, or sorcery, practices. In the Old Norse Heathen psychology, the soul was seen as a composite, having various parts, as we have discussed in other traditions. Two aspects of the soul, the *hugr* and the *hamr*, are fundamental to the understanding of the metaphysics of shape-shifting in this context. Hugr refers to a person's temperament, character, desire, or will, and a person with a strong hugr could act magically at a distance. Hamr, on the other hand, was the "skin," or the shape the hugr could take on while traveling outside of the body. A person adept at this practice was known as a hamleypa—literally, a "skin-jumper." According to Catharina Raudvere in her fascinating monograph, *Trolldómr in Early Medieval Scandinavia*, "the ability to change shape and act out of the ordinary body in a new guise was an inborn character or acquired through learning."[54]

For the shaman or European lycanthrope, through the donning of the skin or the use of the ointment, the transformation was complete. We don't know why

◆ *A 1510 engraving of witches preparing ointment to travel to the Sabbath,*
by Hans Baldung.

the medieval werewolves engaged in this bizarre and violent behavior, but we can imagine that they were invoking an aggressive archetype of raw power, like their *úlfhéðnar* and berserkir predecessors. Unlike the ritual transformations of the shamans, and the shape-shifting Scandinavian warriors, their behavior was not sanctioned by their community. I wonder if this is the whole story, and I suspect that there is much more. Like the alraun and its possible connection to the priestesses of ancient Germany, could the flying and lycanthropy ointments be the folk remains of a deep and ancient European shamanic practice? This might be a practice like that of the hamleypa, but one that was gradually severed from its original context and purpose and became a magic later engaged in for pure personal power. Its original intention, in its proper cultural context, may have been to restore and maintain the precarious balance between humankind, the gods, and the natural world.

Perhaps, beyond the destructive goals of warrior magic, one of the original purposes of shape-shifting and soul traveling was something like the *benandanti* heretics of northern Italy described by Carlo Ginzburg in his book *Ecstasies: Deciphering the Witches' Sabbath*. The benandanti, according to Ginzburg, were shaman-like visionaries who fought the forces of evil and chaos in trance states to preserve the crops from spiritual harm. More germane to the present topic, however, are the werewolves of Livonia, in Germany, who like the benandanti, Ginzburg tells us, did battle against the devil and sorcerers on three holy nights every year to preserve the fertility of the fields. If the werewolves failed at their appointed tasks, famine resulted and the crops and harvest failed.[55] One wonders, too, if the visionary night battles of these ecstatic warriors really did help their crops thrive, if these agrarian seers could see the harmful forces that assailed their growing crops and communities on the imaginal plane and stop them before they became manifest.

Having investigated the botany, horticulture, and historical use of several of the most fascinating troublemakers of the nightshade family of plants, we might ask: What is their relevance to practical geurgy, to the art of the cunning farmer? For one thing, some of these plants are rare in cultivation, so preserving them as specimens is important work. They are extremely important as medicines, with powerful effects on the body that can be extremely effective remedies for common ailments such as asthma and diarrhea, which can be very serious if left untreated. Such uses of these plants should obviously not be attempted except by trained professionals. However, observing the fragile state of our world today, with a very uncertain future for civilized humanity, it seems to me like the knowledge of how to propagate and use these strong medicines may again be essential knowledge in the future. It would be important to have a seed stock as well. Tobacco use also isn't going to go away in the event of a new dark age, and those who know how to grow it would be at an advantage. Of course, beyond the medical and economic uses of these plants, you should remember the spiritual benefits of these potent plant allies for magical practice. We must cultivate the friendship of their dark and alluring genii as offering plants and as specimen plants.

17
PLANT MAGIC

In this chapter, we will be continuing our exploration of magical plants. In the last few chapters, we discussed various plants that induce an altered state of consciousness and how magical traditions have made use of them. Here, we will be investigating nonpharmacological use of plants in the work of the cunning farmer, plants as purely *materia magica*, spell ingredients, and the like. Why would magical practitioners use cinquefoil (*Potentilla reptans*) for eloquence, as the medieval miscellany known in its English translation as *The Book of Secrets of Albertus Magnus* recommends? What is it about this unassuming little weed that grows in the rocky and eroded soil around the edges of my yard that makes it useful for Mercurial work in some traditions or for money-drawing spells in other traditions? Is it a folk knowledge of some obscure pharmacological properties the herb possesses, or is it the form the plant takes, having a five-fingered, hand-shaped "palmate" leaf that indicates its ability to "grab" things and bring them to the spell-worker? In some cases, it is the shape or form of the plants or its color that indicates its medicinal or magical use.

OCCULT VIRTUE IN BOTANICAL *MATERIA MEDICA*

Ancient people believed, for instance, that walnuts were good for the brain because of their resemblance to that organ, or that beets were good for the blood because of their brilliant red juice. Many times the theory is correct, and the plant in question is indeed helpful to the organ or other anatomical structures or fluids it resembles. The idea that the outward form of a plant signifies its medicinal or magical virtue is known as the "doctrine of signatures." Although widely discredited by skeptics today, the idea is an ancient one, having been mentioned by authorities in antiquity such as Pliny and Dioscorides. As German mystic Jakob Böhme said, "Everything as it is inwardly (its innate virtue and quality) so it is outwardly signed, therefore the greatest understanding lies in the signature, wherein man may not only learn to known himself, but also may learn to know the essence of all essences; for by the external form of all creatures . . . the hidden spirit is known." He continues, "Everything has its mouth to manifestation; and this is the language of nature, whence everything speaks out of its property, and continually manifests, declares, and sets forth itself for what it is good and profitable."[1]

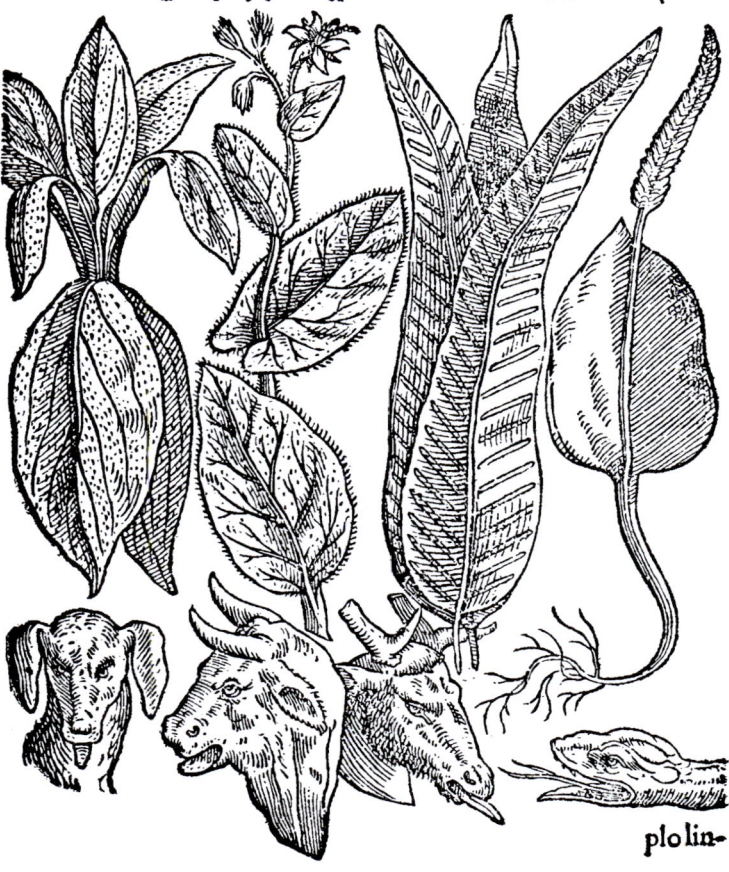

◆ *The doctrine of signatures illustrated. Facsimile by Lucien Marcus Underwood after the 1591 work of Giambattista della Porta.*

In the premodern, enchanted worldview that is still prevalent among mystical philosophers such as Böhme and Paracelsus, the doctrine of signatures showed the Divine Providence at work in the world. The intelligence of the creator showed itself in creation, there for those with an eye to see. Indeed, the biblical Book of Genesis states, "Behold, I have given you every plant yielding seed which is upon the face of the Earth, and every tree with seed in its fruit; you shall have them for food."[2] And by extension, in the postlapsarian world, not only food but medicine and magic too.

Although widely sneered at today, the doctrine of signatures has been studied by the modern ethnobotanist Bradley Bennett. In a 2007 paper in the journal *Economic Botany*, he examines the issue critically and concludes that, although there is no way of testing whether or not the doctrine ever led to the discovery of medicinal uses of any plants, it is a post hoc attribution—that is, it is applied to medicinal plants after the fact, as a mnemonic device. This is very useful in the preliterate societies that depend on plants for medicine.[3] How those preliterate societies discovered the properties of the plants they use remains an open question. There are some who believe that the plants themselves may have taught humans how to use them.

From the doctrine of signatures, we move to the doctrine of cosmic sympathy. In the Neoplatonic and Hermetic understanding of the cosmos that underpins the Western magical tradition, we see the idea of chains of sympathy that extend throughout the whole universe, known in Greek as *serai* (chains or series), that link the gods with the archetypal world, the planetary spheres, the sublunar realm, and all physical reality. Through these chains, divine intellectual influence descends and ascends the various levels of being, and by making use of these connections, the theurgist can bring higher influences to bear in his or her material affairs, summon divine entities, or ascend spiritually to higher levels.

There is a divine intelligence that binds the entire cosmos together in all of its parts, from the furthest galaxies to the smallest terrestrial herbs. This binding influence, in the Hermetic view, is not random but exhibits an archetypal order. To quote the Hermetic treatise *Asclepius*:

> The heavens, a perceptible god, administer all bodies whose growth and decline have been charged by the Sun and Moon. But god, who is their maker, is himself governor of heaven and of soul itself and of all things that are in the world. From all these, governed by the same god, a continuous influence carries through the world and through the soul of all kinds and all forms throughout nature.[4]

Matter is a receptacle for the divine influence, which is arranged in chains of influence from the stars and constellations. It is important to keep that in mind when the Hermetic texts use the word *God* and even the male pronoun that the word refers to a nongendered first principle that is beyond duality. God, says the *Asclepius*:

> dispenses and distributes his bounty—consciousness, soul, and life—to all forms and kinds in the world, so the world grants and supplies all that mortals deem good, the succession of seasons, fruits emerging, growing and ripening, and other such things. And thus seated atop the summit of the highest heaven, god is everywhere and surveys everything all around. For there is a starless place beyond heaven and remote from bodily objects. The one who dispenses [life] whom we call Jupiter [Zeus, the demiurge], occupies the place between heaven and earth. But Jupiter Plutonius rules over the earth and sea, and it is he who nourishes mortal things that bear fruit. The power of these gods invigorates crops, trees, and soil, but the power of other gods will be distributed through all things that are.[5]

We see here the Hermetic notion of a divine that is at once transcendent and immanent. Three divine beings are mentioned in this passage: a total transcendent divine principle that is unnamed and exists outside of the cosmos, beyond space and time; a hypercosmic demiurge; and Jupiter Plutonius, the god of the material realm, who administers the gods who control the growth of plants and crops and all other material affairs. The linking of the disparate parts of the cosmos by the human intellect connects the links of the chain so the divine influence can flow.

The lowest plants and stones are connected to the gods, as Proclus, the Neoplatonist—here quoting Thales, the pre-Socratic philosopher—said: "All things are full of gods, then: things in earth are full of super celestials; and each chain continues abounding up to its final members."[6] Proclus states that the ancient priests observed the sympathies of the universe, "the sympathy that all visible things have for one another and for the invisible powers—and have also framed their priestly knowledge. For they were amazed to see the last in the first and the very first in the last; in heaven they saw things acting causally and in a heavenly manner, in the earth heavenly things in an earthly manner."[7] Proclus describes the Sun-seeking motion of the heliotrope (perhaps *Valeriana officinalis*) and the Moon-seeking of the *selenotrope*, the identity of which is unclear. Proclus says beautifully, "All things pray according to their own ability and sing hymns, either intellectually or rationally or sensibly, to the heads of entire chains [the gods]. And since the heliotrope is also moved toward that to which readily opens, if anyone hears it striking the air as it moves about, he perceives in the sound that it offers to the King the kind of hymn a plant can sing."[8] The love and longing, the eros, of lower things for higher is the foundation of prayer in all things; plant and human alike seek their divine source in the spheres. The properties of the luminaries and planets are distributed throughout all of the kingdoms of the sublunary world, in plants, in minerals, and in animals. It is not our place here to discuss the latter two kingdoms, but Proclus discusses theurgic rites, wherein the theurgist ritually mixes the material receptacles in order to attract the divine influences, calling forth the power of the god.

Iamblichus, another of the theurgic Neoplatonist philosophers of late antiquity, states the theory succinctly: "The theurgic art in many cases links together stones, plants, aromatic substances and other such things that are sacred, perfect, and godlike, and then from all these composes an integrated and pure receptacle." He adds, "The sacrifice of such material rouses up the gods to manifestation, summons them to reception, welcomes them when they appear, and ensures their perfect reception."[9] This is the rationale for the selection of *materia magica* in all types of spellwork: the linking of harmonious substances to attract spiritual forces congenial to the goal of the operator.

Love in the Hermetic Universe

During the Renaissance, the astrologer and philosopher Marsilio Ficino, the genius responsible for translating the works of Plato and the entire Hermetic corpus, as well as many other Greek philosophical works that had been lost to the Latin West since antiquity, reintroduced and developed the theory of Platonic astral magic in several of his works, notably his treatise on astrological medicine and magic, *Three Books on Life*. The ancient doctrine of magic is presented succinctly in the third book, where he states, "Down from every single star there hangs its own series of things proper to it down to the lowest." For Ficino, it is love, eros, desire that attracts these influences, the celestial to the earthly receptacle. He consistently uses the language of sexuality to describe the work of the magician. "Yes," he writes, "everywhere nature is a sorceress . . . in that she entices particular things by particular foods, just as she attracts heavy things by the power of the earth's center, light things by the power of the Moon's sphere, leaves by heat, roots by moisture, and so on. By means of this attraction the wise men of

◆ *Preface to Marsilio Ficino's* Three Books on Life *(1489)*
with a portrait of Ficino.

India testify, the world binds itself together; and they say that the world is an animal that is masculine and at the same feminine throughout and that it links itself in the mutual love of its members and so holds together."[10] The entire cosmos is infused with the mystery of sexual desire and polarity in Ficino's version of the animate cosmos. This universe of the Platonists and the Hermeticists, in which eros is the glue which holds all things together, is a far cry from the cold dead matter of materialist science. "Orpheus called the very nature of the cosmos and the cosmic Zeus masculine and feminine. So eager is the world for the mutual union of its parts," continues Ficino.[11] This interplay of gendered opposites may fall uneasily on modern ears; polarity, however, is nearly a universal concept in many of the spiritual systems of the world. The ancient Chinese Taoist treatise, the Tao Te Ching, says, "The ten thousand things (all of the various parts of the cosmos) carry yin and embrace yang. They achieve harmony by combining these forces."[12] The sacred marriage of Heaven and Earth is also a potent symbol of the erotic interaction of cosmic polarities that gives rise to life itself, the *logos spermatikos* falling on the fertile matrix from which all becoming arises.

The living things of the sublunar realm receive the planetary logoi passing from the source through the planetary spheres, through the medium of the celestial pneuma, or spirit, as do all things in unequal measure. Some of them accumulate more of one planet's influence than the others, becoming, in effect, a symbol of that planet, a conduit for the virtues of that planet. The plants also retain influences from the four elements, with one element or elemental quality predominating. Thus, in this system, a plant can be cold or hot, moist or dry, as well as solar, lunar, venusian, martial, jovial, or saturnine, depending on its nature and qualities. This is the doctrine of occult virtue as it pertains to plants. As Henry Cornelius Agrippa puts it in the *First Book of Occult Philosophy*, "There is, therefore, a wonderful virtue, and operation in every herb, and stone, but greater in a star, beyond which, even from the governing celestial intelligences everything receiveth, and obtains many things for itself, especially from the Supreme Cause."[13] Remember, "all things are full of gods" through the pneuma that binds the influences of the world soul to matter. It remains for the operator to combine the proper influences to bring forth the desired result.

The theurgist or the magician compounds the *materia*, the herbs, stones, and bones, seeds and incense, colored candles and poetic invocations, and the star daimons, representing the gods themselves. The celestial influences are attracted as if by erotic desire to the sympathetic material receptacle. Whatever influence the worker wants to attract, the theory is the same: Venusian *daimones* are attracted to sweet-smelling herbs, incense, roses, and beautiful poetry; Saturnian ones to foul smells and poisonous, soporific plants, black candles, and dirgelike music; Martial spirits are invoked with fiery and pungent herbs, chilis and black pepper, red candles and warlike music. We can attract the spirits of the Sun with gold and yellow candles, frankincense, marigolds and sunflowers, the spirits of the Moon with silver and white or purple candles, selenite, white poppy seed, and camphor. Jupiterian daimons are attracted to amethyst and tin, the color blue, lignum aloes, storax and benzoin, and Mercurial ones to the color orange, tin, agates, topaz, mastic, cloves, and cinquefoil. A full list of correspondences may be found in Agrippa's *Three Books of Occult Philosophy*, particularly in the first one.

Rather than digress into specific recipes for invoking all of the planets, I would like to keep the focus here on examining the rationale for using certain plants for magical purposes, as spirit conduits, or as spiritual entities in their own right. So this section is not meant to be exhaustive but more of a guide to further study. Agrippa lists all of these correspondences in the first book, but

♦ *Compounding herbal medicine, from an Arabic manuscript (c. 1224) of Dioscorides's* De Materia Medica *(a Greek text from the first century BCE).*

examples of the specific incense and plants assigned to the planets are in order. Some are listed as belonging to several planets, and other authorities—such as (pseudo) Albertus Magnus, Paracelsus, and Culpeper—list other attributions. But since we are discussing magic and not medicine (although the two subjects are deeply intertwined; indeed they are not two subjects, in the period we are discussing) we will stick with Agrippa. According to Agrippa (paraphrased from Henry Cornelius Agrippa, *Three Books of Occult Philosophy*), starting with the luminaries:

The Sun

Magical works pertaining to the Sun are those which involve the acquisition of wealth and influence and fame.

Herbs of the Sun include the marigold, the peony, celandine, balm, ginger, gentian, dittany, vervain, bay tree, cedar, ash tree, palm trees, ivy, vine, mint, mastic, zedoary, saffron, balsam, amber, musk, lignum, aloes, cloves, cinnamon, calamus, pepper, frankincense, marjoram, and libanotis (identified with rosemary), and lemon balm.

The Moon

Magical works pertaining to the Moon are those involving intuition, dreams, psychic powers, growth of plants, and water.

Herbs of the Moon include hyssop, rosemary, *agnus castus*, the olive tree, and poppies.

Saturn

Magical works of Saturn pertain to restriction and constriction, destruction, withering, decay, cursing, binding, and, paradoxically, construction and farming.

Herbs of Saturn include the daffodil (probably asphodel), dragonwort, rue, cumin, hellebore, mandrake, opium, "and those things which stupefy" (which would place *Datura*, henbane, and belladonna all under the influence of Saturn, as other authorities do), the fig tree, pine tree, cypress, and mullein.

Jupiter

Works pertaining to Jupiter include increasing wealth, happiness, honor, benevolence, prosperity, religion, and a "glorious life," according to Agrippa.

Herbs of Jupiter include basil, bugloss, mace, spike, mints, mastic, elecam-
pane, violet, darnel, the poplar tree, the oak, aesculus (horse chestnut), the
holm tree, the beech tree, hazel (also often attributed to Mercury), the service tree, the white fig tree, the pear tree, the apple tree, the vine (again),
the plum tree, ash, the dog tree, the olive tree, all types of "corn" or grain,
and sage.

Mars

Works pertaining to Mars include those of "boldness, courage and good fortune in wars, and contentions," as well as "making a man powerful in good
and evil so that he be feared by all."

Herbs of Mars include hellebore, garlic, euphorbium, ammoniac, radish, laurel, wolfsbane, scammony, "and all such as are poisonous, by reason of too
much heat, and those that are beset round about with prickles, or touching
the skin burn it, prick it, or make it swell, as cardis, the nettle, crowfoot,
and such as being eaten cause tears, as onions, *ascolonia* (perhaps shallots),
leeks, mustard seed, and all thorny trees." One could easily place red chilis
and tobacco here, but in Agrippa's time they had not yet been brought to
Europe from the Americas.

Venus

Works pertaining to Venus are those of love, sex, friendship, and beauty, as
well as making one "pleasant, jocund, strong, cheerful, and [bringing]
beauty."

Herbs of Venus include vervain, violet, maidenhair, valerian, thyme, gum
ladanum, ambergris (not an herb), sanders (sandalwood), coriander, "and
all sweet perfumes and delightful and sweet fruits, as sweet pears and
pomegranates, which the poets say in Cyprus, was first sown by Venus.
Also the rose of *Lucifer* [a Latin epithet of Venus, as the Morning Star] was
dedicated to her, also the myrtle tree of Hesperus."

Mercury

Works pertaining to Mercury include those that "conferreth knowledge, eloquence, diligence in merchandising and gain; moreover [those that] beget
peace and concord, and [those that] cure fevers," as well as those for healing and astrology or the occult in general.

Herbs of Mercury include hazel, "five-leaved grass" or cinquefoil, herb mercury, fumitory, pimpernel, marjoram, parsley, and "such as have shorter and less leaves, being compounded of mixed natures."[14]

This is just a brief sample of planetary attributions of magical plants. There are many, many more. For example, in his book of kabbalistic correspondences, *Liber 777*, Aleister Crowley has several tables giving the attributions to the ten Sephiroth of the kabbalistic cosmological diagram known to most magicians, the Tree of Life and twenty-two paths on the Tree of Life. He lists various "plants, real or imaginary" in table 39 and "vegetable drugs" in table 43.[15] The attributions of the plants of those seven of the ten Sephiroth that correspond to the planets don't match any of the older correspondences, and, not surprisingly for those who are aware of his public persona, include all of the more naughty plants, the intoxicating ones, placed prominently in various Sephiroth, whereas older sources place them all mostly in Saturn due to their cold and soporific nature.

An extremely helpful, recently published resource on the subject of magical plants is the second volume of David Rankine's *Grimoire Encyclopaedia*, which collates the magical uses of herbs as *materia magica* and incense in most of the major historical books of magic of the last two millennia. The work is a monumental piece of scholarship and is a fascinating tool for the comparison of systems of magical practice. A brief glance at the entries for a few of the various herbs quickly reveals the diversity of uses each of these herbs were put to, which may be surprising, considering the Neoplatonic theory of innate meaning and inherent sympathy we discussed earlier. This brings us to the next approach to the historical use of magical plants.

THE SPECIALIZED NATURES OF PLANTS

Although plants may exhibit planetary virtues and receive the influences of the stars, they also have a unique and individual genius, an indwelling spirit. Although two plants may be attributed to the same planet, their virtue is their own and flows from the land whereon they grew, as well as the heavens. Albertus Magnus, the medieval philosopher and alchemist, believed, according to Lynn Thorndike:

> The properties of plants are produced by the combination of five virtues: that of the element which preponderates in the composition of the plant, the cooperating virtue of the other elements which are mixed with it, the virtue of the

◆ *Botanical illustrations, from a Persian manuscript (c. 1595) of*
Dioscorides's De Materia Medica.

proportion in which they are mixed, the influence of the stars, and the virtue of the vegetable soul. "*The virtue of the place (where the plant grows) and the virtue of the surrounding air are also effective* [emphasis mine], but they do not enter into the plant's nature so essentially as the aforesaid five virtues." "Its specific form," upon which its occult virtues largely depend, is given to the plant by the motion of the heavens, especially by the movement of the planets through the circle of the zodiac, and their position in relation to the fixed stars. Plants receive this influence at the time of their formation, when vapors, potentially seminal and formative, ascend from the depths of earth and meet the dewy air as it descends.[16]

The plant is thus an individual, a product of its type, as well as its circumstances, an ensouled being with whom the practitioner must enter into a relationship in order to be most effective. One can go to a store and buy a bottle of herbal capsules for various ailments, but how much more effective must the virtue of the plant be when planted, cultivated, and harvested with care and attention to the

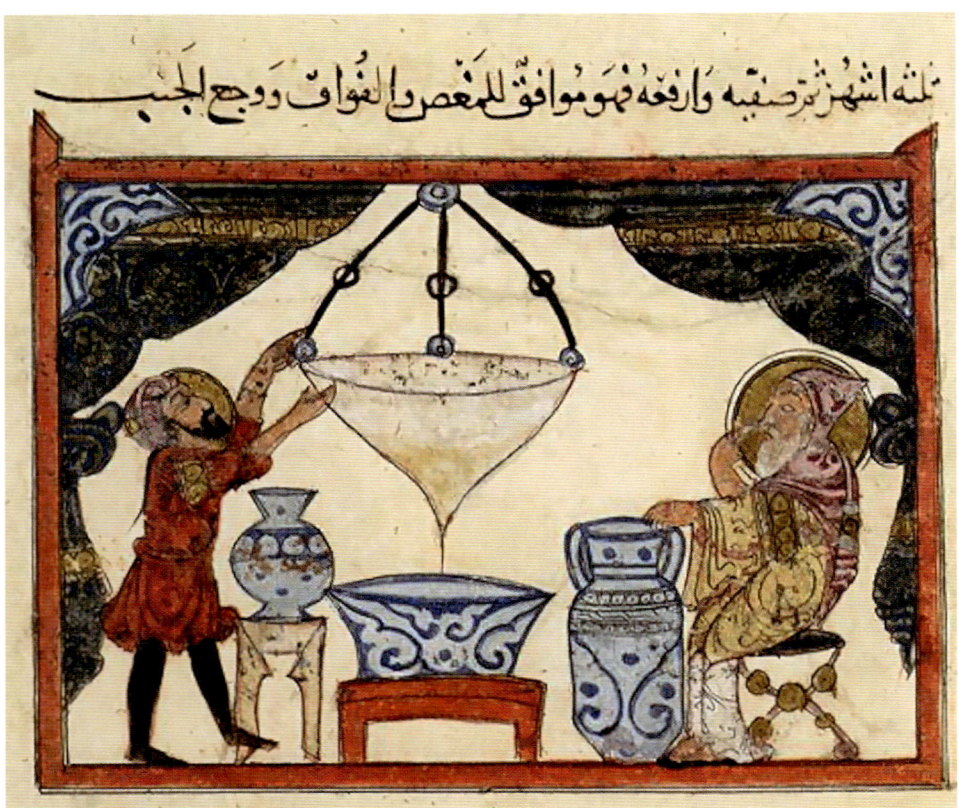

✦ *Doctors preparing medicine, from an Arabic manuscript (c. 1224) of Dioscorides's* De Materia Medica.

movements of the celestial forces whose influences it receives and transmits? Or, working with wild plants, when they are harvested with the same in mind, the celestial rhythms of the planetary movements, as well as being harvested with kindness and respect?

The Neoplatonic system, with its chains of stellar and planetary influences, places the plant within a cosmological framework that is enchanted and animist. All beings are endowed with spirit and soul by participation in the world soul and receiving the enlivening pneuma. This learned system of magic and theurgy required scholarly knowledge and advanced training in the mysteries of astrology in order to predict the optimal moments to perform rituals, when the planetary influences were strongest. By and large, this knowledge would have been unavailable to common rural folk for most of history. Indigenous and shamanic traditions of folk magic practice had to rely on the easily observable motions of the Sun, Moon, planets, and stars, as well as an intimate relationship with the plant people themselves. Prayers and offerings to the spirit of the plant and visionary techniques of communication with the spiritual entities within the plant can also bring about a level of magical efficacy.

Animist cultures have an expanded notion of personhood that applies to all objects in the natural world, not just plants and animals but also rocks, the sky, weather phenomena, and rivers, streams, and other water features. People within those cultures live in relationship with their environment as members of a community of beings. Their magical practice flows from this connectivity with the natural world, and this is evident in their attitude of reverence and kinship. Indigenous cultures view plants, animals, and humans as relatives, as deserving of care and respect, even as humans consume them as food or in other ways.

Plants are also seen as having a spiritual and metaphysical dimension in Indigenous cultures. According to Nancy Turner in her study of plant use among the First Nations peoples of western Canada, *Ancient Pathways, Ancestral Knowledge*:

> As entities of the spirit world, plants have innate powers to influence human destiny and fortune. This influence can be negative or positive, depending upon the actions and attitudes of the people who encounter them. Thus, plants can bring good fortune and success in hunting, fishing, root digging, berry picking, basketry, or canoe making. Plants can also protect people against evil forces, illness, or even death. On the other hand, if people neglect to show appropriate respect and appreciation in their actions, and if they disregard certain taboos or

constraints, plants can cause serious harm to them. The same is true of animals, rivers, and mountains.[17]

The plant spirits are honored with ritual action, prayer, and reverence, in order to keep the relationship on positive terms.

Greek mythology retained a heritage of animist plant spirits into the classical period. The tales of tree spirits such as dryads and hamadryads, nymphs, and mortals transformed into plants represent, I believe, a remnant tradition of vegetal animism dating from a more shamanistic phase of culture. Both the pre-Hellenic people of Greece and the Indo-European ancestors of the Hellenes, in earlier times, before the advent of cities, must have lived in small tribal groups engaging in pastoralism and small-scale agriculture, living closer to the land. The spirits of place would have figured more in their religious and spiritual life. A brief survey of Greek myths turns up a variety of plant spirits and tree nymphs who would have been—we can presume by observing cultures in similar situations, even today—given reverence and offerings in exchange for their gifts, and they would have made appearances in the visionary states of the shamans and healers of the tribe, as teachers and healing allies.

Robert Graves cites numerous examples of plant spirits in his classic work *The Greek Myths*. First we have the tree spirits, oak nymphs, known as *Dryad*, *Mēliae*, apple tree nymphs, *Caryatids*, nut tree nymphs, *Meliae*, ash nymphs. We also have Daphne, daughter of the river Peneius and beloved of Apollo, transformed into a laurel. Apollo seduced Dryope, an oak nymph. Dionysus was, according to some versions of his myth, born to the oak goddess Dione, and fathered by Zeus. Male mortals were transformed into the flowers narcissus and hyacinth.

The bawdy tales of seduction in classical mythology hide an esoteric meaning that relates back to the cosmic erotic attraction already discussed, the Love of Empedocles, primal force of the universe, which binds all things together. Apollo and Zeus and their various liaisons with nymphs are the mythic memory a patriarchal, civilized, urban culture had of its more primal, animist past. The sexual union of Zeus, the rain and thunder god, and Apollo, the Sun god, with the nature spirits, the nymphs and dryads, gives rise to all life in its various forms, exactly as described by Ficino. Love binds, and Strife sunders all that has been bound, in the eternal play of being and becoming. This is the great secret of the Pagan mystery cults, played out in many of the most basic myths and cleverly hidden in plain sight.

◆ The Dryad, *by Evelyn De Morgan, c. 1884–85.*

PLANT MAGIC ACROSS CULTURES

Communicating directly with the plant intelligences as allies or familiar spirits is a part of shamanic practice, and there is a growing body of literature around these practices among neo-shamanic practitioners in the West. I believe that this was a part of ancient Pagan and Heathen cultures of western and northern Europe as well and that some of these practices may have survived into the early modern period before being eliminated first by witch hunts and then secularism. Folkloric evidence is extremely scarce, owing to the fact that the practitioners were rural people lacking conventional education and didn't keep written records, unlike the learned herbalists, alchemists, and magicians of the Middle Ages and Renaissance. The evidence is nevertheless there, written between the lines in myths, folktales, and the records of witch trials.

James George Frazer in *The Golden Bough* lists various examples of spirits of vegetation from around the world. It is not our concern here to find examples of gods of vegetation in general, but rather the genii of particular plants with which the magician may interact. The lore of the alraun, the spirit of the mandrake that becomes a familiar spirit for a magician, is the type of interaction we are seeking evidence for. Surely if the mandrake had its devotees, then the oak tree, the ash, the cedar, or the humble herbs of the field must have had their share of interaction with the Heathen healers and shamans who are the spiritual ancestors of the cunning farmer. For people whose survival depended on plants and the plant world for food, medicine, and other needs (a situation that still hasn't changed), harmony with the spirit forces of the plant world is of paramount importance. That is why I believe it is important to make the case today for a reevaluation of our relationship with plants and with all of nature, as well as to show how for Western people these are ancestral practices, not just quaint practices of tribes from the other side of the world.

Plant spirits play a role in Germanic and Norse mythology, where they are known in English as *plant wights*, from an Anglo-Saxon word meaning a person or a being but having the connotation of a spirit. The plant wights are a subset of land wights or *landvættir*, as they were known in Old Norse, where they formed an important part of the animistic Heathen religion of northern Europe, which comes down to us in fragmentary form in sources such as the Icelandic Eddas and Sagas, as well as Germanic and Anglo-Saxon poetry.

In Nicholaj De Mattos Frisvold's excellent and highly recommended book *Trollrún*, the author mentions the importance of plant wights in the magical practice of the animist Scandinavian Heathens, saying that the cunning folk of the North knew "how to commune with and use the spirits of wood, stones, plants and beasts to affect the world."[18] The northern Europeans, like the Indigenous Americans mentioned above, practiced "speaking to crops and animals, reciting poems and giving offerings to trees, stones, houses and plants . . . because everything was ensouled and held a consciousness. The more the practitioner generated a sympathetic energy, the more he or she would generate a bond that made the *trolldom* effective."[19] Echoing the opinion of Albertus Magnus, that the land itself left an imprint on the virtues of the plants that grew on it, De Mattos Frisvold says, "It was the *vættir* that gave a particular place, plant, stone, waterfall, pond, mound, and so forth its particular energy or vibration."[20] The genius loci is an integral part of the genius of the plant. Veneration and reverence to both is important to the practice of the cunning farmer.

Remnants of this tradition of European animist vegetal shamanism can perhaps be seen in names of "witches' devils" quoted by Jacob Grimm in his book *Teutonic Mythology*. Grimm noted that the names of the devils, or familiar spirits, from the records of German witch trials bore a resemblance to the names of elves and kobolds from Germanic folklore. He said that such names "may afford us a welcome glimpse into the old elvish domestic economy itself. Some are taken from healing herbs and flowers, and are certainly the product of an innocent, not diabolic fancy." The names listed include *Wolgemut* (oregano), *Schöne* (daisy), *Wiegetritt* (plantain), *Blümchenblau, Peterlein* (parsley), oak, linden, pear, rue, elder, and others.[21] These names remind one of the fairies from *A Midsummer Night's Dream*, like Peaseblossom and Mustardseed. Whether or not the German witches were preserving and continuing a practice of incorporating the plant wights into their magical and healing practice as familiar spirits we will likely never know, but there are practitioners preserving these methods of consulting with the spirits of the plants themselves, up to the present day. It is of course regrettable and shameful that these presumably harmless members of what appears to have been a medieval version of Shamanic folk healing ran afoul of the forces of repression and were dragged before the office of the Inquisition.

These intuitive practitioners—heirs to the animist shamanic traditions that were once common throughout the whole world, when pre-urban humanity lived in close contact with the environment—saw the world very differently from the modern materialist understanding common today. As Rudolf Steiner said in a 1923 lecture:

When spiritual vision is directed to the plant-world, we are immediately led to a whole host of beings, which were known and recognized in the old times of instinctive clairvoyance, but which were afterwards forgotten and today remain only as names used by the poet, names to which modern man ascribes no reality. To the same degree, however, in which we deny reality to the beings which whirl and weave around the plants, to that degree do we lose understanding of the plant-world. This understanding of the plant-world, which, for instance, would be so necessary for the art of healing, has been entirely lost to present-day humanity.[22]

Applying the "instinctive clairvoyance" Steiner spoke of to the spiritual potencies of the plant world allowed the magi and the healers of the past to uncover their occult properties as *materia magica*. These occult properties are distinct from, but sometimes related to, the pharmacological and physical properties of the plants in question. For instance, the solanaceous herbs we discussed in the last chapter were believed to be under the government of the planet Saturn and to have an effect on the organism that was cooling in the extreme. Because of their toxic and cooling properties they were assigned to the greater malefic, Saturn. Their nonpharmacological use as ingredients in spells most often stems from their Saturnine virtues, such as in the construction of a ring to draw down the power of the fixed Behenian star in the left wing of the constellation the Crow, *Ala Corvus Sinistra*, which is of the nature of Saturn and Mars, for example. Ficino recommends the use of henbane and burdock, or sorrel—henbane, perhaps because it is of the nature of Saturn as well, though burdock and sorrel are of the nature of Venus, which would seem to be contrary to the nature of the star.[23] These correspondences are traditions of great antiquity, perhaps dating back to the Hermetic books spoken of by Iamblichus in *De Mysteriis*. Translated many times and copied by many scribes, there have been many opportunities for errors to creep in, but perhaps it is instinctive clairvoyance that has made those decisions in the long course of the development of the tradition.

Many love spells call for the use of the herb vervain because of its long-standing assignment to the planet Venus and history of magical use. The plant was considered sacred to the Druids and was in demand as medicine through the sixteenth and seventeenth centuries, when it was included in the herbals of Gerard and Parkinson. The identity of the Druidical vervain is not clear, but it has been identified with *Verbena officinalis*. According to Eleanour Sinclair Rohde, "In Druidical times libations of honey had to be offered to the earth

✦ *A page from a sixth-century Anglo-Saxon manuscript copy of the fourth-century herbal* Pseudo-Apuleius.

from which it was dug, mystic ceremonies attended the digging of it and the plant was lifted out with the left hand. This uprooting had always to be performed at the rising of the dog star and when neither the sun nor the moon was shining."[24] The reputation of vervain endured through the centuries as *materia magica*, which can be demonstrated by its use in love spells such as this one from the grimoire of Arthur Gauntlet, the sixteenth-century cunning man:

> To gain the love of a man or woman:
> Go to the herb vervain when it is flowered near the full Moon and say to it +* The lord's Prayer + then say + in the name of the Father + vervain, I have sought thee In the name of the Son I have found thee + In the name of the Holy Ghost + I will gather thee + Then say + I charge thee Vervain, by the virtue of Our Lord Jesus Christ and by the holy names of God + *Helion* + *Heloy* + *Adonoy* + and by all of the holy names of God that when I carry thee in my mouth that whosoever I may touch that thou make them obedient to me And to do my will in all things + fiat + fiat + fiat + Amen +[25]

*A reminder that in grimoires the "+" symbol is usually understood to mean that one is to make the sign of the cross in the air over the object indicated—in this case, the vervain sprig.

This "love" spell is definitely a bit coercive and belongs comfortably in the ancient tradition of "erotic binding spells" that date from antiquity, but the central role of the occult virtue of the Venusian herb vervain is clear.

Other spells call for herbal materia of a hot or Martial virtue for maleficia in order to cause uncomfortable heat in the victim by magical means. One such spell is the famous "hotfoot" powder of the African-American Hoodoo tradition, of which there are many local variations. Most involve the use of powdered red cayenne pepper, *Capsicum annuum,* often in combination with graveyard dirt and other arcane ingredients, which are sprinkled in the footprints or in the proximity of someone the conjuror wants to be rid of.

Early twentieth-century American anthropologist Newbell Niles Puckett, in his problematic but nevertheless fascinating collection of African American folklore, *Folk Beliefs of the Southern Negro,* recorded many specific practices of root doctors and conjurors throughout the southern United States. Unfortunately, they were interspersed with the kind of patronizing and blatantly racist commentary typical of the anthropology of the day. However, as with James George Frazer, Franz Boas, and others of this era, many traditional magical practices would not have been recorded and passed down to posterity were it not for their fieldwork and scholarship. Puckett records many magical uses of red pepper among his informants for driving away the unwanted as well as for counter-magic: "Red pepper is another effective and widespread charm for making and breaking hoodoo, combining sharpness of taste with the fetish color. Red pepper

✦ Gris-gris, *a painting showing some of the elements of Hoodoo conjure, including the mojo bag, many of which derive from botanical ingredients, by Charles Gandolfo, founder of the New Orleans Historic Voodoo Museum. Permission by Jerry Gandolfo.*

in the shoes, (some say put in the heel of the shoe) or hung over the door is a sure counter-charm." Puckett gives this example of red pepper counter-magic from an informant in Georgia, who upon feeling sensations of pain starting in his foot, and then spreading over his whole body and his head, said, "I . . . knew I was conjured. I looked for the cunjer, found a little bag under my front doorstep, containing graveyard dirt, some nightshade (Jimson weed) roots [note its non-pharmacological use as a source of malefic Saturnian virtue], and some devil's snuff [most likely puffball spores, *Lycoperdon* spp.]; took the bag, and dug a hole in the middle of the public road where people walked, and buried the bag, and sprinkled red pepper and sulphur in my house."[26]

Moving continents and centuries to Anglo-Saxon England, many examples of historical use of herbs and magical practices can be found in the *Lacnunga*, or leechbooks, previously introduced in chapter 14, manuscripts that recorded many medical recipes that combine magical methods and prayer with herbal remedies. The *Nigon Wyrta Galdor*, the famous Nine Herbs Charm, combines an incanta-

✦ *The sacred herb verbena. From a sixth-century Anglo-Saxon manuscript copy of the fourth-century herbal* Pseudo-Apuleius.

tion that calls upon the Pagan god Woden and perhaps Christ, but the manuscript is not clear. The charm dates from a time period in which the Christian religion was in the process of consolidating control, so the invocation of Pagan deities in a Christian charm is interesting. The astute reader may recognize the Old English word *galdor*, which is the same as the Norse and carries the same meaning; namely, magic or incantation. To magical practitioners of all times, the effectiveness of the operations is more important than doctrinal purity, and some syncretism is to be expected. Woden, or as he is known in the Old Norse of the Eddas, Óðinn, is the master of magic, who "hung on a wind-battered tree, nine long nights"[27] and learned the secrets of spells and runes to cure and to harm. The fascinating text of the Nine Herbs Charm from Felix Grendon's 1909 book *Anglo-Saxon Charms* is worth quoting in full to give us the flavor of the plant magic of the period. These charms stand on their own as invocations to the various herbs in particular:

Remember, Mugwort, what you revealed,
What you prepared at Regenmeld.
Una, you are called, eldest of herbs.
You avail against three and against thirty,
You avail against poison and against infectious sickness,
You avail against the loathsome fiend that wanders through the land.

And you, Plantain, mother of herbs.
Open from the east, mighty from within.
Over you carts creaked, over you queens rode.
Brides exclaimed over you, over you bulls gnashed their teeth.
Yet all these you withstood and fought against:
So may you poison and infectious sicknesses resist
And the loathsome fiend that wanders through the land.

Stime this herb is named; on stone it grew.
It stands against poison, it combats pain.
Fierce it is called, it fights against venom,
It expels malicious [demons], it casts out venom.
This is the herb that fought against the snake,
This avails against venom, it avails against infectious illnesses,
It avails against the loathsome fiend that wanders through the land.

Fly now, Betonica, the less from the greater,

The greater from the less, until there be a remedy for both.

Remember, Camomile, what you revealed,

What you brought about at Alorford:

That he nevermore gave up the ghost because of ills infectious,

Since Camomile into a drug for him was made.

This is the herb called Wergulu.

The seal sent this over the ocean's ridge

To heal the horror of other poison.

These nine fought against nine poisons:

A snake came sneaking, it slew a man.

Then Woden took nine thunderbolts

And struck the serpent so that in nine parts it flew.

There apple destroyed the serpent's poison:

That it nevermore in house would dwell.

Thyme and Fennel, an exceeding mighty two.

These herbs the wise Lord created.

Holy in heaven, while hanging [on the cross].

He laid and placed them in the seven worlds,

As a help for the poor and the rich alike.

It stands against pain, it fights against poison.

It is potent against three and against thirty.

Against a demon's hand, and against sudden guile.

Against enchantment by vile creatures.

Now these nine herbs avail against nine accursed spirits.

Against nine poisons and against nine infectious ills.

Against the red poison, against the running poison,

Against the white poison, against the blue poison.

Against the yellow poison, against the green poison.

Against the black poison, against the blue poison.

Against the brown poison, against the scarlet poison.

Against worm-blister, against water-blister,

Against thorn-blister, against thistle-blister,

Against ice-blister, against poison-blister.

If any infection come flying from the east, or any come from the north,

Or any come from the west upon the people.

Christ stood over poison of every kind.

I alone know [the use of] running water, and the nine serpents take heed [of it].

All pastures now may spring up with herbs.

The seas, all salt water, vanish,

When I blow this poison from you.

Mugwort, plantain which is open eastward, lamb's cress, betony, camomile, nettle, crab-apple, thyme and fennel, [and] old soap; reduce the herbs to a powder, mix [this] with the soap and with the juice of the apple. Make a paste of water and of ashes; take fennel, boil it in the paste and bathe with egg-mixture, either before or after the patient ap-plies [sic] the salve. Sing the charm on each of the herbs: three times before he brews them, and on the apple likewise; and before he applies the salve, sing the charm into the patient's mouth and into both his ear and into the wound.[28]

SYNCRETISM IN PLANT MAGIC

The Anglo-Saxon leechbook tradition existed in a liminal zone between the old Saxon Heathen religion, which had much in common with the Norse traditions we have discussed, incorporating an animist approach to the land and living things, and the Catholic Christian traditions of continental Europe, where the Platonic approach we also discussed was the intellectual substrate. Both traditions are useful together when they can work in harmony, and a syncretic approach underpins the attitude of cunning folk and folk magicians of all kinds. The land and the spirits themselves put their own particular form on the practice, which exists necessarily within a cosmological framework that aims to take in the whole spiritual cosmos, at least as a provisional model. Each part of the dual-faith syncretism provides a necessary emphasis on a different understanding of the cosmology. Christian and Heathen are not opposites to the magician but complements. This is seen in traditions from Scandinavian trolldom and its Anglo-Saxon cousins to the Greco-Egyptian Magic of the Greek Magical Papyri, in which Jewish and Christian angels and names of God combine with Greek and Egyptian deities. This is never going to be accepted by the Orthodox or the gatekeepers of either tradition, but the wise know the One is beyond all names and outside all paradigms.

In late antiquity Egypt of the Greek Magical Papyri we find an invocation to be said while gathering herbs for use in magic, found in PGM IV, which beautifully calls upon the powers of the Greek and Egyptian gods and invokes the daimon of the plant:

> The invocation for him [the daimon] which he speaks over any herb, generally at the moment of picking, is as follows:
>
> "You were sown by Kronos, you were conceived by Hera, you were maintained by Ammon, you were given birth by Isis, you were nourished by Zeus the god of rain, you were given growth by Helios and dew. You [are] the dew of all the gods, you [are] the heart of Hermes, you are the seed of the primordial gods, you are the eye of Helios, you are the light of Selene, you are the zeal of Osiris, you are the beauty and glory of Ouranos, you are the soul of Osiris' daimon, which revels in every place, you are the spirit of Ammon. As you have exalted Osiris, so exalt yourself and rise just as Helios rises each day. Your size is equal to the zenith of Helios, your roots come from the depths, but your powers are in the heart of Hermes, your fibers are the bones of Mnevis, and your flowers are the eye of Horus, your seed is Pan's seed. I am washing you in resin as also I wash the gods even [as I do this] for my own health. You also are cleaned by prayer and give us power as Ares and Athena do. I am Hermes. I am acquiring you with Good Fortune [Agathe Tyche] and Good Daimon [Agathodaimon] both a propitious hour and on a propitious day that is effective for all things.[29]

Although in the full text a rubric is given for a harvesting ritual—where offerings of incense and a libation of milk are given—there is no reason why any herb or plant couldn't be consecrated with this prayer at a later date, before preparation or ritual use, for instance.

In many cultures and many times, shamans, magicians, and healers have relied on their relationship with plants and their spirits in order to accomplish their various practical goals, be they healing or other sorcerous intentions. These practitioners envision a world of living intelligences and intimate connections, an animate network of subtle influences working on reality beneath the outer surface of things. Once the veneer of solidity is peeled away, the plants are seen to be more than just machines taking in light and water and putting out economically useful organic matter in return. They are souls and spirits, our relatives, sharing our lives and our world. As a farmer, I know that this relationship isn't easy, that it is a give and take, but our green friends are more than generous with the sustenance, warmth, healing, and wisdom they give us. We are nothing without them.

18

AGRICULTURAL MAGIC

Pray to Zeus of the land and to hallowed Demeter to make
Demeter's grain ripen, as you begin plowing at the very start.[1]

HESIOD, *WORKS AND DAYS*

In this chapter, I hope to introduce the basics of a philosophy of agriculture, which is essential to understanding where we all come from, culturally speaking. Agriculture is the foundation of all of the civilizations of the world and the soil from which all of the great religious systems of the world have sprung. No matter how far removed the sophisticated urban philosophers of antiquity were, they were still not very far removed from the farmers who grew their food. The cosmological systems they invented, with which they tracked the movements of the stars and luminaries, were used to time the humble agricultural operations upon which all life depended. The rising of the constellations reliably marked the agricultural year and the movement of the wandering planets across the sky indicated the auguries of the year to come. In the colder northern climates, the stations of the Sun and Moon helped ancient people to plan the timing of both the agricultural operations and the great ritual festivals that ensured the blessing of the divine ones on the crops and herds. Many of the shamans and cunning people we have discussed in these pages actually were farmers. Far from being only the concern of rural folk, a basic understanding of agriculture and the soil is necessary for an embodied spirituality and esoteric philosophy. We all come from the Earth, and to Her we all return. My intention in this chapter is to make this topic relevant for all readers and not just the ones with green thumbs. Whether you raise a few houseplants in pots or tend acres of crops, you still eat, and so the soil is literally the foundation of your life.

Earth, dirt, clay, soil—it is called by a multitude of names in many tongues—is a mixture of minute particles of silicate- and alumina-based minerals in combination with organic matter in varying amounts. It covers the ground in most places where humans live. No substance in the world is more taken for granted than the humble skin of our sacred mother. This ubiquitous substance has mirac-

✦ *Wheat harvest on the author's farm. Photo by Esmee McKee.*

ulous properties, for from it springs vegetative life. Indeed, we are literally made from mineral substances originating from the soil in which the food we eat was raised. It is possible to tell by the various levels of mineral isotopes in bone fragments what part of the world they originated from, due to differences in the mineral content of soils in various parts of the world.

But the soil is not just minerals. It is also teeming with life; a teaspoon of soil contains billions of microorganisms. The living creatures contained within the soil are essential to its function as a substrate for the growth of plants. Bacteria, fungi, protozoa, and tiny invertebrates of all kinds live out their tiny little lives unnoticed by larger creatures like us who depend on them for the health of the soil we need to produce our food. This is animism in its most literal form. Even the dirt is literally alive in this vibrant cosmos. The soil is an organ of the Earth, and has a vitality all its own. What is alive is also dead, dying, and being reborn in a constant state of transformation. Plants and animals fall to the ground and are broken down by the teeming crew of recyclers, to be physically reborn anew in the bodies of other generations of living things. The cycle of matter embodied in the life cycle of the soil is a physical analogue of the metaphysical life of the sparks of consciousness that enliven more complicated beings such as ourselves.

It has been mentioned in past chapters that the ancient agricultural peoples of the world took great comfort in the continuing cycle of the life of the soil and of crops, seeing them as a symbol of the immortality of the soul, which like the seed is cast into the ground and is reborn as a plant. But even the stubble and the straw of the old plant is reborn in future plants, the body as well as the soul having its own form of eternal life as raw material for new life. To reiterate, the element of earth is the starting and ending point for all physical life; from it we are born and into it we are absorbed. The acknowledgement of our dependence on the physical as embodied entities is essential for the re-sacralization of nature that we are promoting in these pages, in contrast to systems of spiritual thought that seek to deny the physical in favor of the spiritual. The physical world is infused with spirit and saturated with the divine. There is no separation. If we have spoken of the world as fallen, we were speaking metaphorically. It is we who have fallen, by allowing our perception of the divinity all around us to become clouded.

As cunning farmers it is the medium of our art, with which we create, or more properly co-create, because the farmer does nothing unaided by the spiritual life of the land and the cosmos. To paraphrase St. Paul, the farmer plants and waters, but growth comes from the divine realm. It is our task to prepare the land to receive the seed, to make it fertile, to put all of the elements in motion, and to wait for the gods to do their part. We are, after all these thousands of years, still at the mercy of the vagaries of the weather, insects, animals, and all of the other random events of happenstance that fate and fortune can

✦ *Offerings to the Germanic god Freyr to begin the season on the author's farm. The author is driving the tractor. Photo by Esmee McKee.*

throw our way. These forces were known to the ancient peoples of the world as divine beings: Demeter and Gaia, who preside over the seed, the growth of plants, and the fertility of the crops; Zeus, who controls the weather and the rain; Pan, the lord of the animals; the Moirae, the Fates; and Tyche, fortune herself. These deities or the forces they represent are known to all farming cultures by many names and represented by a multitude of different images.

The ancient Roman agronomist Marcus Porcius Cato (234–139 BCE) begins his treatise *De Agricultura* with an invocation "of those twelve divinities who are the tutelaries of husbandmen [farmers]":

> First: I call upon Father Jupiter and Mother Earth, who fecundate all the processes of agriculture in the air and in the soil, and hence are called the great parents.
>
> Second: I invoke the Sun and the Moon by whom the seasons for sowing and reaping are measured.
>
> Third: I invoke Ceres and Bacchus because the fruits they mature are most necessary to life, and by their aid the land yields food and drink.
>
> Fourth: I invoke Robigus and Flora by whose influence the blight is kept from crop and tree, and in due season they bear fruit (for which reason is the annual festival of the *robigalia* celebrated in honour of Robigus, and that of the *floralia* in honour of Flora).
>
> Next: I supplicate Minerva, who protects the olive; and Venus, goddess of the garden, wherefore is she worshipped at the rural wine festivals.
>
> And last: I adjure Lympha, goddess of the fountains, and Bonus Eventus, god of good fortune, since without water all vegetation is starved and stunted and without due order and good luck all tillage is in vain.[2]

So, having paid our duty to the gods, we will proceed to an outline of the basics of cultivation.

A BRIEF PRIMER IN AGRICULTURAL MAGIC AND FARMING TECHNIQUES

During my career as a farmer, I have had the privilege of turning several raw pieces of land into fertile gardens and productive homesteads and farms, and the first two steps were tillage and manure (in one of them clearing the few trees that grew there was necessary before tilling could begin). From where I stand now, as a devotee of the spirits of the land, the first step should have been to make offerings to the genii loci, the spirits of the place, an act Cato would have definitely approved of. No self-respecting Roman farmer would ever have embarked upon clearing forest for farming or setting up a farmstead where there had been none previously without the obligatory prayers and sacrifices. In two of the homesteads I started, my house would be surrounded at night by packs of coyotes howling loudly and eerily in the darkness, perhaps voicing the displeasure of the land spirits at our sudden intrusion on their sanctum.

I started out with no farm machinery at first, except a twenty-year-old Troy-Bilt rototiller and a chainsaw, along with shovels, rakes, and hoes. Armed with these basic tools, we cleared a few garden spots, burned the brush and the stumps, and hauled endless pickup truck loads of manure from a neighbor's mule barn to our woodland homestead. I proceeded to till the hard clay soil and added enough manure to turn it into black loam. We planted medicinal herbs and vegetables, along with the obligatory *Brugmansias*, *Daturas*, and tobacco plants that have always been my favorites. The woodland clearing with a small cabin at the center became a paradise, the kind of place you might find in a fairy tale.

And it was fertile! All of that manure, I learned, was the key to organic farming. A forest is an example of the power of humus to produce luxuriant plant growth. Forest soils receive a continuous supply of leaves as well as the detritus from fallen trees and trunks and gradually build up a layer of soil that is rich in organic matter, or humus. The fallen vegetation mixes with the bodies of a multitude of insects, worms, invertebrates, and even higher animals, and the result, over centuries, is the formation of extremely rich and productive soil. Organic farmers and gardeners seek to imitate this process and speed it up by applying large amounts of compost, or well-rotted animal manure, and mixing it with the topsoil. This allows farmers to improve marginal, eroded, and otherwise unproductive soils quickly.

In natural soil systems, worms and other burrowing animals mix the decaying plant material and humus into ever deeper layers of soil, thus deepening the zone of friable and loose soil in which roots can spread, in turn allowing the organic matter from the decaying roots themselves to enter into ever-deepening layers. Natural soil systems are thus a positive feedback system in which the processes of life stimulate the growth of ever more living things. Another ancient Roman agronomist, Columella (4–70 CE), mentioned this in his classic agricultural handbook *De Re Rustica*, which has been influential for nearly two thousand years. Columella wrote that other authorities had noticed that wooded land brought under cultivation is initially highly productive and soon becomes less so, though he wrongly likened the drop in productivity to the aging process in animals. He says the exhaustion of the soil is not like the debility that accompanies old age. Rather, he writes, the recently forested land has received:

> leaves and herbage of many years, which it has kept producing naturally, fattened, so to speak, with more plentiful nourishment, it more readily satisfies the requirements for bringing forth crops and supporting them. But when the roots of the plants, broken by mattocks and ploughs, and when the trees, cut down by the axe, cease to nourish their mother with their foliage; when the leaves which fell from bushes and trees in the autumn season and which were spread over her are presently turned under by the ploughshare and mixed with the subsoil, which is usually thinner, and are used up, the result is that the soil, being deprived of its old-time nourishment, grows lean. It is not, therefore, because of weariness, as very many have believed, nor because of old age, but manifestly because of our own lack of energy that our cultivated lands yield us a less generous return. For we may reap greater harvests if the earth is quickened again by frequent, timely, and moderate manuring.[3]

The perceptive farmers of the ancient world noticed the tendency of forested land to build soil and of tilled land to become exhausted by agriculture, and the power of additions of manure to cultivated land to restore it. The tendency of modern agriculture, by contrast, is not to imitate the example of the forest and to build soil with regular additions of humus but rather to add back only the major nutrients needed by the crops in the form of chemical fertilizer, thereby exhausting the land while keeping it productive at the same time. As Rudolf Steiner, founder of the influential biodynamic system of natural farming, noted, in all types of farming we are exploiting the land. We are removing forces from the Earth that must be replaced, and that is why we manure. But we must add back living forces, not dead chemicals. The quality of our food and our health has suffered as a result of our willful neglect of both the spiritual and physical health of our agricultural lands.

Many today can understand the need to preserve wild nature, and many of us find it easy to perceive the animate spirits of place in a forest or other wilderness place, but it is our agricultural land that feeds us, clothes us, and gives us life by allowing the growth of the plants that we depend on directly for our life support.

These lands—the province of the benign gods and goddesses of the soil, their hosts of spiritual helpers, and the plant devas of our agricultural crops—need our devotion and support too. We owe them.

One of the first revolutions in agriculture was the use of tools. First were hand tools like the mattock (a mattock, for those who don't know, is a heavy-duty version of a hoe, a tool with a long handle and a blade at the end turned at a right angle). Then some enterprising inventor came up with the idea of mounting a mattock blade on a longer shaft, attaching two long handles, and pulling the whole thing with the help of draft animals, like oxen. This implement is known as an ard, which is a light-duty, primitive type of plow, still in use in some parts of the world. Heavier plows requiring more draft animals were subsequently invented, which allowed for higher crop yields and required still greater social organization, as the number of draft animals and the cost of the machinery itself placed it beyond the reach of the ordinary peasant. Modern plows, called mold-board plows, have a curved, wave-like shape that inverts and crumbles the soil while slicing it. Tillage, whether by plow, mattock, or shovel, allows the farmer to create a clean seedbed of bare soil in which to place the seed, and this allows the crop seedlings to gain an advantage over the hardier and more vigorous weeds that compete with the crop for light, nutrients, and water. Plowing allows the grower to incorporate large additions of organic matter to the body of the soil, rapidly improving its physical structure and fertility. Excessive working of the soil can also damage its structure and burn up organic matter by introducing too much oxygen to the system. Everything must be in balance.

◆ *Medieval plow team, c. 1000 CE.*

In recent years, there has been a trend among some growers to move away from tillage in an effort to avoid damaging the intricate web of living beings in an agricultural soil. I tend to follow an older model of agriculture, in which tillage, compost, soil-restoring cover crops, and periodic rest from tillage all play a role in maintaining soil fertility. Yes, tillage is destructive to some soil microorganisms and invertebrates, but it is beneficial to others, especially when it is coupled with the addition of a food source, like composted manure or a lush green crop of clover. The rich smell and explosion of the population of earthworms a week or two after an addition of organic matter is evidence of this. The ancient authors—Columella, Varro, and company—all recommended planting legumes, members of the bean family, which were known to improve yields for the following crops due to their ability to "fix" atmospheric nitrogen. The ancient agronomists were unaware of the presence of colonies of symbiotic bacteria on the roots of the legume plants, but they were aware of the ecological benefit; as Cato said, "Lupines, field beans, and vetches manure corn land."[4] The patient farmers of the ancient world, whose animistic spirituality and attention to ecology I endeavor to imitate, noticed subtleties that are lost on many modern agriculturalists. I once suggested to the teacher of a USDA-sponsored class on agricultural economics that I was taking that he follow the crop of peas in his home garden with his late sweet corn to take advantage of the nitrogen left behind by the peas for the benefit of the corn. He informed me that he didn't bother with such things because he used chemical fertilizer. Traditional farmers who worked without chemical quick fixes had to rely on the intricate workings of nature in order to have successful crops.

Traditional agriculture normally followed a crop rotation that allowed soil to rest and recover organic matter and humus under several years in a legume crop intended for livestock feed, such as clover or alfalfa. This was followed by the crops with the highest nutrient demand, such as corn (maize),* *Brassicas* (members of the cabbage family), or tobacco, then crops requiring less nitrogen, eventually returning to the legume sod for several years. In this way, the farm produced its own feed for animals and its own fertilizer from those animals, as well as crops for sale, all with minimal inputs of purchased nutrients. For thousands of years, farmers practiced systems of crop rotation informed by traditional knowledge of what crops thrived in their region, the weather patterns of those

---

*When discussing *Zea mays*, the plant known to Americans as corn and to the rest of the English-speaking world as maize, I will alternate between the two terms.

◆ *Plough Monday festivities in Yorkshire, England, from George Walker's* The Costume of Yorkshire *(1814).*

regions, and the particular soils with which they were working. They knew what work had to be done on which land and in which season, and in this they were guided by the festivals and holidays interspersed throughout the year that marked the passing of time.

Ancient Greek farmers, not having paper calendars or almanacs to tell them when to perform important farming operations, observed the heliacal rising of various stars and constellations, principally Arcturus, Orion, and the Pleiades. A star's heliacal rising is when it can first be seen right before the sunrise, having just come out from "under the beams" of the Sun. The Sun, observed daily, is seen to move westward against the background of the fixed stars, through the constellations of the zodiac. Each day, different stars are revealed rising in the east right before dawn. For example, the fall plowing in Ancient Greece commenced at the rising of Arcturus and the setting of the Pleiades, at that time in September. Appropriately enough, Acturus is the brightest star in the constellation Boötes, also known as the Plowman.[5]

In northern Europe, specifically in England, the agricultural year in the Middle Ages commenced on Plough Monday, which was the first Monday after the Feast of Epiphany, in early January. This occasion was marked by festivities and mild extortion of gifts and "largesse" by the plowmen and other agricultural workers, not too different in spirit from trick or treat. The trick, in this case, would be having one's yard plowed up and turned to mud, according to historian Dorothy Hartley.[6] Plowing, by horse or oxen, is a slow business, and in many cases the work wouldn't be finished until March, in time for the land to dry and mellow a bit for seedbed preparation and sowing of spring wheat, oats, and barley.

In my area, winter is usually too wet to do any plowing, so we usually wait until the ground warms in spring. We use tractors here nowadays, which are heavier and move faster, causing more compaction to wet soils. So we prefer to do our plowing in the autumn, when the soil is a bit drier. Fall plowing allows the soil to warm up and be exposed to the repeated freezing and thawing of winter, breaking clods and giving the soil a smooth, mellow consistency. The beneficial effects to agricultural soils of repeated freezing and thawing didn't go unnoticed by the perceptive Roman farmers. The Roman poet Virgil advised starting plowing early enough to allow the land to freeze several times, writing in his *Georgics*, a series of four agricultural poems offering advice to Roman farmers:

> *In early spring tide, when the icy drip*
> *Melt from the mountains hoar, and Zephyr's breath*
> *Unbinds the icy clod, even then 'tis time;*
> *Press deep your plough behind the groaning ox,*
> *And teach the furrow-burnished share to shine,*
> *That land craving farmer's prayer fulfills,*
> *Which twice the sunshine, twice the frost has felt;*
> *Ay, that's the land whose boundless harvest-crops*
> *Burst, see! The barns.*[7]

Virgil here notes the curious fact that the plow share, rusty and dull when the work begins, emerges from the work gleaming with a mirror-like polish. Indeed, allowing the bare soil to absorb the rain and snow, the Sun and the cold, making sure that organic matter is added often, and erosion is prevented by selecting level sites for fall tillage, seems to enhance the fertility as Virgil promises, allowing for reduced soil compaction and increased vitality. The soil warms quickly as well, allowing for earlier planting and faster growth of crops.

Gardeners without plow, tractor, or oxen can still approximate the effects of plowing, albeit more slowly, but also more gently, with a shovel, either a flat-headed or round-headed spade. The method is simple to describe but hard work to be sure. If you aren't used to physical work, take it slowly at first. First, you dig a trench, throwing the pieces of sod or dirt onto a tarp outside of the trench. Next, you dig a second trench, parallel to the first, throwing the sod grass-side down into the first trench (the dirt chunks go upside down if you aren't turning over sod.) The process is continued until the desired area is turned. Then you rest your sore back and wait until the soil mellows and the grass and weeds die, perhaps a month or two. This isn't a method for someone looking for instant results, but it is a time-honored method of making a garden without powered machinery. When the soil is ready, it will have to be leveled out with rakes and hoes before planting, an operation farmers call "secondary tillage." Farmers use implements known as harrows, which used to be as simple as dragging a bush or tree behind the team of oxen and now are huge steel affairs coming in many shapes and variations. In all cases, the clods are broken up and the ground is leveled out to remove high spots and low spots in the field (or garden spot).

Having prepared a fine and mellow seedbed, we move to the next operation: sowing, a very important operation in which the precious seed is actually committed to the ground. The seed is precious, because it is the assurance of future survival. No matter how hungry a peasant family got in the winter, the seed corn would only be eaten in the most dire circumstance. It was selected from the best of the crop in hopes of passing on the characteristics of the parent plants to the progeny, a process known as selective breeding, a remarkably effective technology from which all of the varieties of crop plants known to humanity, selected by generations of our peasant farmer ancestors, have come down to us. This is a beautiful example of the mutuality that exists between the generous plant genii and the humans who lovingly tend them. With the concentration and modernization of agriculture, many of these traditional peasant crop cultivars, known as landraces or heirloom varieties, have been lost to agriculture forever because they are not as profitable to modern agriculture as modern varieties, as they are not as high yielding or have a shorter storage life.

Planting on My Farm

On our farm we grow, or have grown, nearly every kind of crop plant that can be grown in our region. Many of these crops, such as wheat and corn, are not very profitable economically and can be bought much more cheaply than they can be grown, due to economies of scale and government subsidies that ensure that larger operations with huge equipment can produce these crops nearly at a loss. Sometimes modern farmers need to grow thousands of acres and borrow millions of dollars against the crop just to make a very modest middle-class income, after all costs

✦ The Harvesters, *a 1565 painting by Pieter Brueghel the Elder, showing the harvesting process for what appears to be rye, or a very tall variety of wheat, in exquisite detail. It is a window into the life of farmers in the sixteenth century.*

of production are paid, at great financial risk. It sounds like madness to me, but if it wasn't for the efforts of the farmers who take these risks and work long hours behind the driver's seats of massive tractors and combine harvesters, we wouldn't be able to produce enough food to feed the world's burgeoning population, because that is how the system is set up. This is also why we primarily raise vegetables on our farm, which are much more profitable at our small scale than grains.

We do grow grains on a small, subsistence scale, mostly for our own use. We sell a bit of cornmeal, which we grind in an antique stone grist mill. I have planted small crops of wheat and barley and reaped and threshed them with hand tools, a backbreaking and time-consuming labor, which really taught me firsthand of the hard work our ancestors had to do in order to earn their bread. We milled the wheat into flour for bread and malted the barley and brewed amazing beer. The bread and beer, being so hard won and fresh from the Earth herself, were the most amazing I have ever tasted. But I can only recommend this on a very limited scale, just for the experience.

THE MAGIC OF DEMETER: WORKING THE LAND

Because we are concerned here about gleaning (sorry for the pun) effective praxis from history, in this section we will concentrate on historical examples, folklore, and mythology pertaining mostly to those staple crops on whose survival most people depended: in this case, grain, wheat, barley, and, in North America, maize. Sowing these crops in modern times involves machinery, but in ancient times the small grains were broadcast by hand, an operation that requires a good sense of rhythm and coordination if it is to be done right. The seed must be spread evenly and not too thickly on the surface of the field—not enough seed and yields suffer and weeds can quickly overpower the growing crop, while using too much seed is wasteful of a precious resource and causes the plants to grow thin and spindly due to too much competition. The seed must be quickly covered by passing over it with a harrow or, on a small scale, a rake. This is the time when the crop's future is most vulnerable, due to birds who eat the seeds, heavy rain (which can wash it out of the ground or harden the ground so it can't emerge properly), and cold weather, which, coupled with excessive dampness, can cause the precious seed to rot.

This is precisely why ancient farmers all over the world practiced agricultural magic to ensure the success of the crop at the time of sowing. In the latter half of the nineteenth century, Scottish folklorist Alexander Carmichael lived among the Gaelic-speaking people of the Hebrides in Scotland and collected hundreds of songs, prayers, and incantations in use among the common folk. He translated them into English and published *Carmina Gadelica*, c. 1900. In describing the worldview of his informants, who lived close to nature as crofters and fisherfolk, Carmichael said, "The Celtic missionaries allowed the pagan stock to stand, grafting their Christian cult thereon. Hence the blending of the Pagan and Christian religion in these poems, which to many minds will constitute their chief charm."[8] He recorded this prayer that accompanied the ritual to consecrate the seed corn:

I will go out to sow the seed,
In the name of Him who gave it growth;
I will place my front in the wind,
And throw a gracious handful on high.
Should a grain fall on a bare rock,
It shall have no soil in which to grow;
As much as falls into the earth.
The dew will make to be full.

Friday, day auspicious,
The dew will come down to welcome
Every seed that lay in sleep
Since the coming of cold without mercy,
Every seed will take root in the earth,
As the King of the elements desired,
The braird will come forth with the dew.
It will inhale life from the soft wind.

I will come round with my step,
I will go rightways with the sun,
In name of Ariel and the angels nine,
In name of Gabriel and the Apostles kind.

Father, Son, and Spirit Holy,
Be giving growth and kindly substance
To every thing that is in my ground.
Till the day of gladness shall come.

The Feast Day of Michael, day beneficent,
I will put my sickle round about
The root of my corn as was wont;
I will lift the first cut quickly,
I will put it three turns round
My head, saying my rune the while,
My back to the airt of the north.
My face to the fair sun of power.

I shall throw the handful far from me,

I shall close my two eyes twice,

Should it fall in one bunch

My stacks will be productive and lasting,

No Carlin will come with bad times

To ask a palm bannock from us.

What time rough storms come with frowns

Nor stint nor hardship shall be on us.[9]

Carmichael tells us that:

The preparation of the seed-corn is of great importance to the people, who bestow much care on this work. Many ceremonies and proverbs are applied to seedtime and harvest. . . . All these preparations are made to assist Nature in the coming Spring. Three days before being sown the seed is sprinkled with clear cold water, in the name of Father, and of Son, and of Spirit, the person sprinkling the seed walking sunwise the while. The ritual is picturesque, and is performed with great care and solemnity, and, like many of these ceremonies, is a combination of Paganism and Christianity. The moistening of the seed has the effect of hastening its growth when committed to the ground, which is generally begun on a Friday, that day being auspicious for all operations not necessitating the use of iron.[10]

An interesting item in this particular incantation is the use of the word *rune*. Carmichael is apparently using the word *rune* to translate the Gaelic *rann*, which also means rhyme, verse, or stanza, but I wonder if this is a local translation that comes from Norse influence on Hebridean culture.

We can see that the mystically minded Hebridean farmers of the nineteenth century were leaving nothing to chance, placing this powerful intention on the seed and calling upon the powers of the angels and the apostles, sowing while envisioning the harvest months ahead. These were the true cunning farmers, working their magic even as they worked their land, enriching their lives and giving strength to their crops with these deeply moving incantations.

In ancient Rome, the day of sowing was observed with prayers and ceremonies also; the Feriae Sementiuae, the festival of sowing, was celebrated in January. The poet Ovid gives us this prayer for the occasion, writing, "Propitiate Earth and Ceres, the mothers of the corn, with their own spelt and flesh of teeming sow. Ceres and Earth discharge a common function: the one lends to the corn its vital force, the other lends it room." The prayer itself, which functions almost as a fertility-generating charm, boosting the vital power of the crop with the help of the goddesses, is as follows:

> Partners in labour, ye who reformed the days of old and replaced the acorns of the oak by food more profitable, O satisfy the eager husbandmen with boundless crops, that they may reap the due reward of their tillage. O grant unto the tender seeds unbroken increase; let not the sprouting shoot be nipped by chilly snows. When we sow, let the sky be cloudless and winds blow fair; but when the seed is buried, then sprinkle it with water from the sky. Forbid the birds—pests of the tilled land—to devastate the fields of corn with their destructive flocks.
>
> You too, ye ants, O spare the sown grain; so shall ye have a more abundant booty after the harvest. Meantime may no scurfy mildew blight the growing crop nor foul weather blanch it to a sickly hue; may it neither shrivel up nor swell unduly and be choked by its own rank luxuriance. May the fields be free from darnel, that spoils the eyes, and may no barren wild oats spring from the tilled ground. May the farm yield, with manifold interest, crops of wheat, of barley, and of spelt, which twice shall bear the fire.
>
> These petitions I offer for you, ye husbandmen, and do ye offer them yourselves, and may the two goddesses grant our prayers.[11]

Ceres was the Roman goddess of grain, often equated with Demeter but not identical with her. According to Georges Dumézil, Ceres was worshipped in Latium long before being absorbed into the cult of Demeter and her mysteries. Dumézil also mentions tantalizing hints of a specific Roman priest, the Flamen Cerialis, who we can presume is in charge of the cult of the grain goddess, but he gives us no specifics. Also mentioned are the *semones*, the spirits of the sowings, mentioned only in one source, the so-called "Song of the Arval Brethren." The Arval Brethren were an order of priests dedicated to Bona Dea, the Good Goddess, who were charged with performing agricultural rites during the Ambarvalia, a spring agricultural festival.[12]

With the crops sown and blessed and petitions for successful crops done, the farmers' attention is turned to other pressing matters until harvest season, which

◆ *The author harvesting maize, or corn, by hand in the fall of 2023. We grow a variety known as Reid's Yellow Dent, an open-pollinated variety dating from the 1850s, which makes excellent cornmeal.*

occurs at different times around the world. In my region, wheat and barley ripen in June and early July, which isn't ideal, because that is also the rainiest time of year here and is yet another reason why maize is the staple grain in this part of the United States, rather than wheat or barley. Maize, by contrast, is ready to harvest in the autumn, September or October, when the weather is dry. In our region, yields of maize are also up to three or four times higher than yields of wheat, so maize is the major grain grown here.

Cultivating Maize in the Americas

Maize was selectively bred by Indigenous people over a period of several thousand years from a grass that grows wild in Mexico called teosinte. Tropical flint corn began to be grown in eastern North America around 200 CE, and a more adaptable and productive variety known as eastern flint corn began to be grown around 800 to 1000 CE. Maize cultivation gradually spread north from its original home in Mexico farther into North America as well as south into South America, becoming the staple crop of the cultures archaeologists refer to as the Mississippian and Fort Ancient peoples but whose names have been lost to history. These people grew the famous "Three Sisters" crops of Indigenous

✦ *Eighth-century Mayan ceramic sculpture of the young maize god.*

◆ Green Corn Dance (Minatarrees), *an 1861 painting by George Catlin.*

America: maize, beans, and squash. The three crops are very suited to the climate of eastern North America, and the addition of beans to a diet of maize adds a source of the amino acid lysine, which the corn lacks. The extreme productivity of these crops allowed for the growth of large villages and cities across the North American continent, all the way from Mexico to what had once been the largest city in North America before European contact, the Mississippian culture site known as Cahokia in the present-day state of Illinois.[13]

The modern crops of maize that grow on the Eagle Creek bottomland below my farm were likely preceded by the maize crops of the Mississipian and Fort Ancient peoples. Today, the corn is sown and harvested without song or ceremony, planted and harvested by machine operators working hundreds of acres a day, in many cases not even owning or having any relationship with the land. In contrast, if we can judge from the ethnological literature and current practices of present-day Indigenous people, the ancient Kentuckians who worked this land before it was taken from them treated every aspect of the cultivation of the maize that gave them life as an occasion for ceremony, much like the Scottish crofters and Roman farmers we have discussed. It continues to amaze me as I research these pages how much the people of the world throughout history who have lived in relation with the living land as a spiritual entity have in common with one another. It also makes me sad to think of those little corn plants: unloved and not sung to, treated as a source of industrial raw material, genetically modified and sprayed with toxins, a far cry from the sacred maize of the Indigenous American people.

The native people of North America did (and do) attend every aspect of the cultivation of maize with prayers and ceremony. There are many examples of these in the literature, but for the sake of brevity we will detail one example from anthropologist Charles Hudson's classic book, *The Southeastern Indians*. According to Hudson, the most sacred occasion for the Indigenous people of southeastern North America was the Green Corn Ceremony, which was celebrated in one form or another by all of the native cultures living in the region. It was celebrated on the occasion of the first Full Moon after the appearance of mature ears of corn in the milk stage, something like our modern sweet corn but a bit starchier and more al dente. Southerners in the United States, as well as people familiar with the Mexican version known as *elote*, will likely know about "roasting ears," harvested when green from the cornfield and grilled over an open fire. This delicious treat, which is only available once a year, is much anticipated. The community assembled to express their gratitude and thanksgiving for the success of the corn crop. The occasion involved communal feasting and fasting, construction of public buildings, and, according to Hudson, most likely included the enlarging of the earthen mounds for which the region is famous. A new fire was often kindled, and the first fruits of the harvest were ceremoniously committed to the flames. People were given the opportunity to turn over a new leaf and leave behind bad habits, much as we do on the New Year. The central role of this festival in the lives of Native Americans tells us much about how important the maize crop would have been to them, not only for food, but for the life it gives to the whole culture.[14]

Harvesting Cereal Crops in Europe

Historically, ancient and medieval Europeans had similar festivities to celebrate the harvest of their own staple cereal crops. In his excellent book, *The Stations of the Sun*, historian and folklorist Ronald Hutton details the folklore of the harvest season in the British Isles. In Ireland and Scotland, Hutton tells us, the first day of August represented the beginning of autumn and the commencement of the harvest season. The holiday is mentioned in one of the earliest of the Irish sagas, *The Wooing of Emer*, part of the great cycle of tales of the hero Cú Chulainn, known as the *Táin Bó Cúailnge*. In the story, the hero tries to win the daughter of a chieftain, a girl named Emer. She sets before him a number of nearly impossible tasks, among them staying awake for a whole year, listing the fourfold division of the Irish calendar by the festivals that make the seasons, staying awake from "Samhain, when the summer goes to its rest, until Imbolc, when the ewes are milked at spring's beginning; from Imbolc to Beltine at the summer's beginning and from Beltine to Brón Trogain, earth's sorrowing autumn."[15]

✦ Summer, *a 1624 painting by Pieter Brueghel the Younger.*

Other medieval sources refer to the holiday as Lughnasadh, the festival of the Irish god Lugh, who is often described as the Irish Sun god but, according to Professor Hutton, is nothing of the sort. Neither is he a god of the harvest, but "the patron of all human skills, with an especial interest in kings and heroes."[16]

According to Irish folklorist Máire MacNeill, the festival involved:

> a solemn cutting of the first of the corn of which an offering would be made to the deity by bringing it up to a high place and burying it; a meal of the new food and of bilberries of which everyone must partake; a sacrifice of a sacred bull, a feast of its flesh, with some ceremony involving its hide, and its replacement by a young bull; a ritual dance-play perhaps telling of a struggle for a goddess and a ritual fight; an installation of a head on top of the hill and a triumphing over it by an actor impersonating Lugh; another play representing the confinement by Lugh of the monster blight or famine; a three-day celebration presided over by the brilliant young god or his human representative. Finally, a ceremony indicates that the interregnum was over, and the chief god in his right place again.[17]

Here we see a similar ceremony of feasting and sacrifice of the first fruits as we saw among the Indigenous Americans, the grateful response of agricultural people to the forces of nature that give them life.

Professor Hutton devotes a few pages to a critique of the reconstruction of the festival quoted above, saying the evidence is scant for a pan-Celtic harvest festival on August first. He does, however, mention that there was an Anglo-Saxon holiday called Lammas on the same date that marked the beginning of the harvest season, for which there is ample evidence from medieval texts. Lammas, which in Old English is *hlaef-mass* (loaf mass), was the feast of first fruits, during which the Anglo-Saxon people gave thanks for their harvest. According to Hutton, "It was clearly by then the custom to reap the first of the ripe cereals and bake them into bread which was consecrated at the church upon that day."[18] The Anglo-Saxons at this point in history were officially Christian and were perhaps following the biblical injunction from the Book of Leviticus to offer the first fruits of the harvest to God, before anyone was permitted to eat them. As we have seen in other chapters, the Anglo-Saxons were enthusiastic about the use of charms, and the first loaf of Lammas was no exception. The blessed bread was broken in four pieces, crumbled, and sprinkled in the four corners of the harvest barn to prepare it to receive the harvested sheaves.[19]

After the blessings and sacrifices were made, the harvest began in earnest. Harvesting small grains by hand is a relatively simple procedure. In most places, a knife with a curved blade called a sickle was held in the right hand, and the wheat or barley stalks were grasped in the left while the razor-sharp blade sliced easily through the hollow stems, leaving them as long as possible for use as straw, an important material. The reaper kept cutting until their left hand was full of the cut grain and then laid the stalks on the ground in neat piles, all facing the same direction, to be gathered up and tied together in sheaves by other workers, a process known as binding. The sheaves were all gathered up and placed carefully in piles strategically made to shed water in case of rain, known as shocks or stooks. When all the grain was cut, the workers then loaded all of the shocks on wagons and either brought them to a barn or stored them in carefully made thatched stacks near the farmstead.

The work of the grain harvest is tedious, but it has a ritual quality. There is a unique feeling of timelessness that comes from looking out over the standing grain, the patch growing smaller as the weary hours roll on, the hot Sun beating down, the neat rows of sheaves nested in the stubble where the waving grain had been, the promise of resting one's sore back and enjoying a cold drink and a large meal when the work is finished. One loses a bit of one's individual identity at that moment, becoming for a little while just another laborer in the harvest, doing as our ancestors had done, a job that has changed little in twelve thousand years and is still carried on this way even today in places where there is little money for farm machinery. To participate in the grain harvest with hand tools is to participate in one of humanity's oldest rituals; there are no religious ceremonies with that long of a pedigree. At the end of the harvest, the workers are as relieved that the job is over as they are grateful that there will be enough to eat for the coming year. Prayers of gratitude are offered to whatever deities the community worships, and the harvest is finished for another year.

Alexander Carmichael gives us this account of the harvest and these humble reaping prayers, as well as an intriguing divination practice, from the Gaelic-speaking crofters of the Hebrides in *Carmina Gadelica*:

The day the people began to reap the corn was a day of commotion and ceremonial in the townland. The whole family repaired to the field dressed in their best attire, to hail the God of the harvest. Laying his bonnet on the ground, the father of the family took up his sickle, and, facing the sun, he cut a handful of corn. Putting the handful of corn three times sunwise round his head, the man raised the "Iollach Buana," reaping salutation. The whole family took up the strain and praised the God of the harvest, who gave them corn and bread, food and flocks, wool and clothing, health and strength, and peace and plenty. When the reaping was finished the people had a trial called "cur nan corran," casting the sickles, and "deuchain chorran," trial of hooks. This consisted, among other things, of throwing the sickles high up in the air, and observing how they came down, how each struck the earth, and how it lay on the ground. From these observations the people augured who was to remain single and who was to be married, who was to be sick and who was to die, before the next reaping came round.

Carmichael also includes a "Reaping Blessing":

On Tuesday of the feast at the rise of the sun,
And the back of the ear of corn to the east,
I will go forth with my sickle under my arm,
And I will reap the cut the first act.

I will let my sickle down,
While the fruitful ear is in my grasp,
I will raise mine eye upwards,
I will turn me on my heel quickly,

Rightway as travels the sun.
From the airt of the east to the west,
From the airt of the north with motion slow
To the very core of the airt of the south.

I will give thanks to the King of grace
For the growing crops of the ground,
He will give food to ourselves and to the flocks
According as He disposeth to us.

James and John, Peter and Paul,
Mary beloved the fulness of light.

On Michaelmas Eve and Christmas,
We will all taste of the bannock.[20]

Interestingly, airts are the cardinal directions, associated with the winds, or airs. According to F. Marian McNeill in *The Silver Bough*, Scottish magic "was always a very practical affair: it had to bring rain to the crops, fish to the nets, a loved one to the lover's arms, a child to the barren womb; to protect from misfortune, avert the evil eye, or bestow victory in battle." She adds, "There are two elements in the magical act—the Spell and the Rite. The spell is the uttering of words according to a formula; the rite is the accompanying set of actions by which the spell is conveyed to the object it is desired to affect."[21] The rune mentioned in the consecration of the seed, above, is in this case what McNeill refers to as the Spell, and the actions and orientation to the direction indicated by the Sun and the northern airt is the Rite. It will be remembered that the north, which the Scottish paterfamilias performing the ritual turns his back to, is the direction of cold, dark, and misfortune. The fact that the whole family turned up to the field in their best attire shows us the reverence they paid to their crops and the Creator who provided them.

James George Frazer records multiple instances of harvest customs from all over Europe in which the last sheaf of grain cut by the harvest crew is the subject of various ceremonies, in which it is often referred to by feminine epithets, such as the Corn Mother, the Old Woman, Cailleach among Gaelic speaking Scots, the Hag or Wrach in Wales, sometimes the Maiden, sometimes the Bride. Often, the unfortunate laborer who cut the last sheaf was subject to hazing and pranks. The last strip of the crop was often left uncut, because no one wanted the misfortune attendant upon being the one to cut the Hag. It is this Old Woman, the personification of famine, referred to in the incantations above as the Carlin, that the crofters sought to avoid by magical means. In other districts, the last sheaf was made into a doll or dressed in a girl's dress and carried by a young woman who was herself the Harvest Queen.[22] Frazer believed these customs to be an example of an animistic belief in a Corn Spirit, present in the last sheaf, a vegetation deity surviving in folk customs until modern times. This view was extremely influential but fell into disfavor in the twentieth century, later interpreters believing that these customs had a more practical intent, encouraging competition among the crews and preventing children from trespassing in the standing crop. Ronald Hutton in *Stations of the Sun* declares the "theory of belief in an animating corn spirit is effectively discounted."[23] Discounted by academics perhaps, but those who have seen the play of the wind in a waving crop of barley or wheat—which flashes almost silver in the breeze, as if something unseen were running in it—or noticed how the last strip of standing grain almost crackles with vitality as it is being reaped or cut by the combine, clouds of insects and frightened rabbits that had sought refuge in the ever-dwindling strip of grain flying out in all directions, will know for certain that there is a spirit of the grain.

PRAYERS TO THE EARTH GODDESS

The pragmatic Anglo-Saxon magical herbalists and farmers would not hesitate to pray to Woden, or Mother Earth, alongside Christ and the Virgin. Several of the manuscripts contain prayers to Mother Earth, which antiquarian and folklorist Hilda Ellis Davidson referred to as "an isolated piece of evidence for a pre-Christian goddess in England."[24] Here is a passage from the Æcer-Bōt, a ritual for blessing the land before plowing, which incorporates a petition to the Earth Goddess:

> *Erce, erce, erce*, mother of Earth,
> May the Almighty, the eternal Lord, grant you
> Fields flourishing and bountiful.
> Fruitful and sustaining,
> Abundance of bright millet-harvests,
> And of broad barley-harvests.
> And of white wheat-harvests.
> And all the harvests of the earth!
> Grant him, O Eternal Lord,
> And his saints in Heaven that be,
> That his farm be kept from every foe,
> And guarded from each harmful thing
> Of witchcrafts sown throughout the land.
>
> Now I pray the Prince who shaped this world,
> That no witch so artful, nor seer so cunning be
> [That e*er] may overturn the words hereto pronounced.
>
> Then drive forth the plough and make the first furrow.
>
> Then say:
>
> "All hail, Earth, mother of men!

---

*Square brackets in the source material likely denote a lacuna in the original text.

Be fruitful in God's embracing arm,
Filled with food for the needs of men."

Then take meal of every kind, and have a loaf baked as big as will lie
in the hand, and knead it with milk and with holy water, and lay it under
the first furrow. Say then,—

"Full field of food for the race of man,
Brightly blooming, be you blessed
In the holy name of Him who shaped
Heaven, and earth whereon we dwell.
May God, who made these grounds, grant growing gifts,
That all our grain may come to use!"
Then say thrice, "*Crescite, in nomine patris sitis benedicti*. Amen"
and Paternoster thrice.[25]

This quotation from a more elaborate ritual shows us how important ceremonial observance in the operations of agriculture was to the Anglo-Saxon people in the period that this manuscript was written, probably during the eighth century CE. The people of the culture that wrote these manuscripts were nominally Christian. But in their private observance, on the land, they saw fit to honor the older chthonic powers when engaging with the spirits of land and plants. The country people still lived among the elemental forces of nature, one poor harvest or injury away from starvation and ruin. The warmth of the summer and the coming cold and dark of the winter very powerfully symbolized the battle between gods of light and of darkness, the Aesir and Jotuns. This became Christianized into the battle between Christ and the Devil, but the substance of their lives remained unchanged regardless of the names they heard in the Sunday sermon.

The Saxon healer magicians also observed great ceremony in the harvesting and administering of their medicinal herbs, owing to the great reverence they had for the spiritual forces indwelling the plants, according to Rohde, who writes: "Some of the most remarkable passages in the manuscripts are those concerning the ceremonies to be observed both in the picking and in the administering of herbs. What the mystery of plant life which has so deeply affected the minds of men in all ages and of all civilisations meant to our ancestors, we can but dimly apprehend as we study these ceremonies. They carry us back to that worship of earth and the forces of Nature which prevailed when Woden was yet unborn."[26]

Additional evidence that the common people were still following the old ways can be gained from the content of ecclesiastical pronouncements against the Heathen customs of gathering herbs with incantations. The practice of harvesting plants for medicine and magic with attendant prayers is also recorded in sources as diverse as the Greek Magical Papyri and the ancient Roman *Precatio Omnium Herbarum*. Gathering herbs with prayer and intention is a nearly universal practice in all times and places. Speaking to the spirit of the plant and the divinities who preside over them, and commending their use to the purpose for which they were gathered, puts the magician or healer in intimate connection with the animate force of the plant itself. Here is one of the prayers to Mother Earth that was intended to accompany the harvesting of medicinal herbs, from a twelfth-century herbal manuscript:

Earth, divine goddess, Mother Nature, who generatest all things and bringest forth anew the sun which thou hast given to the nations; Guardian of sky and sea and of all gods and powers; through thy power all nature falls silent and then sinks in sleep. And again thou bringest back the light and chasest away night and yet again thou coverest us most securely with thy shades. Thou dost contain chaos infinite, yea and winds and showers and storms: thou sendest them out when thou wilt and causest the seas to roar; thou chasest away the sun and arousest the storm. Again when thou wilt thou sendest forth the joyous day and givest the nourishment of life with thy eternal surety; and when the soul departs to thee we return. Thou indeed art duly called great Mother of the gods; thou conquerest by thy divine name. Thou art the source of the strength of nations and of gods, without thee nothing can be brought to perfection or be born; thou art great, queen of the gods. Goddess! I adore thee as divine; I call upon thy name; be pleased to grant that which I ask thee, so shall I give thanks to thee, goddess, with due faith.

Hear, I beseech thee, and be favourable to my prayer. Whatsoever herb thy power dost produce, give, I pray, with goodwill to all nations to save them and grant me this my medicine. Come to me with thy powers, and howsoever I may use them may they have good success, and to whomsoever I may give them. Whatever thou dost grant, may it prosper. To thee all things return. Those who rightly receive these herbs from me, do thou make them whole, goddess, I beseech thee; I pray thee as a supplicant that by thy majesty thou grant this to me.

> Now I make intercession to you, all ye powers and herbs and to your majesty, ye whom earth, parent of all, hath produced and given as a medicine of health to all nations and hath put majesty upon you, be I pray you the greatest help to the human race. This I pray and beseech from you, and be present here with your virtues, for she who created you hath herself promised that I may gather you with the goodwill of him on whom the art of medicine was bestowed, and grant for health's sake good medicine by grace of your powers.
>
> I pray, grant me through your virtues that whatsoe'er is wrought by me through you may have in all its powers a good and speedy effect and good success and that I may always be permitted with the favour of your majesty to gather you with my hands and to glean your fruits: so shall I give thanks to you in the name of that majesty which ordained your birth.[27]

This prayer itself is a combination of two prayers from Ancient Rome, the *Precatio Terrae* and the *Precatio Omnium Herbarum*, with some of the more problematic (to the Christian scribes who transcribed it) Pagan references to Apollo, Pluto, and the gods removed. It is interesting that the scribes were not troubled by adoration to Mother Earth; her obvious goodness, virtue, and universal appeal overrode the fact that this is still a Pagan invocation of the kind forbidden by other contemporary churchmen. Thus it seems that in our sources we do not find evidence for the survival of Pagan practice but rather tantalizing hints that people were still interacting with the same spiritual forces of land and fertility, conversing with the same plant spirits, and operating with the same instinctive clairvoyance as in earlier times. The Latin prayer was preceded by a multitude of Heathen versions in the vernacular, recited by generations of herbalists and cunning folk whose words are lost to history. It is not hard to imagine what they would have said, addressing their own petitions to the gods of healing and the spirits of the place and the plants and making their humble and heartfelt offerings.

TOWARDS A RE-ENCHANTED AGRICULTURE

In my writing, I realize, I have given voice to some of the older schools of folklore and anthropology, ones that have fallen into disfavor in modern scholarship, and this is because I believe, though Frazer may have overstated the pervasiveness of the animist hypothesis, that he was nevertheless onto something. Although there may or may not have been Pagan fertility cults surviving into the nineteenth and twentieth centuries in the customs of rural people, the aliveness and intelligence in nature, its spiritual presences, is there for anyone with eyes to see, including people from every generation. "Paganism" and animist modes of understanding do not have to survive from the distant past in an unbroken tradition; they are reinvented anew in each child who looks at nature with wonder and curiosity, sprung from our innate human nature.

I have provided a few examples of sacred agriculture, one that honors the gods, the Earth, and her spirits, as well as the creative Source of all things, and keeps them in mind in all of our actions in relation to the way we earn our bread. To me this is not idle speculation; all of these practices that I describe here are simple ways to put this philosophy into practice, not necessarily by slavishly imitating the past, but by adopting, in spirit, the attitudes embraced by the cultures we have examined. These ancestral practices, which are now lost to our peril, kept us in balance with nature, mitigating with reverence the worst of our hubris. Many of the actual examples I have discussed here are from cultures like the ancient Romans', which were rapidly falling away from their animistic heritage but hadn't forgotten it completely, as our culture seems to have.

Agriculture, in spite of the advances in the field of chemistry and the ever-increasing scale of mechanization, remains very much an art as much as it is a science. So much comes down to the farmer's knowledge of and intimacy with the land and the workings of their craft. Every farm, every region, every field, and even each part of the field is different. Experience and practice, gained through trial and error—and failure, which is the best teacher—show each person the way to work their own land, whatever piece of it they are given to tend. The modern methods of massive machinery with GPS-guided seed drills and fertilizer applicators have replaced countless farmers living in intimacy with their land. The process of mechanization has been going on for a long time, beginning with the invention of the ard so many millennia ago. Where I live, the number of full-time farmers is dwindling every year due to several factors, including depressed farm economics, increased concentration of land in fewer and fewer hands, and the fact that farming can be a terrible career, with long hours, low returns, and a great deal of financial and personal risk. Sometimes, it seems like intimacy with the Earth means being either muddy or dusty much of the time and having a sore back and knees, something fewer people want to put up with in their daily lives.

I don't want to idealize the farming life for you, reader. To be sure, to some extent, it is better that many have been freed from the obligation to work the land in order to gain their sustenance. It has allowed us to have all of the gifts of higher culture. It begs the question, however: Have we gone too far? The sense of reverence toward the land and its spiritual forces, the ecological sensitivity, the intimate contact with the elements and familiarity with nature and its moods have been lost to most, replaced with a hollow consumerism and a need to be constantly entertained by electronic media. With eight billion of us sharing this planet, most living in cities and many in megacities, how can we regain what has been lost? It seems unclear what interest a dweller in a megacity, who spends most of their time interacting with electronic media and eating food produced by large-scale agriculture and processed in factories by machines, whose interaction with nature is distant and abstract, can have in reinvigorating and re-enchanting their relationship with that same nature.

So, this is where we are, in the twenty-first century, the heirs of a long history of culture and conflict. It is a fascinating place to find ourselves, with the knowledge of the ages literally at our fingertips, but, for many of us, a disconnection from the Earth that gives us life. If you, like me, have been born with a hunger for nature, a love of dirt and growing things, then this disconnection will be very unsettling and will move you to prioritize seeking that connection and finding modes of living that allow for intimate contact with the land and its presences. There are many opportunities to exercise a green thumb in this world, from internships and caretaking opportunities, to community gardens, to backyard farming, to joining a community-supported agriculture program run by cunning farmers, to, if you can afford it, buying some rural land. Whatever you do, get out of Babylon if you can. Megacities are no place for souls like ours to live.

19

SACRED FORESTS AND SPIRITUAL PRACTICES WITH TREES

Let us pause for a moment to take a walk into the ancient forest that once covered the land in times past. We halt at the edge of the clearing and pass through a nearly impenetrable hedge of brambles and low bushes that mark the boundary between the wood and the field. We ask permission from the forest spirits to enter their realm. We know the gods and land wights give signs through the happenings of nature and the movements of animals, so we wait for a few moments for a signal. As if to answer, a gentle breeze trembles the leaves around us and the cheery song of a wren calls to us from a nearby tree. Taking this as an affirmative answer, we proceed down the steep slope into the wooded valley. It is late afternoon, and the Sun is sinking fast. We will not have much time before the lowering dark surrounds, leaving us alone in the cold night. We notice the dappled light shining down from the canopy onto the dim forest floor, the towering trunks of the ancient trees. We hear the rustle of the leaves as the deer move quietly through the hidden paths. The squirrels bark at our clumsy trespass, and the raucous calls of crows and jays warn all that we are in their domain. The rich smell of the damp fallen oak leaves and the aromatic scent of the cedars fill the air. Following the ancient path, we cross a dry creek bed and start up the hill on the other side. Carefully, we climb the steep rocky path from the valley until we reach a circular clearing in the midst of a grove of ancient white oak trees.

In the midst of the clearing stands a great and venerable oak tree, its spreading canopy of branches bare of leaves. Under the tree, a short distance from the thick trunk, is a large flat stone, which serves as an altar table. With humble reverence, we approach the sacred table, and with bowed heads, we unshoulder our packs and take out the humble offering we have brought for the gods of field and forest as well as Mother Earth. We place an earthenware jug of ale on the table, as well as a loaf of bread and another jug of milk. We are here to give thanks for the gifts of the forest today, because we have had enough firewood and food from successful hunting and medicine from forest herbs to survive through the long winter, and we are grateful. We voice our prayers and pour our libations onto the soft earth, saying the ancient oration:

We call you, Ancient One

We call you, Wild One

We call you, the Green One

We call you, Hair-Covered One

We call you, He Who Roams the Wild Wood

We call you, He Who Guards the Boundaries of the Land.

We call you here, Ward, Defender, Bringer of Bounty, Lord of the Animals

Mighty One Who Roars Through the Land.

Wind Bringer

Rain Bringer

Deer Hunter

He of the Oak and Ash

Come to us!

Be present here in the Hallowed Grove,

Where from ancient times your people have called you.

Accept these gifts which we offer you today from the bounty you have given us.

Meat and milk from your herds.

Beer and meal from your harvest.

Bless us, we ask, mighty one,

With good health and good fortune.

Enough game, healthy herds,

Wood for homes and hearths,

And a bountiful harvest.

As we wait in silence for a sign that our offer was appreciated, a huge stag runs into the clearing and stops for a moment to stare at us, clouds of steam blowing from his nostrils. He gives us a second look, snorts loudly, and runs off again. We comment on the auspiciousness of this sign and slowly walk home in silence to our farmstead through the darkening forest, feeling blessed by the gods.

This brief imaginative exercise gives us a hint of the role forest and sacred groves played in the religious lives of the people who lived in them. In this chapter, we will be exploring the role of the forest and trees in the life and practice of the cunning farmer. The woodland is a feast for the senses, in whatever season. I have enjoyed some of the finest hours of my life simply wandering in the woods among the trees, just to see what I would encounter and enjoying the peace and sacredness of the place. In this I am not unique. The forest and its trees have been places of encounter between humans and numina, or divine presences, since our ancestors first learned to perceive them, in the shadowy depths of mythic time.

✦ *A massive and majestic American white oak tree in winter near the author's farm.*

In this chapter, we will discuss the role of the forest as a sacred place and home of land spirits, in history and folklore, and as a source of some of the materials necessary for life on the land, namely firewood, lumber, game, medicinal plants, and materia magica. We will also investigate the rich symbolism of the tree in animist cosmologies, as both axis mundi and tree of life. We will close by discussing some practices to help connect us with the spirit presences of the woodland trees.

DEITIES OF THE FOREST

From earliest times, trees have been considered dwelling places for spiritual beings—numina, as the ancient Roman called them, dryads and hamadryads to the Greeks. These were the beings responsible for the feeling you get, while walking in the woods, that you aren't completely alone. As Ovid said, "Under the Aventine there lay a grove black with the shade of holm-oaks; at the sight of it you could say, 'There is a spirit here.'"[1] I have felt this feeling many times when walking through the woods, the sense of presence and sacredness amid the stillness and the sounds of bird song and the wind roaring through the branches. In ancient times, many of the cultures labeled "Pagan" revered trees either as images of the divine or habitations of spirits, as sites for communicating with the gods by receiving oracles or offering sacrifices.

In addition to lesser spirits of the forest, gods were seen to inhabit the forests as well. Ancient Roman traditional religion did include the cult of a god of the forest, whom they called Silvanus, a word meaning "the one who manages the forest," often associated with but not identical to Mars. He was worshipped, according to Cato, in the forest. He was associated with sacred trees and groves throughout the Roman world and was the embodiment of "trees, shrubs, and branches." He presided over the liminal zone between the wild, uncultivated spaces and the land being brought under cultivation. He later became associated with agriculture and pasturage. He is frequently pictured as a man, old or rugged, having a beard and described as *horridus*, or hairy, and holding a falx, a sickle, and a tree branch, and pictured with a dog at his side. Silvanus was the guardian of the limes or boundaries of the farm and had much in common with the domestic deities, the lares, we have discussed in other chapters.[2] The cult of Silvanus was syncretized with that of other deities including Pan and Faunus, blending his attributes with those of the other gods, although there were major differences. Faunus and Pan, though sometimes conflated with Silvanus, had fairly different cults, attributes, and personalities in practice, sometimes showing a decidedly

◆ *A rustic statuette of Silvanus, the Roman god of the forest. Image by Wellcome Images, London.*

dark side. Faunus was noted for his sexual aggression and was active after dark, causing nightmares. His home is in the ancient primeval forest and uncultivated regions. St. Augustine, perhaps conflating Silvanus with Faunus, accuses Silvanus of endangering women in childbirth and threatening agriculture unless appeased with a sacrifice. Faunus represents, in my opinion, the darker, more transgressive side to the spirit of the forest, an amoral animal force of the wild wood. He is the forest after dark, far from the hearthfire of the home, the source of panic, the fear which comes from Pan.[3] In his guise as Faunus and Pan, the more kindly and dignified Silvanus shows his overtly sexual side. In iconography, Pan is often pictured *in flagrante delicto* with nymphs and even goats, whereas Silvanus is often naked but never ithyphallic or in amorous embrace.[4]

In the *Interpretatio Romana*, the system whereby the ancient Romans equated the gods of conquered foreign people with their own, Roman gods were identified with the gods of the nations of the Celtic and other peoples. In Britannia,

♦ *A votive relief of Silvanus blended with the attributes of Pan, from Croatia, c. second to third century CE (in the Archaeological Museum in Split). Image by Carole Raddato.*

Silvanus was identified with two Celtic deities, Coccidius and Vinotonus. Peter Dorcey offers this interesting information about a Silvanus inscription from near Hadrian's Wall, "At a number of sites near the Wall altars depict a naked or semi-nude, phallic horned deity, the so-called 'Horned God' of the local British tribe, the Brigantes. He is sometimes called Mars, sometimes Mercury. At Moresby he is equated with Silvanus. This is the only instance in which Silvanus is ithyphallic or theriomorphic."[5] Through Silvanus, foreign gods of forest and grove were introduced to the Roman pantheon all over the Roman territory.

The Bigfoot Myth

The whole Silvanus/Faunus/Pan complex is reminiscent of the modern American phenomena of Bigfoot, a hairy, sometimes-benign, sometimes-frightening humanoid dweller in the primeval forest, capable of inducing panic in those who trespass on his silvan sanctuaries. When you read the account later in this chapter about the Druidical grove from Lucan's *Pharsalia*, notice that even the priest—the Druid himself—feared the approach of the Lord of the Grove. The link between the European god of the forest and the elusive North American cryptid was the subject of a 1983 book by Jim Brandon called *The Rebirth of Pan: Hidden Faces of the American Earth Spirit.*[6] Brandon's book, which is very speculative and a bit far-out but intriguing nonetheless, discusses some of my favorite themes of Earth energies and the Hopewell and Adena sacred sites, tying them to the Bigfoot mystery and the Pan myth. I have not personally seen Bigfoot, but I have heard some loud knocking in the woods at night. Not too long ago, on an evening walk in the woods to visit my sacred spot, I heard an eerie, humanlike noise, somewhere between a scream and a howl, uncomfortably close by, followed by a similar call in answer, farther away. I heard several frightened deer fleeing the source of the noise down the hill from me. I imitated the first call and was answered once, but then nothing more and decided to go home. Whatever was calling wasn't a coyote; I'm very familiar with their calls and howling. Perhaps the forest numina were afoot in the darkening woods that evening.

THE WORLD TREE IN SHAMANISM AND NORTHERN MYTHOLOGY

Imagine a towering tree, the kind you might find in an old-growth forest, roots sunk deep in the dark earth, crown extending into the heavens, towering above you. It appears to be holding up the sky. It is a living pillar making the connection between heaven and Earth. If you could climb the branches, you could ascend to the very heavens. If you could enter within and descend down the column to the roots, you could enter the underworld itself. The branches are like levels of the cosmos, ascending higher and higher, worlds apart, cosmic realms where gods and spirits ascend and descend like earthly birds, down to the dark world of shadows below in which we pass our lives.

As discussed in earlier chapters, this tree is the very center of the cosmos, the world axis of the shamanic cosmology. The symbol of the world axis is central to the shamanic cosmology that ultimately underpins many of our magical

♦ The Ash Yggdrasil, *by Friedrich Wilhelm Heine, 1886.*

systems, at least symbolically. In many of these shamanic cultures, the axis is a tree. The tree is the cosmos. Its roots are the underworld, the world of the dead; its branches are the divine realm; the heavens above are in the realm of the Creator; and the ground on which the tree stands is this world, the middle realm. The idea of the tree as divine symbol of the cosmos, an *imago mundi*, and home of the gods was central to many of the Pagan cultures we have also discussed in this work. The symbol of the tree is of such paramount importance that it plays an important role in mythologies and religions worldwide. World trees feature in shamanic traditions on every continent, many containing similar features of cosmic levels, ascended by visionaries. The tree as sacred symbol is an ancient and nearly universal motif in world mythology. Scholar of mythology Mircea Eliade detailed many of them in his book *Shamanism: Archaic Techniques of Ecstasy*, to which I refer the interested reader.

The world tree of central and north Asian shamanism seems to have entered Norse and Germanic mythology at some point as Yggdrasil, the World Ash that supports the cosmos, whose trunk, roots, and branches frame the Nine Worlds of Norse cosmology. Yggdrasil exemplifies the archetype of the sacred tree. As the "Voluspa," the Eddic poem that tells of the creation and destruction of the gods and the cosmos, says of it:

> *I know an ash tree*
> *named Yggdrasil*
> *a high tree, speckled*
> *with white clay;*
> *dewdrops fall from it*
> *upon the valleys;*
> *it stands, forever green,*
> *Above Urth's well.*[7]

The tree has three great roots, which reach through three of the worlds. Watering the root that passed beneath Jotunheim, the land of the giants, was the spring of Mimir, which gave to the one who drank them wisdom and understanding. A second root passed into the land of the Æsir, the gods, and was watered by the Well of Urd; this was the Well of Fate. The third root passed into Hel, the land of the dead, and was watered from the spring Hvergelmir. The Well of Fate was the dwelling place of the three Norns, the Norse Fates, who controlled the destiny of all things, spinning the thread of life. Their names are Urðr, Fate;

Verðandi, Being; and Skuld, Necessity. The Norns water the roots of Yggdrasil with the clear cool water of Fate from the well and coat the trunk with wet white clay, giving it life, and the tree returns the water to the Earth as dew and rain.[8,9]

This image of the goddesses of fate watering the world tree with the waters of the Well of Urðr is rich in symbolic meaning. The water of fate, its essence, is the substance that sustains the cosmos itself, the very sap of the world tree. Fate is inseparable from existence itself. This is the concept the Anglo-Saxons called *wyrd*, the Old English word for Urðr, the underlying fate pattern of all beings. No one can avoid it because it is inescapable. As the anonymous poet who wrote the Old English poem "The Wanderer" said, *"Wyrd bið ful aræd,"* meaning "Destiny is all," or "Fate is inexorable." I find the fatalism of Northern mythology strangely comforting. If fate is woven into the fabric of the universe, there's no use struggling against it. One must just get down to the business of living the best life one can under the circumstances one has been given. Stop blaming God, or the gods, or other people for what is one's allotted fate, as if things could have been different.

♦ *The three Norns—Urðr, Verðandi, and Skuld—seated beneath Yggdrasil. By Ludwig Burger, 1882.*

Beneath the tree, a great serpent and multitudes of smaller ones gnaw continually at the roots, and a great eagle perched on top of the tree makes perpetual war with the serpent. Four stags continually eat the tree's foliage. As the *Prose Edda* says, "Yggdrasil suffers hardships more than people realize. Stag bites above, at the sides it rots, Niðhöggr eats away at it below." This shows us that the cosmos is continually in peril from the forces of disorder and decay, or at least that chaos and entropy are part of the life of the cosmos and have their place in the divine order.[10]

Yggdrasil supports the Nine Worlds, home to the various supernatural and natural creatures of the Cosmos. Starting from the worlds below the roots of the tree, the worlds are Helheim, the world of the dead; Jotunheim, the world of the giants, or Jotun; Niflheim; the shadowy underworld; Darkalfheim, the world of the gnomes or dark elves; Midgard, Middle Earth, our world; Muspelheim, the world of fire; Alfheim, world of the elves; Vanaheim, world of the Vanir gods; and Asgard, world of the Æsir gods, at the top of the tree.[11] A rainbow bridge, Bifrost, linked Asgard to the lower realms. This schema is not canonical, and I have seen it arranged differently in different secondary sources. It is interesting that Nicolas De Mattos Frisvold, in discussing the names of the worlds of Norse cosmology, sometimes translates the Eddic word *heima* as "farmsteads" instead of worlds, saying, "The medieval Norsemen were using the material world for a reference for understanding the other side."[12]

Oðinn's Sacrifice on Yggdrasil

In Scandinavian Paganism, Yggdrasil was an image of the spiritual cosmos, connecting the world of the gods and other spiritual beings of Norse mythology with the human realm. It was also intimately connected with the myth of Oðinn, the All-Father, god of wisdom and magic. The word *Yggdrasil* translates as "the steed of Ygg," which is an epithet of Oðinn, which means "the terrible one."[13] At the Well of Mimir, beneath the tree, Oðinn sacrificed his eye for one drink of the water of wisdom and in return was gifted with power in the art of *seiðr*, a type of magic that is difficult to explain briefly but was probably a form of spirit possession and was usually considered to be the province of women. Oðinn sacrificed himself on the tree, hanging for nine nights, pierced by a spear, and gained the power to use the runes, power in another of the magic arts. The parallels to shamanic initiatory ordeals are obvious and were noted by Mircea Eliade.[14] Oðinn ascends the tree in ecstatic flight, between life and death, learning the ways of fate, returning to life, possessed of new magical power.

For the shamans of the Norse tradition, spirit travel enabled the traversing of the levels of the world tree, often in animal form, allowing one to converse with the beings there and gain knowledge of the geography of the nine realms. It is to be remembered that the cosmological and eschatological poem, the "Voluspa," literally meaning the "Vision of the Seeress," is written as a vision by a visionary, a *volva* in Old Norse, given to the god Oðinn, of the beginning and end of the cosmos. It is probable that the content of this and most other cosmological myths and tales of imaginal travel have their origin in the visions of seers, in soul travel, anabasis and katabasis in visionary states, and that these tales and myths form a map of the shifting geography of the imaginal realm.

SACRED TREES IN THE NEAR EAST

As we discuss sacred trees and groves in places as far removed as Assyria, ancient Palestine, Gaul, Britain, Germany, and Scandinavia, we will see similar themes in all of them. All bear witness to a pattern of repression by conquering nations and factions who destroy the symbols that give meaning to the conquered culture in an attempt to sever its link to the land that gives it life. Trees are a representation of these cultures' cosmologies and the dwelling places of their divinities. Destruction of their sacred groves and holy trees is the first step in cultural genocide. In an unfortunate drama that has been played out innumerable times throughout history and is still being played out today, genocide begins with ecocide. Destroy the sacred trees, cut the sacred groves, ban the ancient rites, slander the old ways with a propaganda campaign, preferably by erasing all counternarratives, and impose the conqueror's cultural traditions by force.

The biblical tradition of the ancient Hebrew people had its own version of the sacred tree myth, which stems from the biblical Book of Genesis. The Bible tells that the creator god Yahweh Elohim "planted a garden in Eden in the east; and there he placed the man he formed. And out of the ground Yahweh Elohim made to grow every tree that is pleasant to the sight and good for food, the tree of life also in the midst of the garden, and the Tree of Knowledge of Good and

♦ Assyrian Tree of Life, from 865–860 BCE (in the British Museum). King Ashurnasirpal appears twice in front of the sacred tree, accompanied by protective spirits. Image by Sanjar Alimov.

Evil."[15] A river flowed out of the Garden of Eden, which became four rivers—the Gihon, the Pishon, the Tigris, and the Euphrates—which flowed from the Tree of Life at the center of the world, like a mandala, a central tree with four rivers watering the whole world. In the familiar tale, Adam and Eve eat the fruit of the Tree of Knowledge but are expelled from the garden for their disobedience before they have a chance to eat the fruit of the Tree of Life. It is possible that this story is related to Assyrian tales of the Tree of Life, a legendary cedar or date palm known as the *kiškānû.*

Assyriologist Archibald H. Sayce related the story of the kiškānû with both the myth of Tammuz and the Garden of Eden. Tammuz was the son of Ea, known to the Sumerians as Enki, and the goddess Dav-kina, and his original home was the "garden" of Edin, or Eden, which, according to Sayce, in Babylonian tradition, was located near Eridu, twelve miles from the city of Ur in what is now Iraq. This ancient hymn preserved in a cuneiform tablet tells of the tree of Tammuz in Edin:

1. (In) Eridu a stalk grew over-shadowing; in a holy place did it become green;

2. its root ([*sur*]*sum*) was of white crystal which stretched towards the deep;

3. (before) Ea was its course in Eridu, teeming with fertility;

4. its seat was the (central) place of the earth;

5. its foliage (?) was the couch of Zikum (the primæval) mother.

6. Into the heart of its holy house which spread its shade like a forest hath no man entered.

7. (There is the home) of the mighty mother who passes across the sky.

8. (In) the midst of it was Tammuz.

9. [This line is missing.]

10. (There is the shrine?) of the two (gods).[16]

Interestingly, Sayce explicitly mentioned a similarity between the tree of Tammuz and Yggdrasil, writing the following:

> The description reminds us of the famous Ygg-drasil of Norse mythology, the world-tree whose roots descend into the world of death, while its branches rise into Asgard, the heaven of the gods. The Babylonian poet evidently imagined his tree also to be a world-tree, whose roots stretched downwards into the abysmal deep, where Ea presided, nourishing the earth with the springs and streams that forced their way upwards from it to the surface of the ground. Its seat was the earth itself, which stood midway between the deep below and Zikum, the primordial heavens, above, who rested as it were upon the overshadowing branches of the mighty "stem." Within it, it would seem, was the holy house of Dav-kina, "the great mother," and of Tammuz her son, a temple too sacred and far hidden in the recesses of the earth for mortal man to enter.[17]

The similarity between the two trees must be purely archetypal, as far removed in time and space as the cultures that created them are. But we do see similar tree symbolism in much nearer central Asia, the home of the Indo-European cultural ancestors of the Germanic Norse.

Sayce's identification of the kiškānû with the biblical Tree of Life has fallen out of favor with scholars, but it is interesting: a sacred tree in a garden with divine beings, in the center of the world, sounds much like the Edenic trees as well. Again, the link may be the archetypal universality of the tree motif, rather than cultural influence. However, after the first siege of Jerusalem in 597 BCE and the fall of the Kingdom of Judah, many Judean people were brought to Babylon by their conquerors and were held captive there for nearly sixty years, until 538 BCE,[18] which would have been several generations, during which time many would have absorbed a large amount of Babylonian culture. Some parts of the creation narrative in the Book of Genesis were believed to have been redacted and compiled from earlier sources during the captivity,[19] so it is possible that Babylonian myths of a primal garden of the gods and sacred trees with magical properties were influences on the narrative. It is also probable, owing to the close proximity of Judah, Israel, and Mesopotamia, and to other linguistic and cultural ties, that the neighboring cultures shared many mythological traditions in common.

Assyriologist Simo Parpola, in a paper titled "The Assyrian Tree of Life: Tracing the Origin of Jewish Mysticism and Greek Philosophy," asserted that the source of the famous cosmological diagram known as the Tree of Life, the Otz Chaim, which is central to the medieval Jewish mystical system known as the Kabbalah, originated with the Tree of Life motif central to the Mesopotamian worldview. Parpola tells us that the Assyrian tree "symbolized the divine world order" and that "it could also be projected on the king to portray him as the perfect man."[20] The kabbalistic tree also symbolizes the divine order. It describes a graded system of divine emanations from a transcendent source to the material world. The divine emanation passes through ten levels, known as Sephiroth, in the creation of the material realm, which is sustained by pulses of divine efflux moment to moment. Many systems have been formulated to equate these Sephiroth with gods, angels, and planetary intelligences, in a manner highly influenced by and influential on Gnosticism and Neoplatonism. Parpola adds his own nuance by graphing the Babylonian gods on the tree, in the Sephiroth he deems appropriate to their own powers and attributes, in a way that seems both natural and appropriate.[21]

Parpola asserts that "the emergence of Kabbalah as a doctrinal structure can now be reliably traced to the first century A.D. The renowned rabbinical schools of Babylonia were the major centers from which the Kabbalistic doctrines spread to Europe during the high Middle Ages."[22] Placing the development of this influential system of Jewish mysticism well into antiquity presents a very good case for Babylonian origins to this "branch" of the myth of the sacred tree, one with its roots in the supercelestial realm of the ultimate transcendent.

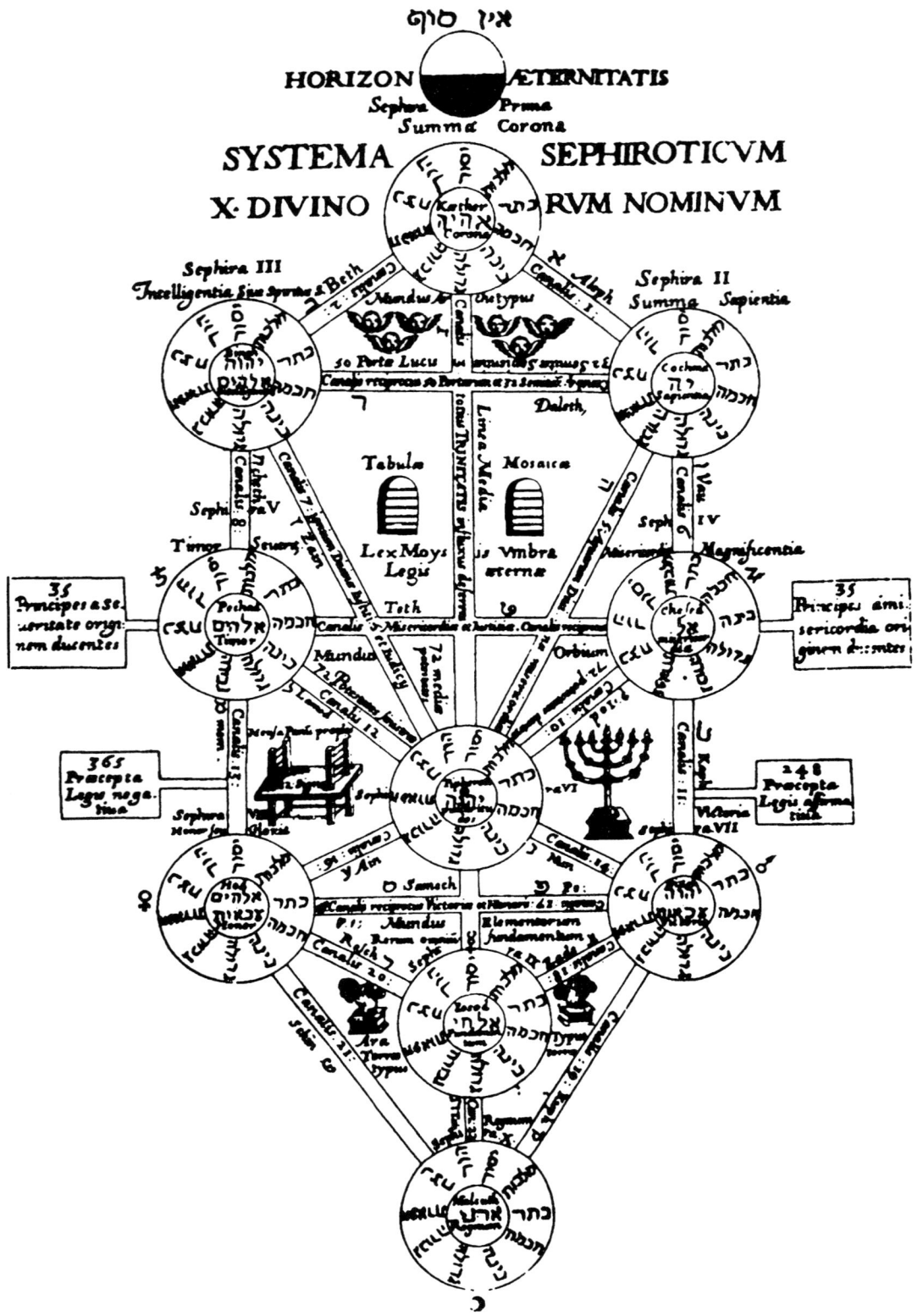

✦ *The kabbalistic Tree of Life as pictured by the Renaissance mystic*
Athanasius Kircher in 1652.

The Goddess Asherah

Another part of the story of Middle Eastern and Jewish myths of the sacred tree is the presence in the biblical tradition of a being or object called Asherah, who is at times described as either a goddess or a sacred pole or tree of some sort, and in all likelihood is both. She may be associated with the Babylonian traditions of the sacred tree described above and be related to Dav-kina, the consort of Ea and mother of Tammuz. Some historians have proposed that Yahweh or perhaps El, who was subsumed into the Yahweh cult, are forms of Ea/Enlil, and Asherah has been thought to be the consort of Yahweh, so there is actually a chance that these are yet two more branches of the same tradition.

In ancient Palestine, the traditional religion of the ancient Israelites, before the Yahweh cult rose to ascendancy, involved the worship of a pantheon of divinities, including the goddess Asherah. Asherah was a Canaanite fertility goddess, divine consort of the chief god El, who was later absorbed by the Yahweh cult.[23] The cult of Asherah seems to have involved trees, groves, and wooden pillars, set up in mountaintop sanctuaries throughout Israel and Judah. Indeed the plural *asherim* is used mostly to refer, scholars say, to wooden poles and carved cult

✦ An Asherah figurine from Israel (in the Israel Museum). The goddess is pictured giving birth to twins, the gods Shahar and Shalem. Image by עליון-פעמי.

objects and sometimes live trees representing the goddess. On the basis of several inscriptions that mention "Yahweh and his Asherah," some authorities believe that she was worshipped as a consort of Yahweh in southern Palestine. The Old Testament mentions the word *AShRH* forty times in nine books. Many of the mentions record attempts by the leaders of the Yahweh cult to wipe out the cult of the goddess.

Ancient Palestine was a region of considerable competition among various cults and factions. For example, King Solomon became polytheist in his old age, worshipping "Ashtoreth the goddess of the Sidonians and Milcom the abomination of the Ammonites."[24] Scholars believe that Ashtoreth and Asherah, despite the similarity in their names, are actually different goddesses. Solomon also built a temple, "a high place for Chemosh the abomination of Moab and for Molech the abomination of the Ammonites, on the mountain east of Jerusalem. So he did for all his foreign wives who burned incense and sacrificed to their gods."[25] Under his successor, his son Jeroboam, the polytheism continued in the kingdom of Judah: "And Judah did what was evil in the sight of Yahweh, and they provoked him to jealousy with their sins they committed, more than that of their fathers had done. For they built for themselves high places and pillars and Asherim on every high hill and under every green tree."[26] In many of the verses in the Old Testament that mention Asherah, she is consistently equated with a living tree. Biblical scholar Tilde Binger in her book exploring the goddess, *Asherah: Goddesses in Ugarit, Israel and the Old Testament*, after reviewing the textual evidence in the Hebrew scriptures, concludes, "It is possible that the goddess was not only represented by a statue or some kind of idol in the cult, but was seen as being present in certain specimens of a particular tree: the oak or the terebinth. If this is the case—which it might very well be . . . the connection is made. The Elah [another Hebrew word associated with the goddess in several places in the Old Testament] is the goddess, who is also Asherah, and is the tree."[27]

In order to extirpate all trace of Canaanite polytheism from the region of Palestine, the Yahweh cult instituted a campaign of cultural genocide, which included desecration of the high places, images, and, yes, the groves of the cults of the old gods, as Yahweh tells Moses in the Book of Exodus: "Behold, I will drive out before you the Amorites, the Canaanites, the Hittites, the Perizzites, the Hivites, and the Jebusites. Take heed to yourself lest you make a covenant with the inhabitants of the land whither you go, lest it become a snare in the midst of you. You shall tear down their altars, break their pillars, and cut down their Asherim (for you shall worship no other god, for Yahweh, whose name

✦ Sophia as the goddess of wisdom. Statue from the Library of Celsus in Ephesus, Turkey. Image by Joseph Kranak.

is Jealous, is a jealous God)."[28] Lest we think that the cult of Asherah was an obscure nature religion of a small minority of Israelites, a later text, giving credence to the theory that Asherah was the consort of Yahweh, records the removal of the cult objects of Baal and Asherah from the Jerusalem temple, the religious center of the Kingdom of Judah and of the Yahweh cult. Josiah burned, crushed, and defiled all of the sanctuaries of the traditional polytheistic religion and rounded up their priests: "Josiah put away all of mediums and wizards and all of the teraphim and idols and all of the abominations that were seen in the land of Judah and in Jerusalem."[29]

The goddess of the tree was not so easily banished from the spiritual life of her people, however. Archetypes of the divine feminine have continued to creep into Jewish mysticism and esotericism over the millennia. The Jewish Platonist philosopher Philo of Alexandria speaks of the logos, the divine reason, whose father is God and whose mother is Sophia, Divine Wisdom.[30] Philo uses this metaphor in several of his works, so it probably forms an important part of his metaphysics, as indeed feminine principles do in the metaphysics of other Middle Platonists. In the mystical writings of the Kabbalah, the Shekinah, the divine presence of God, is often spoken of as a feminine being—the "Wife of the King" or the "Bride of the Sabbath"—and, according to noted Kabbalah scholar Gershom Scholem, explicitly identified with the Edenic trees.[31] Archetypes, indeed goddesses, are very hard to kill.

Perhaps it is unfair to judge these accounts in the light of the modern virtues of tolerance and respect for one's neighbor. However, those values were taught first in the Western world by the same prophetic tradition that opposed the Asherah cult, the same values that were spread by a famous descendant of the people of these tales, the Teacher of Nazareth, whose lessons, unfortunately, have gone unheeded by the most zealous of his followers down throughout history, even to the present day. The words in the Old Testament, giving the land of many Indigenous tribes of Canaan to a tribe of invaders and giving a moral warrant for genocide, are still bearing foul fruit in the same land where they were first uttered, three millennia later. And as we will see, these exhortations to ethnic cleansing and not the message of love of one's neighbor are the ones that were brought to northern and western Europe and later the whole world by Christian missionaries, at the point of a sword. And the blade of the axe, for we will also learn that among the first acts of the misguided prophets of imperial Christianity was to fell the beloved trees and groves of the Heathen tribes of the North.

SACRED TREES IN THE ROMAN AND POST-ROMAN WORLDS

The axe-wielding Christian zealots of the Dark Ages were preceded centuries earlier by their Roman Pagan predecessors, who, while waging a war of conquest in Celtic Gaul and Britain, encountered the ancient groves sacred to the traditional Celtic religion and its oak-priests, the Druids. Groves and oak trees were sacred to the Roman traditional religion as well, being descendents of a common Indo-European heritage that held that the oak tree was sacred to the Thunder God in all of his various guises in the many countries where he was honored. In Rome, the Thunder God was Jupiter; in Greece, Zeus; in Gaul and Britain, among the Celtic nations, he was known as Taranis; and by the peoples of the Germanic tribes in modern Germany, England, and Scandinavia, he was known as Donar, Thunor, and Thor, respectively. In all of these cultures, sanctuaries were made for him in sacred groves of oak trees. During the Roman Civil War of 49–45 BCE, Roman legionaries under the command of Julius Caesar discovered a Celtic temple grove in a forest near Marseille, in modern-day France, while cutting timber for the construction of military fortifications during a siege of the fortified city of Massilia.[32] The poet and historian Lucan (Marcus Annaeus Lucanus, 39 CE–65 CE) gave the following vivid account in his epic historical poem, *Pharsalia*:

> A grove there was, untouched by men's hands from ancient times, whose interlacing boughs enclosed a space of darkness and cold shade, and banished the sunlight far above. No rural Pan dwelt there, no Silvanus, ruler of the woods, no Nymphs; but gods were worshipped there with savage rites, the altars were heaped with hideous offerings, and every tree was sprinkled with human gore. On those boughs—if antiquity, reverential of the gods, deserves any credit—birds feared to perch; in those coverts wild beasts would not lie down; no wind ever bore down upon that wood, nor thunderbolt hurled from black clouds; the trees, even when they spread their leaves to no breeze, rustled of themselves. Water, also, fell there in abundance from dark springs. The images of the gods, grim and rude, were uncouth blocks formed of felled tree-trunks. Their mere antiquity and the ghastly hue of their rotten timber struck terror; men feel less awe of deities worshipped under familiar forms; so much does it increase their sense of fear, not to know the gods whom they dread. Legend also told that often the subterranean hollows quaked and bellowed, that yew-trees fell down

and rose again, that the glare of conflagration came from trees that were not on fire, and that serpents twined and glided round the stems. The people never resorted thither to worship at close quarters, but left the place to the gods. For, when the sun is in midheaven or dark night fills the sky, the priest himself dreads their approach and fears to surprise the lord of the grove.[33]

This account is embellished by Roman ethnocentrist sentiments and political propaganda, but it nevertheless gives us a glimpse into the emotional power the sacred site had on the Roman soldiers who were set on desecrating it, as well as the Druid priests who served there. The fear the soldiers must have felt upon encountering the numina of these foreign gods in their own temple is palpable in Lucan's description. The Romans were an extremely religious people, and the soldiers hesitated due to reverence. They feared to desecrate the grove, which was indeed the last part of the forest, according to Lucan, to fall under the axe. Until urged on by Caesar himself, the disciplined Roman soldiers refused to fell the trees as ordered. Lucan continues his tale:

> This grove was sentenced by Caesar to fall before the stroke of the axe; for it grew near his works. Spared in earlier warfare, it stood there covered with trees among hills already cleared. But strong arms faltered; and the men, awed by the solemnity and terror of the place, believed that, if they aimed a blow at the sacred trunks, their axes would rebound against their own limbs. When Caesar saw that his soldiers were sore hindered and paralyzed, he was the first to snatch an axe and swing it, and dared to cleave a towering oak with the steel: driving the blade into the desecrated wood, he cried: "Believe that I am guilty of sacrilege, and thenceforth none of you need fear to cut down the trees." Then all the men obeyed his bidding; they were not easy in their minds, nor had their fears been removed; but they had weighed Caesar's wrath against the wrath of heaven. Ash trees were felled, gnarled holm-oaks overthrown; Dodona's oak, the alder that suits the sea, the cypress that bears witness to a monarch's grief, all lost their leaves for the first time. . . . The peoples of Gaul groaned at the sight; but the besieged men rejoiced; for who could have supposed that the injury to the gods would go unpunished?[34]

Several important attitudes toward trees and groves are revealed in this account: first, the Roman soldiers, even Caesar, the Flamen Dialis, the High Priest of Jupiter, hesitated to desecrate the groves, because of their own tradition of sacred groves; and second, the inhabitants of the besieged city rejoiced, expecting that the gods would avenge the sacrilege. As the poet goes on to say, "Fortune often guards the guilty," and Caesar's bloody war of conquest to fund his political ambitions continued apace. We will see the connection between deforestation, genocide, and sacrilege repeated over and over again throughout history, continuing to the present day, as the sacred forests of Indigenous people all over the world continue to fall to fund the appetite of capitalism for endless expansion and raw materials.

The Celtic nations of Western Europe conducted ceremonies and worship in woodland clearings with open sky all over the territories in which they lived. These ritual sites were called *nemeton*, or *drunemeton*, the latter word incorporating the Indo-European root *dru*, meaning oak.[34] In the account of Lucan, only the priests, the Druids, were permitted to enter the hallowed grove to perform the ancient rites. Whatever those hallowed rituals may have been were lost to history because the Druids famously did not commit their teachings to writing. In inquiries of this type, so many questions as to the actual practices of the ancient Pagan nations will remain unanswered, mostly due to the fact that history, as it is often said, is written by the victors. We have the records of the Roman conquerors and later hostile witnesses in the form of Christian apologists, aside from a handful of vague references of Greek geographers, such as Strabo, to tell us about the religion of the ancient Celts. Other sources include Irish myths, recorded by monks long after the fact, that give a record of the doings of gods and heroes but tantalizingly little about the practices of the priestly class of Druids. We know more about the Germanic people whose confrontation with the imperial faith of the new religion was more recent.

In February of 313 CE, with the Edict of Milan, the Emperor Constantine legalized Christianity and the union of Catholic Christianity with Roman imperialism was begun. As Christianity began to spread in the territories of the Roman Empire, an increasingly intolerant attitude took hold in the newly ascendant religion, an attitude markedly different from the sentiments of the Teacher. Continuing the venerable tradition begun by their Yahwist spiritual forebears in ancient Palestine, the evangelists of the New Jerusalem began to vigorously and violently oppose traditional polytheism in the lands of the collapsing Roman Empire, as it was also suppressed among the Germanic tribes that came to power in the former imperial territories of the West.

As noted earlier, the Germanic-speaking tribes of the north of Europe venerated trees and groves, in much the same way as their Celtic, Roman, and Greek neighbors, as physical symbols of their deities, as if they were infused with the power of the gods themselves. The groves where they grew in ancient majesty were the sacred temples of those who, according to Roman historian Tacitus, "do not think it in keeping with divine majesty to confine gods within walls or to portray them in the likeness of any human countenance. Their holy places are woods and groves and they apply the names of deities to that hidden presence which can only be seen with the eye of reverence."[35]

The Dangers of Idealizing the Past

It has been said that Tacitus was one of the first historians to advance the concept of the "noble savage," an idealized view that more "primitive" peoples are somehow more virtuous and vigorous, brave and pious than their decadent "civilized" counterparts. Reading him, I can see this is true, and I confess that my own view can be seen to fall under the same prejudice from time to time. First, let me say that I believe that people have been much the same in all times and in all places and that I don't want to advance an overly idealized view of the polytheists of ancient Palestine, the Celts, the Norse, or anyone else. History is not often a simple matter of heroes versus villains; all the parties involved were motivated by complex factors, usually political as well as religious. As with other turning points in history, that which was admirable about the old ways gets discarded in the embracing of new ways. Looking back from the vantage point of centuries later, we wish to recall the admirable ways that we have forgotten in the heedless march of progress. As we discussed, the cosmic tree is watered by the Well of Wyrd, of Fate, and this world has to be as it is right now, for whatever reason only the Fates know.

The Judeans, Romans, and Christians all made important contributions to the world we live in today, many of them good and valuable ones, necessary to the world, which is in some ways safer and more peaceful than the world of the Middle Ages and antiquity. This is not the place to pass judgment on them. New ways become old, civilizations rise and fall, and many of yesterday's conquerors have become the conquered and the downtrodden. New conquerors have arisen, new ways that make the systems of domination of the three groups of past conquerors we have examined pale in comparison. The Heathen and polytheist cultures we have discussed, while living closer to the natural environment and having a strong spiritual tie with the land, were no angels either. They practiced slavery, large-scale warfare, and even human sacrifice, and we certainly shouldn't make them models for how to live in peace with our neighbors. I admire their enchanted and animist worldview and the deep spirituality that comes down to us in what survives of their literature. Personally, in my heart, though, I stand against all systems of domination and wish to see a world where we integrate the best practices of our ancestors with those of the present day in order to regain our connection to nature.

Between the last centuries of the first millennium and the first century of the second, Western Europe lost what remained of its Indigenous tradition, and indeed its sacred groves, due to the spread of Christianity at the point of the sword and the blade of the axe. The Frankish and Anglo-Saxon kingdoms had converted to Christianity by the seventh century, leaving the Germanic nations outside the Frankish fold, with Frisia, Saxony, and Scandinavia still retaining their traditional Heathen ways. In the year 724 CE, an Anglo-Saxon missionary named Wynfrith of Crediton cut down a massive oak dedicated to Thunor at Geismar in Hesse, in what is now Germany, and used the wood of the fallen tree to build a church on the site. Posterity remembers Wynfrith as St. Boniface, apostle to the Germans.[36] The story of Boniface's cutting of the sacred oak of Thunor may have been inspired by the fourth-century story of the felling of a sacred pine tree by St. Martin of Tours. In both cases, a miraculous wind intervened, showing the power of the Christian God over those of the Pagans. As in the case of the felling of the Druidical grove in Massilia, the native Pagans would have been surprised by the fact that those who desecrated their sacred precincts went unpunished by the gods.

As we have seen, these trees were living symbols of divinity and of the entire cosmology of the Pagans. When they were felled so easily, the reaction must have been one of shock and dismay, of the foundations of faith being shaken. Religious historian Carole Cusack's book *The Sacred Tree* has been very helpful to me in the writing of this chapter by emphasizing the paramount importance of sacred trees to the Pagan peoples of Western Europe, as well as the devastating consequences of their loss. According to Cusack, "When conversion to Christianity is equated with conquest and the loss of independence and indigenous traditions, it is surely realistic to explore the likelihood that the conquering, colonizing culture systematically undermined the conquered people and that Christianity was

♦ *St. Boniface fells the Oak of Donar as the tribespeople of the Chatti (a Germanic tribe) look on in astonishment. By Emil Doepler, c. 1905.*

ultimately accepted because the traditional Pagan religion was rendered inoperable and impossible as a means of preserving identity and providing hope for the future."[37] It was a devastating blow to the morale of the native people when their sacred symbols and places of worship were desecrated and destroyed.

In an action that was to have very far-reaching and unintended consequences, the Frankish Emperor Charlemagne began an intensified war on the Heathen inhabitants of Saxony in 772. In the province of Hesse, the Saxons had a sacred wooden pillar called Irminsul that served as a focal point for religious rites, as an axis mundi and representative of the sky god, scholars theorize. As part of his campaign against the Saxon Heathen, Charlemagne ordered the pillar destroyed, igniting a conflict that would last decades, centuries even. The resulting guerilla war of resistance against Frankish imperial Christianity spread far beyond the borders of Saxony and neighboring Frisia and possibly initiated attacks by the Danish Heathen, allies of the Frisians and Saxons, upon Christian settlements as far away as Lindisfarne, and in Britain ultimately ignited the Viking attacks on Christian England and the rest of the British Isles. In 841 and 842, according to a chronicler of the time, "Norsemen and Slavs might unite with the Saxons, who called themselves *Stellinga* (restorers) because they are neighbors, [so] they might invade the kingdom and root out the Christian religion in the area."[38] The conquered people joined forces. As Cusack tells us, "The Frisians

✦ Baum des Totes and Lebens (Tree of Death and Life), *by Berthold Furtmeyr,*
from the fifteenth-century text the Heidelberger Schicksalbuch
(the Heidelberg Book of Fate*).*

and the Saxons, Pagan peoples living on the borders of the Frankish territories, sought to defend their identities through holding firm to their Pagan religion, and attempted to reduce the influence of Christianity in their lands by burning churches and expelling ecclesiastics, and defending their traditional political systems (kingship in the case of the Frisians, elected war leaders in the case of the Saxons)."[39] Ultimately, the revolt was unsuccessful, and by the end of the ninth century most of Saxony had been converted and brought into the Frankish fold—the Holy Roman Empire, as it was then called. Making sacrifices at trees, springs, or groves was punishable by fines, as was not having one's children baptized. Refusing baptism was punishable by death.[40]

The same pattern was repeated in the Christianization of the Scandinavian nations. Missionaries to Scandinavia and recently converted tenth-century kings, such as Olav Tryggvason of Norway and Olav Haraldsson and Harald Bluetooth of Denmark, outlawed Paganism in their realms and instituted systematic repression of the old religion. The former Olav was legendary as a destroyer of Pagan sacred sites, known as *horgs*, or open-air stone altars. By the end of the eleventh century, the process of official conversion was complete and what remained of the veneration of trees and other natural features went underground.[41]

The archetypes of the tree and the grove are universal and all-pervasive symbols of the structure of the spiritual reality underlying our cosmos and point to the presence of the divine in our everyday world. And so it is important to note that the worldview of medieval Christianity had its own version of the sacred tree, which it inherited by way of both its Judean and Pagan spiritual predecessors. The cosmic tree of the earlier shamanic and animistic cultures became the Edenic Tree of Life and Tree of Knowledge of Good and Evil, as well as the cross upon which the world savior Christ was hung, recalling Óðinn's self sacrifice upon Yggdrasil.

FOREST DEITIES, WORLD TREES, AND SACRED GROVES FOR THE CUNNING FARMER

If every tree is somehow partaking of the archetype of the primal world tree, then every tree is a dwelling place of the divine and is in some sense a heirophany, a revelation. How do we as dwellers in the land, as cunning farmers, put in practice our newfound respect for the ordinary trees with whom we share the land?

Trees and forests bring us many gifts, physical as well as spiritual. In addition to being sacred images of divine reality and making, when growing together in their wild state trees provide a uniquely sacred space for ritual, contemplation, or just a peaceful walk. For much of my life, I have lived on the edge of the eastern deciduous hardwood forest, a remnant of a great forest that once covered much of temperate eastern North America. The forest has been my refuge. It has given me firewood to heat my home, as well as the welcome ritual of cutting it. I have sought medicines from the plants that grow there and regularly retreat to the woods that surround our farm to commune with the spirits of the land, much in the way I described in the opening paragraphs of this chapter. On my land, there are great trees, mother trees, much larger than the surrounding trees of the same species. To these I give special reverence and merge my consciousness with theirs, placing my hands on their trunks and attempting to contact the indwelling genius of the mother tree, the hamadryad. I have had powerful, intensely personal contacts with two of these trees on my land: an ancient sycamore that stands guard on the corner of our property, and a large, two-trunked oak that stands over a spring that has a dry-laid stone wall built into it. I don't know how old the wall is, or who built it, but it is a special place.

Trees are, for better or worse, the sources of important materials we use in our everyday lives for building, heating, paper, and more. We can't get by without them. I have had to cut many during my career, mostly for firewood but also for building material, and often because saplings have a knack for coming up in inconvenient places, in fence rows and building foundations. As we noted in the first chapter, the ancient Roman agronomists would not cut a single tree without prayers and offerings to the appropriate gods and spirits. I have always been mindful of the sacredness of trees, and, in my younger days when I first began to harvest trees for use in building projects, I made sure to use an axe to fell them, my reasoning being that the hard work involved was both a limiting factor on my cutting of trees and an offering to whatever spirits I may have offended. These days, I make more formal offerings and make sure to voice a prayer, though I use a chainsaw and mostly cut dead trees now.

I believe in having a relationship of reciprocity with the spirits of the land. They have given me signs and gifts in return. For instance, a few years ago, I went out on the abandoned farm I have written about looking for firewood to cut for the coming winter. I said a prayer to the land spirits to help me find some good and dry dead trees to cut for wood. A large red-tailed hawk appeared and flew along ahead of me as I walked, until he landed in a large dead red oak tree, standing dry on the edge of the woods. I gave thanks, did a ritual of protection, and proceeded to fell the tree. Cutting large trees, especially dead ones, is

✦ *A very large sycamore growing on a corner of the author's farm,*
serving as a guardian of the land.

dangerous work. A protective ritual and prayers are a good idea, as are a hardhat, eye protection, and absolute attention to the job, because there is an element of randomness to the motion of a falling tree, as well as the potential for limbs to become loosened by the vibration of the saw, and the woodcutter has to be ready to move away quickly. It is in these moments of random chance that we are truly in the hands of the gods.

We are allowing steep and eroded sections of the farm to revert to woodland. Such places probably shouldn't have been cleared in the first place. There is more woodland in the eastern United States now than there was a hundred years ago, due to the collapse of the small farm economy; a silver lining there, I suppose. My farm apparently used to be sheep pasture in parts, as sheep used to be economically important in this part of Kentucky. Sheep are, however, notoriously damaging to the soil if not well managed, being able to nibble the grass down to the roots with their close-biting teeth. As a consequence, our hillside pastures are rocky with a thin coating of topsoil, much better suited to the growing of trees than grass. The land where I live wants to revert to forest, as does most fertile agricultural land, so anywhere that isn't mowed for a couple of years turns quickly into a young forest. Cedar trees are the first to grow in the poorer eroded sections, and black locust trees quickly cover the deeper soil. I have lived here for fifteen years, and the steeper hillside pastures have become dense woods of cedar, where light doesn't reach the surface. The locust trees have grown to the size where they can be harvested as fence posts, poles for building, or high-quality firewood. They are a fast-growing and sustainable source of much-needed materials on the farm.

An orchard is much like a tame forest, shady and leafy, but managed for human needs. For the first few years after my wife and I took ownership of the farm, we planted a number of apple and other fruit trees in other sections of the property that were close to the old farmhouse where we used to live. These orchard sites have pretty good soil but are too steep to grow other crops on, so they were perfect for fruit trees. The orchards are thriving and finally yielded a bumper crop of apples last year. It is truly a blessing to enter the shade of the orchard after spending hours working on field crops in the hot sun.

TREES IN SPIRITUAL PRACTICE

First, as I have recommended elsewhere, the importance of making offerings to trees and other forest and land spirits of food and other items cannot be underestimated. It is a practice that changes the one making the offering, because you are making a physical demonstration of goodwill and not just thinking or saying words. You and the forest spirits will both know the difference. They do not need to eat; it's the devotion, the reverence that they feed on. It changes the giver, and that changes the whole relationship.

There are several other practices I have used relating to trees that I have found helpful and meaningful, some that involve being in the presence of actual trees and some that are energy-moving exercises where the practitioner becomes the tree. The former techniques involve entering into the presence of a living tree as a spiritual and energetic entity. The latter techniques involve visualizing oneself as the world tree itself, standing at the center of the cosmos and moving the flowing life between heaven and Earth. Further practices include leaving offerings at particular trees and pathworking that involves traveling to the various worlds of the cosmic tree, using Yggdrasil or the kabbalistic Tree of Life as a model. The interested reader will have to look elsewhere for these methods, as these are a discipline unto themselves.

Working with Tree Energies

To work with tree energy and entities, simply find a tree that you feel drawn to, in a private and safe location. Take some deep breaths. Ground and center. Ask the tree for its permission to work with it.

Pay attention to your feelings. If you feel in any way apprehensive or hesitant, do not proceed. I have not found trees that were unwilling, but your own nervousness could be a sign that the tree is not congenial to further contact. Hold up your hands and feel the life force of the tree flowing up the trunk and out the branches and circulating back to the ground. Place yourself in the flow. Open your mind to the tree. What do you see? What do you feel? Pay attention to impressions that form; they could be the tree attempting to communicate with you. Trees have a very different sense of time than humans; our motions and thoughts are as fast to them as insects are to us. As Treebeard said in Tolkien's *The Two Towers*, "We must not be hasty." The hamadryad of the sycamore appeared to my mind's eye as a bark-covered woman. Even her eyes were bark-covered. She opened them and made eye contact, seeing me, and then closed them again. The dryad of the oak tree by the walled-up spring appeared to me as a leafy woman with red hair. I don't recall any words that passed between us, just a sense of peace and presence. Your impressions may be more or less vivid than mine.

Standing in the stream of the circulating vital force of the tree can cleanse your energy field and calm your emotions. Stand with palms down (the palms have chakras, or energy centers), at the edge of the spread of the branches where the vital force cascades back into the earth like a waterfall. Breathing in deeply, allow the purified qi from the tree to enter the crown chakra and, breathing out fully, allow it to flow back down into the earth through the palms and the feet. Do several circulations of this kind and observe how you feel. Visualize the in-breath as cool and cleansing and the out-breath as hot, removing with it all stress and stagnant life energy from your energy body. Do this for as long as you feel is necessary and enjoyable. At the end of the practice, you should feel cleansed and renewed. It is a good practice to do when you are stressed or run down. This practice comes from the Taoist qigong system of China and can be refined and practiced with any tree or plant, the various trees and plants having their own energetic signatures and benefits.

The archetype of the world tree can be a useful form to use in order to invoke heavenly and earthly vital force into one's energy body. John Michael Greer gives a full ritual, expanding on the Sphere of Protection we discussed in chapter 12, called the Tree of Light working, done after the completion of the Sphere of Protection, the Three Cauldrons working, and the Awakening of the Dragons ritual, all of which can be found both in Greer's *Druid Magic Handbook* as well as *The Dolmen Arch* course handbooks. I highly recommend both of these to the interested reader.

While I have never done it as a stand-alone ritual, the Tree of Light working involves visualizing a seed of light in one's solar plexus that sprouts into an enormous Tree of Light and involves circulating vital energy down from the solar current, the Sun, with the in-breath, down through the leaves, branches, trunk, and into the roots in the center of the Earth, and then pulling the telluric current, the vital energy of the Earth, from the center of the Earth, up through the roots and into the trunk and out the leaves and back into heaven. Repeat the process nine times in all.

❧ Tree of Life Energy Moving

As a stand-alone ritual, here's one based on the Reiki moving meditation, which is similar to t'ai chi and qigong exercises. This is one that I practice daily, as it only takes a few moments. I learned the exercise it is based on during Reiki training, but it wasn't presented as having anything to do with sacred trees.

That is my innovation. In doing the exercise, I realized my hand and arm movements were making the shape of a tree, and the energy circulations were similar to the ones in the Tree of Light. Therefore, let's call this the Tree of Life Energy-Moving exercise. You heard it here first.

To begin: stand with feet slightly apart, knees slightly bent.

Close your eyes. Imagine that you are the Tree of Life, standing in the center of the cosmos. Your roots are in the Earth, drawing up the qi from the center of the Earth. Keeping still, imagine your branches reach up to heaven, drawing down heavenly qi. Your trunk is the cosmic axis, stable and strong. Around you whirl the stars and planets in their orbits.

Stay with this visualization for a few minutes, breathing in and out slowly, relaxing your body and mind.

When you feel ready to proceed, bring your hands up to the level of your heart, in the center of your trunk. Form a ball of light with your hands. The ball should be about eight inches in diameter. Feel and see the ball with your mental senses. Inhaling deeply and slowly, move the ball straight up, as high as you can, pulling energy up your roots from the Earth.

When your arms are fully outstretched, exhale slowly and bring them, fully extended, in a full arc until they are straight along both of your sides, visualizing the heavenly energy of sunlight coming down through your leaves and branches and into your trunk. Inhale slowly, bringing both arms along the body, hands up to the level of your heart again, storing the light in your heart center, completing the inhalation.

Place your hands horizontally, palms facing the earth, elbows and hands at the level of your heart, exhaling slowly, pushing the energy down into your roots, with your hands, grounding yourself, until your arms are fully extended.

Inhale slowly, and bring your arms up slowly along your body, visualizing the golden Earth energy flowing from deep within the center of the Earth, pulling it upward with your arms and breath and forming it into a ball at the level of your heart center. Repeat the process several times until you feel fresh and grounded.

And so we reach the end of our exploration of the sacred tree in history, mythology, and in the actual practice of cunning farmers, who enter into a relationship of mutuality with the living trees on the land on which they live. As noted, trees bring us so many gifts and are an essential part of life on the land in many parts of the world. Trees are living beings with a spiritual presence and a magic that transcends even these precious physical gifts they give us. As we stand in the cool, leaf-dappled shade of the great green beings with whom we share our lives, we can take a moment to pause, breath in the fresh oxygenated air they give us, renewing our souls with the abundant healing life force they exude, and remember that we are in the presence of the holy image of the spiritual cosmos in which we live, a sacred reflection of the Divine reality which constantly enfolds us in its healing presence.

THE CUNNING FARMER IN PRACTICE

We have reached the end of our journey through the four cosmic realms with which the cunning farmer works and have gained valuable insights into the hidden workings of the forces with which we engage in our daily work as caretakers of a part of this holy Earth. We have learned to engage with the animate forces around us, to reinvigorate and re-enchant in our small way the corner of creation that it has been given to us to tend.

I have made the case that we live in a world infused with Divine Intelligence, in which ideas and archetypes emerge from the World Soul, in time with the movements of the stars and planets. These archetypal forces not only shape the thematic contours of our individual lives, they also give form to the spirit of the time. This spirit gives each age its particular character, and is received by each individual according to their circumstance and individual spiritual constitution. Like all of the other ages our world has gone through in the past, ours is marked by its own particular brand of madness, which we cannot see clearly because it is the ocean in which we swim, but it will become clearer upon hindsight. We can make some general observations, however.

When I began this work over three years ago, the world had just witnessed the global pandemic, lockdown, isolation, masking, mass vaccinations, widespread death and sickness, germ paranoia, and all of the other novel insanities of that time. Whatever your memories of that period of our collective history might have been, it probably changed you and made you make a few different life choices about how you wanted to live the rest of your life. It has definitely changed me.

The coming age, one in which all of the outer planets have changed signs and elemental triplicities, promises to be unlike any we have ever seen before.

This will be a time in which technological, cultural, and political change will transform all areas of life, including our relationship with the natural world. The character of the change is as yet unknown. What can be said is that it will leave its mark on all who live through this time, for better or worse.

I hope that this book stands as a statement of defiance against the forces of the wrong kind of change, the kind of change that seeks to tear us away from our roots, both physical and spiritual. The forces of uniformity and conformity, of commercialism, urban ugliness, resource extraction, and religious fundamentalism, of repression and reaction. Let this book be a protest against the forces that despoil our living Mother Earth and desacralize our universe by declaring it to be so much dead matter, material to be used for monetary gain for the oligarchs who increasingly own our world. It is as much a manifesto as a guide for living, a record of a life lived in relationship with nature and spirit.

We are entering a new age, one that was foreseen by the best minds of the twentieth century, people like psychologist Carl Jung and theologian and scientist Pierre Teilhard de Chardin. In this age, much of what was old will be swept aside by the winds of change. As Jung himself said, "We are living in what the Greeks called the *kairos*—the right moment—for a metamorphosis of the gods, of the fundamental principles and symbols."[1] It is my hope that this book becomes a part of that metamorphosis. The ideas presented here are ways of seeing and being that deserve to continue to inform our relationship with our planet, our land, and the living creatures with whom we share it, and ultimately with the Divine Source from which all of this beauty and chaos flows forth.

We are as pilgrims, those of us journeying into this uncertain future from our troubled past. And as pilgrims and travelers from one country emigrating into a new world we bring with us in our patched luggage and steamer trunks much of what made life meaningful in the old world: our songs, stories, myths, and customs, and most precious of all our seeds, by which we would sow our gardens to nourish us in the new land of the future. Let this book be a collection of precious seeds, heirloom seeds, from our past, to be tended in our hearts and minds, watered, nurtured, and protected from the harsh winds and the depredations of pests, until they bear fruit to nourish the next generation. There is much from the past that should be left behind, and unfortunately, probably won't. But the precious esoteric traditions that bring us wisdom, enliven our spirits, and bring us into participation as members of the spiritual community of the cosmos that is our home should be safeguarded as the treasures that they are.

So, as I look back on the journey that has been the writing of this book, I think a few final remarks are in order. We began with the premise that honoring the spirits of the land is the necessary beginning for the work of the cunning farmer. Before we even begin work as farmers, homesteaders, or even urban dwellers, we have to start from a foundation of reverence and love for the place that gives us life. Our first duty is to "consult the genius of the place," to begin the intimate relationship with the land itself. The land gives us life, whether we are urban dwellers or country folk.

I have introduced the concept of farmer or land manager as a sacred calling, one that involves not only the physical work of tending crops, preparing soil, and caring for livestock. Even if you just tend a small garden patch and have some cats, when you work with your own little patch of Earth, you have a role as intermediary between the sacred forces of the cosmos and the imaginal realm and the Earth itself. You are nature's priest or priestess, as the Roman farmer was. We work with the rhythms of the Moon for planning, planting, and other work on the farm. We have learned about harnessing the cosmic forces of creation by using the Moon, the planets, and the stars to improve our life and circumstances with the magical arts. We have explored the spiritual cosmos as a framework for understanding how creative intelligence and soul interact with the material world in which we live.

We have delved into the history and folklore of magical plants and their place in our magical and spiritual practice. Like the trees that symbolize our role as mediators between heaven and Earth, the humble plants and herbs, which the cunning farmers devote their lives to caring for, also bring to bear their own spiritual intelligences and mediate the world soul in their particular ways, as magical materials and medicines for healing. We have learned about their role as spirit teachers and guides and how their influence extends far beyond their pharmacological properties on the physical body. They transmit their subtle influences, which they receive from their own relationship with the heavens and the Earth, in the form of planetary energies, to the magician or the patient, aiding and reinforcing their physical properties.

As we have seen, the web of mind extends far beyond the living personality into the realm of the departed and the underworld, a reality that is ever-present and that interacts with our world in a number of ways. We have explored a mode of seeing that understands our physical existence as rooted in the reality of the Earth and the underworld, and that rises like a tree, heavenward, reaching out to the stars and the transcendent source that sustains us all and that is symbolized by the heavens beyond the heavens, the Empyrean realm in traditional cosmology. Our roots absorb the Earth's energies that sustain our physical vitality, and our leaves absorb the light of the Source that shines down like the Sun on all of us, colored by the tincture of the stars and planets that mediate that uncreated light.

We have envisioned the spiritual cosmos as a tiered structure, and we have added nuance to the traditional three-tiered cosmology of shamanism and Neoplatonism, by claiming that the underworld is not just a place under the earth but yet another plane entirely, outside of space and time, a realm of memory that "underlies" the earth in a metaphysical and symbolic sense, and that the upper realm of the shamans is divided into the celestial realm and the supercelestial realm of the Neoplatonist. In essence, we have attempted to reunite the Neoplatonic worldview of Western magic with the archetypal cosmology of shamanism, providing, in a rough outline, a working hypothetical model for the unique position we find ourselves in as magical earth workers. The overvaluing of the transcendent and supercelestial at the expense of the earthly and chthonic has damaged our ability to live in harmony with the natural world. As the locus of value moved further and further from the earthly, it began to disappear entirely. In more recent centuries, the rejection of the spiritual and transcendent entirely allowed for the unchecked destruction of nature in the pursuit of profit. Further, the transcendent Source of existence itself has also been deeply misunderstood in modern times and has been equated with the apotheosis of the human ego, an irascible projection onto the divine realm of the all-too-human qualities of tribalism and bigotry This twisted god-image is rightly rejected by many, leading to the pervasive atheism and materialism of modern times.

By pursuing a worldview in which the transcendent as well as the chthonic and earthly are both seen as necessary parts of a spiritual organism, I have endeavored, on a personal level, to create a world picture that suits my needs and peculiarities as a seeker of wisdom and a magical practitioner. I believe, however, that I am not unique, and that many of us are striving to create new models in which to conceptualize our journeys of re-enchantment. So, if I can contribute in some small way to the conversation we are having about our place in the world as human beings, and if some of my readers find my views on the matter helpful in placing their lives on our Mother Earth in a larger and more meaningful context, then I will be satisfied in having helped at least a few seekers to find a few more pieces of the puzzle.

In closing, as a statement of a worldview and a memoir, I hope this book will inspire future cunning farmers to get out on the land and begin a re-enchantment revolution, an act of defiance against the forces of technological uniformity and commercialism, against the dehumanization and mechanization of agriculture, against both religious fundamentalisms and materialism. I hope to promote magical thinking as a force for positive change. If we envision and engage with positive thoughtforms we can create through them a new future in which we rekindle the sacred fires and raise the standing stones, chant the holy words and call the divinities of Nature—the Earth goddesses, the Sun gods, the powers of storm and rain, wind and sea, the spirits of the forest and the land, the star beings—back from the forgotten margins where they work behind the scenes, into the light, revered and renewed. Let us carry on the work of countless generations of ancestral spirits of cunning farmers, earth mystics, and shamans to return a sense of the sacredness of the land and of place to the center of our collective lives, and let us begin now.

NOTES

CHAPTER 1. CONSULTING THE GENIUS OF THE PLACE

1. Pope, "Epistles to Several Persons."
2. Cato, *The Complete Works*, 53.
3. Virgil, *The Georgics of Virgil*, 3.
4. Virgil, *The Georgics of Virgil*, 7.
5. Kirk, *The Secret Commonwealth*, 51.
6. William F. Romain, "Werewolf Shamans in the Ancient Woodlands of the Eastern United States," accessed on Academia's website April 2025, 5.

CHAPTER 2. LAND SPIRITS AND SACRED SPACE

1. World Heritage Ohio, *Guide to the Hopewell Ceremonial Earthworks*, 20.
2. Romain, *Mysteries of the Hopewell*, chapter 2.
3. Romain, *Archaeology of the Sacred*, 187.
4. Romain, *Mysteries of the Hopewell*, introduction.
5. Romain, *Mysteries of the Hopewell*, introduction.
6. Romain, *Mysteries of the Hopewell*, appendix.
7. Pirtea, "From Lunar Nodes to Eclipse Dragons," 341.
8. Romain, *Mysteries of the Hopewell*, appendix.
9. World Heritage Ohio, *Guide to Ceremonial Earthworks*, 20.
10. Romain, *Shamans of the Lost World*, 8.
11. Romain, *Shamans of the Lost World*, 37.
12. Romain, *Shamans of the Lost World*, 37.
13. Romain, *Shamans of the Lost World*, 19.
14. Romain, *Shamans of the Lost World*, 20–23.

15. Romain, *Shamans of the Lost World*, 90.

16. Romain, *Shamans of the Lost World*, 91.

17. Romain, *Shamans of the Lost World*, 91.

18. Romain, *Mysteries of the Hopewell*, 203.

19. Kiesel, *Magic Circles in the Grimoire Tradition*, 20.

20. Kiesel, *Magic Circles in the Grimoire Tradition*, 19.

21. Kiesel, *Magic Circles in the Grimoire Tradition*, 20–21.

CHAPTER 3. FIRST PRINCIPLES OF A MAGICAL WORLDVIEW

1. Corbin, "Mundus Imaginalis," 5.

2. Corbin, "Mundus Imaginalis," 6.

3. Lewis, *The Discarded Image*, 95, 166.

4. Tarnas, *Cosmos and Psyche*, 77.

5. Jung, *Synchronicity*, 25.

6. Plotinus, *The Enneads*, 549.

7. Plotinus, *The Enneads*, 549.

8. Plotinus, *The Enneads*, 534.

9. Plotinus, *The Enneads*, 535.

10. Plotinus, *The Enneads*, 535.

11. Plotinus, *The Enneads*, 535.

12. Avelar and Ribeiro, *On the Heavenly Spheres*, 10–15.

13. Avelar and Ribeiro. *On the Heavenly Spheres*, 12–13.

CHAPTER 4. THE MOON, MISTRESS OF MAGIC

1. Crowley, *Magick in Theory and Practice*, xii.

2. Crowley, *Magick in Theory and Practice*, xiii.

3. Agrippa, *Three Books of Occult Philosophy*, 208.

4. Agrippa, *Three Books of Occult Philosophy*, 210.

5. Agrippa, *Three Books of Occult Philosophy*, 206.

6. Ogden, *Magic, Witchcraft, and Ghosts in the Greek and Roman Worlds*, 239.

7. Apuleius, *The Golden Ass*, (Adlington), 7.

8. ní Mheallaigh, *The Moon in the Ancient Greek and Roman Imagination*, 27, 31.

9. ní Mheallaigh, *The Moon in the Ancient Greek and Roman Imagination*, 28.

10. Roebuck, *The Upanishads*, 239.

11. Taylor, *The Mystical Hymns of Orpheus*, 24–28.

12. Mathers, *The Greater Key of Solomon*, 14.

13. Agrippa, *Three Books of Occult Philosophy*, 319.

14. Baker, *The Cunning Man's Handbook*, 272.

15. Needham and Wang, *Science and Civilization in China*, 242.

16. Collins, "Scholasticism and High Medieval Opposition to Magic," 469.

17. Agrippa, *Three Books of Occult Philosophy*, 368.

18. The New Oxford Annotated Bible, Ecclesiastes 3:1–3.

19. The New Oxford Annotated Bible, Wisdom of Solomon 8:17–22.

20. al-Qurtubi, *The Picatrix* (Greer and Warnock), 290.

21. al-Qurtubi, *The Picatrix* (Greer and Warnock), 292.

22. al-Qurtubi, *The Picatrix* (Greer and Warnock), 292.

CHAPTER 5. MOON WORK

1. Agrippa, *Three Books of Occult Philosophy*, 427.

2. National Aeronautics and Space Administration, "Moon Facts."

3. Villavicencio and Virgilio, *The Power of Asteria*, 160.

4. Ptolemy, *Tetrabiblos*, 4.

5. Agrippa, *Three Books of Occult Philosophy*, 366.

6. al-Qurtubi, *The Picatrix* (Greer and Warnock), 66.

7. al-Qurtubi, *The Picatrix* (Attrell and Porreca), 78.

8. al-Qurtubi, *The Picatrix* (Attrell and Porreca), 77.

9. Villavicencio and Virgilio, *The Power of Asteria*, 158.

10. National Radio Astronomy Observatory, "Variability of the Moon's Apparent Motion."

11. al-Qurtubi, *The Picatrix* (Attrell and Porreca), 77.

12. Villavicencio and Virgilio, *The Power of Asteria*, 172.

13. Avelar and Ribeiro, *On the Heavenly Spheres*, 12–15.

14. Villavicencio and Virgilio, *The Power of Asteria*, 172.

15. Villavicencio and Virgilio, *The Power of Asteria*, 172.

16. Avelar and Ribeiro, *On the Heavenly Spheres*, 45–46.

17. Avelar and Ribeiro, *On the Heavenly Spheres*, 67, 70.

18. Avelar and Ribeiro, *On the Heavenly Spheres*, 50.

19. Avelar and Ribeiro, *On the Heavenly Spheres*, 50.

20. McNeill, *The Silver Bough*, 53.

21. McNeill, *The Silver Bough*, 53.

22. McNeill, *The Silver Bough*, 54.

23. Geller, *Melothesia in Babylonia*, 90–93.

24. Wigginton, *The Foxfire Book*, 217–19.

CHAPTER 6. WEATHER MAGIC

1. Authorized King James Bible, Psalm 29:5.

2. Barnes, *Early Greek Philosophy*, 116.

3. Johnson, *Daoist Weather Magic and Feng Shui*, 102.

4. Day, *Psalms*, 40.

5. Peterson, *The Sixth and Seventh Book of Moses*, vii–xviii.

6. Swart, *The Book of Sacred Names*, 70.

7. Day, *Yahweh and the Gods and Goddesses of Canaan*, 91–98.

8. Iamblichus, *On the Mysteries*, 297–99.

9. Swart, *The Book of Sacred Names*, 13–14.

10. Swart, *The Book of Sacred Names*, 13–14.

11. Parke-Taylor, *Yahweh: The Divine Name in the Bible*, 86–87.

12. Parke-Taylor, *Yahweh: The Divine Name in the Bible*, 86–87.

13. Swart, *The Book of Sacred Names*, 106.

14. Swart, *The Book of Sacred Names*, 106.

15. Swart, *The Book of Sacred Names*, 106.

16. Esoterica channel, "Who is Yahweh—How a Warrior-Storm God became the God of the Israelites and World Monotheism," March 3, 2023, YouTube, 43:10.

17. Swart, *The Book of Sacred Names*, 71–72.

18. Johnson, *Daoist Weather Magic and Feng Shui*, 214–15.

19. Frazer, *The Golden Bough,* 72.

20. Crowley, *Magick in Theory and Practice*, 119.

21. Crowley, *Magick in Theory and Practice*, 120–21.

22. Gary, *Traditional Witchcraft*, 157.

23. Peterson, *Secrets of Solomon*, 28.

24. Gary, *Traditional Witchcraft*, 158.

25. Leitão, *Opuscula Cypriani*, 398–400.

26. Aurelius, *The Meditations of Marcus Aurelius*, 299.

27. Augustine of Hippo, *City of God*, 565.

28. Kieckhefer, *Forbidden Rites*, 14.

29. Stratton-Kent, *The Testament of Cyprian the Mage*, 70 ff.

30. Mathers and Crowley, *The Lesser Key of Solomon*, 5.

31. Mathers, *The Greater Key of Solomon*, 80–81.

32. Scaraoschi, "Scaraoschi's Books of Sorcery."

CHAPTER 7. REIKI AND ENERGY TRANSFER

1. Columella, *On Agriculture* (Ash translation), vol. I, book III, 343.

2. Ager, "Roman Agricultural Magic," 286–87.

3. Sheldrake, *The Sense of Being Stared At*, chapter 6.

4. Sheldrake, *The Sense of Being Stared At*, chapter 6.

5. Tompkins and Bird, *The Secret Life of Plants*.

6. Rand, *Reiki*, 17–21.

7. McEvilley, *The Shape of Ancient Thought*, chapter 21.

8. Philosophy and Art Collaboratory, "Qi: Breath and Vital Energy."

9. Posthumus, *All My Relatives*, 208.

10. Brand, *The 72 Angels of Magick*, 2015.

11. Swart, *The Book of Sacred Names*, 16.

12. Swart, *The Book of Sacred Names*, 195 ff.

13. Bilardi, *The Red Church*, 160.

14. Johnson, *Chinese Medical Qigong Therapy*, 50.

15. Rand, *Reiki*, 8–9.

16. Hohman, *The Long Lost Friend*, 70.

17. Tompkins and Bird, *The Secret Life of Plants*, 17–32.

CHAPTER 8. THE COSMOLOGY OF RITUAL

1. Eliade, *Shamanism*, 259.

2. Eliade, *Shamanism*, 260.

3. Agrippa, *Three Books of Occult Philosophy*, 3

4. Agrippa, *Three Books of Occult Philosophy*, 254–61.

5. Gary, *Traditional Witchcraft*, 93.

6. Stratton-Kent *The Sworn and Secret Grimoire*, 84.

7. Scot, *The Discoverie of Witchcraft*, 215–16.

8. Crowley, *Magick in Theory and Practice*, 102.

9. Crowley, *Magick in Theory and Practice*, 104.

10. Kraig, *Modern Magick*, 34–38.

11. Greer, *Circles of Power*, 156–58.

12. Gary, *Traditional Witchcraft*, 124–28.

13. Heath, *Elves, Witches & Gods*, chapter 6.

CHAPTER 9. THEURGY

1. Alighieri, *Paradise*, 152.

2. Alighieri, *Paradise*, 153.

3. Alighieri, *Paradise*, 153.

4. Randles, *The Unmaking of the Medieval Christian Cosmos*, 4.

5. Opsopaus, *Summary of Pythagorean Theology*, 7.

6. Opsopaus, *Summary of Pythagorean Theology*, 49.

7. Majercik, *The Chaldean Oracles*, 59.

8. Iamblichus, *On the Mysteries*, 71.

9. Iamblichus, *On the Mysteries*, 71.

10. Pseudo-Dionysius, *The Complete Works*, 145.

11. Pseudo-Dionysius, *The Complete Works*, 145.

12. Robinson, *The Nag Hammadi Library*, 138.

13. Robinson, *The Nag Hammadi Library*, 135.

14. Ellis, *The Ancient Tradition of Angels*, 32–33.

15. New Oxford Annotated Bible, Acts 17:28.

16. Iamblichus, *On the Mysteries*, 59.

17. Iamblichus, *On the Mysteries*, 59.

18. Iamblichus, *On the Mysteries*, 59.

19. Majercik, *The Chaldean Oracles*, 25.

20. Majercik, *The Chaldean Oracles*, 27.

21. Majercik, *The Chaldean Oracles*, 28.

22. Betz, *The Greek Magical Papyri in Translation*, 51–52.

23. Kieckhefer, *Magic in the Middle Ages*, 118.

24. Scholem, *Major Trends in Jewish Mysticism*, 57.

25. *Hekhalot Rabbati*, 27.

26. *Hekhalot Rabbati*, 18.

27. The Book of Common Prayer, 373.

28. Leitch, *The Essential Enochian Grimoire*, 9.

29. Leitch, *The Essential Enochian Grimoire*, 9.

30. Leitch, *The Essential Enochian Grimoire*, 10.

31. DuQuette, *Enochian Vision Magick*, 30–31.

CHAPTER 10. ASTROLOGICAL MAGIC

1. Thompson, *Reports of the Magicians and Astrologers of Nineveh and Babylon*, LXI.

2. Plato, *The Collected Dialogues of Plato*, 1524.

3. Sheldrake, "Is The Sun Conscious?," 25.

4. Pingree, "The Sabians of Harran and the Classical Tradition," 26.

5. Greer and Warnock, *Astral High Magic*, 8.

6. al-Qurtubi, *Picatrix* (Attrell and Porreca), 38.

7. al-Qurtubi, *Picatrix* (Greer and Warnock), 55.

8. Plato, *The Collected Dialogues of Plato*, 1166.

9. Plotinus, *Enneads*, 3.1.5.

10. Plotinus, *Enneads*, 4.4.32.

11. Plotinus, *Enneads*, 4.4.40.

12. Plotinus, *Enneads*, 4.4.40.

13. Plato, *The Collected Dialogues of Plato*, 1526–27.

14. Cumont, *Astrology and Religion Among the Greeks and Romans*, 64.

15. Cumont, *Astrology and Religion Among the Greeks and Romans*, 65.

16. Saif, "From Ġāyat al-ḥakīm to Šams al-ma'ārif," 305.

17. Saif, "From Ġāyat al-ḥakīm to Šams al-ma'ārif," 305.

18. Saif, "From Ġāyat al-ḥakīm to Šams al-ma'ārif," 304.

19. Saif, "From Ġāyat al-ḥakīm to Šams al-ma'ārif," 306.

20. Saif, "From Ġāyat al-ḥakīm to Šams al-ma'ārif," 306.

21. Cumont, *Astrology and Religion Among the Greeks and Romans*, 20.

22. Greenbaum, *The Daimon in Hellenistic Astrology*, 215.

23. Greenbaum, *The Daimon in Hellenistic Astrology*, 215.

24. Scott, *Hermetica*, 415.

25. Copenhaver, *Hermetica*, 61.

26. Copenhaver, *Hermetica*, 60.

27. Betz, *The Greek Magical Papyri in Translation*, xlvii.

28. Greenbaum, *The Daimon in Hellenistic Astrology*, 164.

29. Irenaeus, "Against Heresies," book 1, chapter 24 in Schaaf, *The Complete Works of the Church Fathers*.

30. Schnieders, *The Books of Enoch*, 2 Enoch 21:7.

31. Kaplan, *Sefer Yetzirah*, 169–70.

32. Kaplan, *Sefer Yetzirah*, 170–71.

33. Agrippa, *Three Books of Occult Philosophy*, 430.

34. Agrippa, *Three Books of Occult Philosophy*, 430.

35. Agrippa, *Three Books of Occult Philosophy*, 431.

36. Agrippa, *Three Books of Occult Philosophy*, 431.

37. Agrippa, *Three Books of Occult Philosophy*, 431.

38. Agrippa, *Three Books of Occult Philosophy*, 499.

39. Plato, *The Collected Dialogues of Plato*, 1209.

40. Ficino, *Three Books on Life*, 351.

CHAPTER 11. EARTH MAGIC

1. Virgil, *The Georgics*, 85.

2. Yeats, *Ideas of Good and Evil*, part II.

3. Yeats, *Ideas of Good and Evil*, part III.

4. Burkert, *Greek Religion*, 199.

5. Campbell, *Primitive Mythology*, 125–27.

6. Campbell, *Primitive Mythology*, 137.

7. Frazer, *The Golden Bough*, 456.

8. Harrison, *Prolegomena to the Study of Greek Religion*, 266.

9. Harrison, *Prolegomena to the Study of Greek Religion*, 267.

10. Sallust, *On the Gods and the World*, 17.

11. Harrison, *Prolegomena to the Study of Greek Religion*, 261.

12. Dorcey, *The Cult of Silvanus*, 81.

13. Burkert, *Greek Religion*, 156–59.

14. Jung, *Alchemical Studies*, 208–9.

15. Jung, *Alchemical Studies*, 238.

16. Lecouteux, *Demons and Spirits of the Land*, 29.

17. Lecouteux, *Demons and Spirits of the Land*, 29.

18. Agrippa, *Three Books of Occult Philosophy*, 567.

19. Silvestris, *Cosmographia*, 108.

20. Agrippa, *Three Books of Occult Philosophy*, 567.

21. Plato, *The Collected Dialogues of Plato*, 525.

22. Larson, *Greek Nymphs*, 9–10.

23. Maria Piranomonte, "Religion and Magic at Rome: The Fountain of Anna Perenna," in *Magical Practice in the Latin West*, ed. Richard Gordon and Francisco Simon (Brill Academic Publishers, 2010), 197.

24. Larson, *Greek Nymphs*, 70.

25. Piranomonte, "Religion and Magic at Rome," 197.

26. Friends of the Wissahickon, "Devil's Pool."

27. Fortune, *Psychic Self Defense*, chapter 7.

CHAPTER 12. WORKING WITH EARTH ENERGIES

1. De Rola, *The Golden Game*, 197.

2. Michell, *The Earth Spirit*, 4.

3. Michell, *The Earth Spirit*, 7.

4. Cowan and Arnold, *Ley Lines and Earth Energies*, 25.

5. Michell, *The Earth Spirit*, 14.

6. Michell, *The Earth Spirit*, 7.

7. Gary, *Traditional Witchcraft*, 62.

8. Gary, *Traditional Witchcraft*, 62–63.

9. Gary, *Traditional Witchcraft*, 64–65.

10. Michell, *The Earth Spirit*, 19–20.

11. Graves, *The Greek Myths, Volume 1*, 27

12. New Oxford Annotated Bible, Genesis 3:4–6.

13. Jung, *Alchemical Studies*, 288.

14. New Oxford Annotated Bible, Genesis 3:17–19.

15. Robinson, *The Nag Hammadi Library*, 164–65.

16. Harrison, *Prolegomena to the Study of Greek Religion*, 332.

17. Homer, *The Homeric Hymns*, 43.

18. Harrison, *Epilogomena to the Study of Greek Religion and Themis*, 428–29.

19. Harrison, *Epilogomena to the Study of Greek Religion and Themis*, 459–60.

20. Campbell, *Occidental Mythology*, 80.

21. Jung, *Aion*, 335.

22. Jung, *Aion*, 372.

23. Jung, *Aion*, 372.

24. Jung, *Aion*, 336.

25. New Oxford Annotated Bible, John 3:14.

26. Riviere, *Tantrik Yoga*, 61–66.

27. McEvilley, *The Shape of Ancient Thought*, chapter 8.

28. Plato, *The Collected Dialogues of Plato*, 1193.

29. P. L. Henry, "The Cauldron of Poesy," *Studia Celtica* 14/15 (1979/1980): 114.

30. Greer, *The Dolmen Arch*, vol. 1, 174–75.

31. Greer, *The Dolmen Arch*, vol. 1, 175.

32. Kingsley, *Ancient Philosophy, Mystery and Magic*, 13–14.

33. Gordon et al., "A Prayer For Blessings on Three Ritual Objects Discovered at Chartres-Autricum," 487–518.

34. I have combined elements from several versions of the Sphere of Protection ritual along with my own wording. The sources used are as follows: Greer, *The Druid Magic Handbook*, chapter 3; Ancient Order of Druids in America, "Introduction to the Sphere of Protection," accessed February 22, 2024.

CHAPTER 13. THE MAGIC OF THE UNDERWORLD

1. Duffy, *Anathema Maranatha*, 57–60.

2. Sturlusson, *Edda*, 24.

3. Johnston, *Restless Dead*, 38.

4. Burkert, *Greek Religion*, 194.

5. Johnston, *Restless Dead*, 42.

6. Burkert, *Greek Religion*, 194–95.

7. King, *The Ancient Roman Afterlife*, 90.

8. Iamblichus, *On the Mysteries*, book II, 83.

9. Luck, *Arcana Mundi*, 168–69.

10. Evans-Wentz, *The Fairy-Faith in the Celtic Countries*, 438.

11. Agrippa, *Three Books of Occult Philosophy*, 575.

12. Agrippa, *Three Books of Occult Philosophy*, 576

13. Johnston, *Restless Dead*, 129–30.

14. William Lee Rand, *Holy Fire III Reiki Master Manual* (International Center for Reiki Training, 2022), 28.

15. Eliade, *Shamanism*, 207.

16. Plato, *The Collected Dialogues of Plato*, 909b, 1464.

17. Betz, *The Greek Magical Papyri in Translation*, 64–66.

18. Johnston, *Restless Dead*, 115.

19. Johnston, *Restless Dead*, 109.

20. Diagoras of Melos, "Derveni Papyrus: A New Translation," 20–21.

21. Johnston, *Restless Dead*, 260.

22. Eliade, *Occultism, Witchcraft, and Cultural Fashions*, 40, 41.

23. Eliade, *Occultism, Witchcraft, and Cultural Fashions*, 41.

24. Luck, *Arcana Mundi*, 231.

25. Eliade, *Occultism, Witchcraft, and Cultural Fashions*, 31.

26. Eliade, *Occultism, Witchcraft, and Cultural Fashions*, 39.

27. Burkert, *Greek Religion*, 195.

28. Plato, *The Collected Dialogues of Plato*, 64.

29. Cumont, *After Life in Roman Paganism*, 4.

30. Cumont, *After Life in Roman Paganism*, 34.

31. New Oxford Annotated Bible, Ecclesiastes 9:10.

32. New Oxford Annotated Bible, Job 7:9.

33. Cumont, *After Life in Roman Paganism*, 70.

34. Eliade, *Occultism, Witchcraft, and Cultural Fashions*, 39.

35. Cumont, *After Life in Roman Paganism*, 75-76.

36. Elliott, *The Apocryphal New Testament*, 185–204.

37. Lilly, *The Center of the Cyclone*, 9.

38. Bernabe and San Cristobal, *Instructions For The Netherworld*, 9–10.

39. Stratton-Kent, *Geosophia*, 46–47.

CHAPTER 14. PRACTICES INVOLVING SPIRITS OF THE UNDERWORLD

1. Evans-Wentz, *The Fairy-Faith in the Celtic Countries*, 75.

2. Evans-Wentz, *The Fairy-Faith in the Celtic Countries*, 280.

3. Evans-Wentz, *The Fairy-Faith in the Celtic Countries*, 438.

4. Pócs, *Fairies and Witches at the Boundaries of South-East Europe*, 16-21.

5. Thompson, *Semitic Magic*, 16-17.

6. Pócs, *Fairies and Witches at the Boundary of South-Eastern and Central Europe*, 23.

7. Spence, *British Fairy Origins*, 75.

8. Adams, *The Power of the Healing Field*, 53, 73, 98.

9. Lewis Bayles Paton, *Spiritism and the Cult of the Dead in Antiquity* (The Macmillan Company, 1921), 212.

10. Thompson, *Devils and Evil Spirits of Babylonia*, 29–37.

11. Thompson, *Devils and Evil Spirits of Babylonia*, 29–37.

12. Singer, "Early English Magic and Medicine," 6.

13. Storms, *Anglo-Saxon Magic*, 223.

14. Heath, *Elves, Witches & Gods*, 246.

15. Heath, *Elves, Witches & Gods*, 90–91.

16. Storms, *Anglo-Saxon Magic*, 143.

17. Hutton, *Queens of the Wild*, 13.

18. Hutton, *Queens of the Wild*, 15.

19. Young, *Twilight of the Godlings*, 260–61.

20. Lecouteux, *The Hidden History of Elves & Dwarfs* (Inner Traditions International, 2018), 14.

21. Morgan, *Sounds of Infinity*, 236.

22. Scot, *The Discoverie of Witchcraft*, 234–35.

23. Connor, *A People's History of Science*, 186.

24. Connor, *A People's History of Science*, 186.

25. Agrippa, *The Fourth Book of Occult Philosophy*, 55–56.

26. Wilby, *Cunning Folk and Familiar Spirits*, 19.

27. Wilby, *Cunning Folk and Familiar Spirits*, 19–20.

28. Wilby, *Cunning Folk and Familiar Spirits*, 20.

29. Authorized King James Bible, Hebrews 11:1.

30. Morgan, *Sounds of Infinity*, 20–21.

31. Joshua Cutchin, *Ecology of Souls* (Horse and Barrel Press, 2022).

CHAPTER 15. ENTHEOGENIC PLANTS AND THE PLANT PATH

1. Muraresku, *The Immortality Key*, 168.

2. Angus, *The Mystery-Religions*, 112–13.

3. Schultes and Hofmann, *Plants of the Gods*, 106–11.

4. Robert Wasson, "Seeking the Magic Mushroom," *Life* 42, no. 19 (May 13, 1957): 100–110, 113–14, 117–18, 120.

5. Pendell, *Pharmako/Gnosis*, 27.

6. Munn, "The Mushrooms of Language," 88–89.

7. Munn, "The Mushrooms of Language," 91.

8. Wasson, "The Hallucinogenic Fungi of Mexico," 38.

9. Pindar, *Odes of Pindar*, 591–92.

10. Otto, "The Meaning of the Eleusinian Mysteries," 20.

11. Wasson et al., *The Road to Eleusis*, 38–44.

12. Robert Graves, "Mushrooms, Food of the Gods." *The Atlantic*, August 1957.

13. Muraresku, *The Immortality Key*, 48.

14. Otto, "The Meaning of the Eleusinian Mysteries," 24.

15. Apuleius, *The Golden Ass* (Graves translation), 280.

16. Jung, *Mysterium Coniunctionis*, 133.

17. Ponzi, "Trinity College M.S. 0.2.48 — The Hermetic Lunatica."

18. Eliade, *The Forge and the Crucible*, 161.

19. Muraresku, *The Immortality Key*, 223–24.

20. New Oxford Annotated Bible, 1 Corinthians 11:20–30.

21. Muraresku, *The Immortality Key*, 225.

CHAPTER 16. BANEFUL PLANTS IN MAGICAL PRACTICE

1. Ott, *Pharmacotheon*, 375.

2. "Tobacco," World Health Organization, updated July 31, 2023.

3. Salmon and Salmon, "Tobacco in Colonial Virginia."

4. Wilbert, *Tobacco and Shamanism in South America*, 6.

5. Ott, *Pharmacotheon*, 373.

6. World Heritage Ohio, *Guide to the Hopewell Ceremonial Earthworks*, 10.

7. Hudson, *The Southeastern Indians*, 351.

8. Hudson, *The Southeastern Indians*, 351.

9. Hudson, *The Southeastern Indians*, 353.

10. Hudson, *The Southeastern Indians*, 353–54.

11. Yronwode, *Hoodoo Herb and Root Magic*, 198–99.

12. Pendell, *Pharmako/Poeia*, 38.

13. Schulke, *Viridarium Umbris*, 401.

14. Disel et al., "Poisoned after Dinner," 51–55.

15. The Other Jamestown, "Algonquian Life and Customs."

16. Schultes and Hofmann, *Plants of the Gods*, 110–11.

17. Kluckhohn, *Navaho Witchcraft*, 63.

18. Castaneda, *The Teachings of Don Juan*, 147.

19. Castaneda, *The Teachings of Don Juan*, 147.

20. al-Qurtubi, *Picatrix* (Attrell and Porreca translation), 213.

21. al-Qurtubi, *Picatrix* (Attrell and Porreca translation), 204.

22. della Porta, *Natural Magick*, 219.

23. Ott, *Pharmacotheon*, 364–65.

24. Schultes and Hofmann, *Plants of the Gods*, 131.

25. Kiernan, "Last Blast for Florida's Teenage Trippers," *New Scientist*, February 4, 1995.

26. Ott, *Pharmacotheon*, 368.

27. New Oxford Annotated Bible, Genesis 30:14–17.

28. Watts, *Dictionary of Plant Lore*, 239.

29. Theophrastus, *Enquiry into Plants*, 261.

30. Watts, *Dictionary of Plant Lore*, 239.

31. Dioscorides, *The Herbal of Dioscorides the Greek*, 624, 627.

32. Watts, *Dictionary of Plant Lore*, 240.

33. Watts, *Dictionary of Plant Lore*, 241.

34. Frazer, *Folklore in The Old Testament*, 381.

35. Frazer, *Folklore in The Old Testament*, 381.

36. Frazer, *Folklore in The Old Testament*, 382.

37. Grimm, *Teutonic Mythology*, 1202.

38. Grimm, *Teutonic Mythology*, 1225.

39. Harner, *Hallucinogens and Shamanism*, 135–36.

40. Schulke, *Veneficium*, 26.

41. Bailey, *The Standard Cyclopedia of Horticulture*, vol. 2, 1628.

42. Murarescu, *The Immortality Key*, 128.

43. Pendell, *Pharmako/Poeia*, 66.

44. Dioscorides, *The Herbal of Dioscorides the Greek*, 615.

45. Schultes and Hofmann, *Plants of the Gods*, 86.

46. Harner, *Hallucinogens and Shamanism*, 140.

47. della Porta, *Natural Magick*, 231–32.

48. Harner, *Hallucinogens and Shamanism*, 145.

49. De Mattos Frisvold, *Trollrún*, 327.

50. Webb and Baby, *The Adena People*, 62–64.

51. Eliade, *Shamanism*, 93.

52. Eliade, *Shamanism*, 459–60.

53. Price, *The Viking Way*, chapter 6.

54. Ankarloo and Clark, *Witchcraft and Magic in Europe*, 102.

55. Ginzburg, Carlo, *Ecstasies*, 153.

CHAPTER 17. PLANT MAGIC

1. Boehme, *The Signature of All Things*, 18.

2. New Oxford Annotated Bible, Genesis 1:29.

3. Bennett, "Doctrine of Signatures," 1.

4. Copenhaver, *Hermetica*, 68.

5. Copenhaver, *Hermetica*, 83.

6. Copenhaver, *Magic in Western Culture*, 92.

7. Copenhaver, *Magic in Western Culture*, 92.

8. Copenhaver, *Magic in Western Culture*, 91.

9. Iamblichus, *On the Mysteries*, 269.

10. Ficino, *Three Books on Life*, 387.

11. Ficino, *Three Books on Life*, 387.

12. Lao Tzu, *Tao Te Ching*, 44.

13. Agrippa, *Three Books of Occult Philosophy*, 39.

14. Agrippa, *Three Books of Occult Philosophy*.

15. Crowley, *777 and Other Kabbalistic Writings of Aleister Crowley*, 10, 13.

16. Thorndike, *A History of Magic and Experimental Science*, 565–566.

17. Turner, *Ancient Pathways, Ancestral Knowledge*, 797.

18. De Mattos Frisvold, *Trollrún*, 66.

19. De Mattos Frisvold, *Trollrún*, 67.

20. De Mattos Frisvold, *Trollrún*, 76.

21. Grimm, *Teutonic Mythology*, 1063.

22. Rudolf Steiner, "Man as Symphony of the Creative World," pt. 3, Rudolf Steiner Archive, accessed February 27, 2024.

23. Ficino, *Three Books on Life*, 279.

24. Rohde, *The Old English Herbals*, 29.

25. Gauntlet, *The Grimoire of Arthur Gauntlet*, 303.

26. Puckett, *Folk Beliefs of the Southern [Omitted]*, 235, 289.

27. Crawford, *The Poetic Edda,* 42.

28. Grendon, *Anglo-Saxon Charms*, 191, 193.

29. Betz, *The Greek Magical Papyri in Translation*, 95.

CHAPTER 18. AGRICULTURAL MAGIC

1. Hesiod, *Works and Days*, 125.

2. Cato, *Roman Farm Management*, 52–53.

3. Columella, *On Agriculture* (Ash translation), vol. I, book II, 107–9.

4. Cato, *Complete Works of Cato the Elder*, 24.

5. Signe and Skydsgaard, *Ancient Greek Agriculture*, 161–62.

6. Hartley, *Lost Country Life*, 51–52.

7. Virgil, *The Poems of Virgil*, lines 43–59.

8. Carmichael, *Carmina Gadelica*, xxix.

9. Carmichael, *Carmina Gadelica*, 245–47.

10. Carmichael, *Carmina Gadelica*, 245.

11. Ovid, *Fasti*, 51–53.

12. Dumézil, *Archaic Roman Religion*, 376.

13. Hudson, *The Southeastern Indians*, 292–94.

14. Hudson, *The Southeastern Indians*, 365–75.

15. Kinsella, *The Táin*, 27.

16. Hutton, *Stations of the Sun*, 302.

17. Hutton, *Stations of the Sun*, 347.

18. Hutton, *Stations of the Sun*, 350.

19. Hutton, *Stations of the Sun*, 350.

20. Carmichael, *Carmina Gadelica*, 249–51.

21. McNeill, *The Silver Bough*, 48.

22. Frazer, *The Golden Bough*, 463–77.

23. Hutton, *Stations of the Sun*, 358.

24. Davidson, *Roles of the Northern Goddess*, 61.

25. Grendon, *Anglo-Saxon Charms*, 175.

26. Rohde, *The Old English Herbals*, 36.

27. Singer, Charles, "Early English Magic and Medicine," 32–33.

CHAPTER 19. SACRED FORESTS AND SPIRITUAL PRACTICES WITH TREES

1. Ovid, *Fasti*, Book III: 143, 295.

2. Dorcey, *The Cult of Silvanus*, 16–25.

3. Dorcey, *The Cult of Silvanus*, 36–42.

4. Dorcey, *The Cult of Silvanus*, 42.

5. Dorcey, *The Cult of Silvanus*, 56.

6. Brandon, *The Rebirth of Pan*.

7. *The Poetic Edda*, 6.

8. Davidson, *Gods and Myths of Northern Europe*, 26.

9. De Mattos Frisvold, *Trollrún*, 168.

10. Sturlusson, *Edda*, 19.

11. De Mattos Frisvold, *Trollrún*, 43–49.

12. De Mattos Frisvold, *Trollrún*, 43–49.

13. Cusack, *The Sacred Tree*, 155.

14. Eliade, *Shamanism*, 380.

15. New Oxford Annotated Bible, Genesis 2:8–9.

16. Sayce, *Lectures on the Origin and Growth of Religion*, 238.

17. Sayce, *Lectures on the Origin and Growth of Religion*, 238–39.

18. Schama, *The History of the Jews*, 428.

19. David M. Carr, "Genesis: Date of Composition and Literary History," in The New Oxford Annotated Bible, Revised Standard Edition (Oxford University Press, 1977): 7–8.

20. Simo Parpola, "The Assyrian Tree of Life: Tracing the Origin of Jewish Mysticism and Greek Philosophy," *Journal of Near East Studies* 52, no. 3 (July 1993): 166–67.

21. Parpola, "The Assyrian Tree of Life," 175–81.

22. Parpola, "The Assyrian Tree of Life," 174.

23. Day, *Yahweh and The Gods and Goddesses of Canaan*, 60.

24. New Oxford Annotated Bible, 1 Kings 11:5.

25. New Oxford Annotated Bible, 1 Kings 11:7-8.

26. New Oxford Annotated Bible, 1 Kings 14:22–24.

27. Binger, *Asherah*, 139.

28. New Oxford Annotated Bible, Exodus 34:11–14.

29. New Oxford Annotated Bible, 2 Kings 23:24.

30. Philo of Alexandria, *The Works of Philo*, 331.

31. Scholem, *Kabbalah,* 111–12.

32. Cusack, *The Sacred Tree*, 65.

33. Lucan, *Pharsalia* (lines 399–425), 143–45.

34. Lucan, *Pharsalia* (lines 426–52), 145–47.

35. Tacitus, *The Agricola and the Germania*, 109.

36. Cusack, *The Sacred Tree*, 94–96.

37. Cusack, *The Sacred Tree*, 71–72.

38. Cusack, *The Sacred Tree*, 99.

39. Cusack, *The Sacred Tree*, 108.

40. Cusack, *The Sacred Tree*, 111–12.

41. Winroth, *The Conversion of Scandinavia*, 151

CONCLUSION

1. Jung, *The Undiscovered Self*, 123.

BIBLIOGRAPHY AND RECOMMENDED READING

Adams, Peter Mark. *The Power of the Healing Field*. Healing Arts Press, 2022.

Ager, Britta K. "Roman Agricultural Magic." PhD diss., University of Michigan, 2010.

Agrippa, Henry Cornelius. *Three Books of Occult Philosophy*. Llewellyn, 1993.

Agrippa, Henry Cornelius. *The Fourth Book of Occult Philosophy*. Translated by Robert Turner. Edited by Stephen Skinner. Ibis Press, 2005.

Alighieri, Dante. *The Divine Comedy, Paradise*. Translated by Charles Eliot Norton. Encyclopaedia Britannica, 1952.

al-Qurtubi, Maslama. *The Picatrix*. Translated by Christopher Warnock and John Michael Greer. Adocentyn Press, 2011.

al-Qurtubi, Maslama. *The Picatrix*. Translated by Dan Attrell and David Porreca. Pennsylvania State University Press, 2019.

Angus, S. *The Mystery-Religions*. Dover, 1975.

Ankarloo, Bengt, and Stuart Clark, editors. *Witchcraft and Magic in Europe: The Middle Ages*. University of Pennsylvania Press, 2002.

Apuleius. *The Golden Ass*. Translated by W. Adlington. Revised by S. Gaselee. William Heinemann, 1924.

Apuleius. *The Golden Ass*. Translated by Robert Graves. Farrar, Strauss, and Giroux, 1951.

Augustine of Hippo. *City of God*. Translated by Marcus Dods. Encyclopedia Britannica, 1952.

Aurelius, Marcus. *The Meditations of Marcus Aurelius*. Translated by George Long. Encyclopedia Britannica, 1952.

Avelar, Helena, and Luis Ribeiro. *On the Heavenly Spheres: A Treatise on Traditional Astrology*. American Federation of Astrologers, 2010.

Bailey, L. H. *The Standard Cyclopedia of Horticulture*. Macmillan, 1943.

Baker, Jim. *The Cunning Man's Handbook*. Avalonia, 2013.

Barnes, Jonathan. *Early Greek Philosophy*. Penguin Books, 2001.

Bennett, Bradley C. "Doctrine of Signatures: An Explanation of Medicinal Plant Discovery or Dissemination of Knowledge?" *Economic Botany* 61, no. 3 (2007): 246–55.

Bernabe, Alberto, and Ana Isabel Jimenez San Cristobal. *Instructions for the Netherworld: The Orphic Gold Tablets*. Brill, 2008.

Betz, Hanz Dieter. *The Greek Magical Papyri in Translation*. University of Chicago Press, 1992.

The New Oxford Annotated Bible, Revised Standard Version. Oxford University Press, 1977.

Bilardi, C. R. *The Red Church*. Pendraig Publishers, 2009.

Binger, Tilde. *Asherah: Goddesses in Ugarit, Israel and the Old Testament*. Sheffield Academic Press, 1997.

Boehme, Jacob. *The Signature of All Things*. J. M. Dent and Sons, Ltd., 1912.

The Book of Common Prayer. Church Publishing, Inc., 1986.

Brand, Damon. *The 72 Angels of Magick*. CreateSpace, 2015. Kindle.

Brandon, Jim. *The Rebirth of Pan: The Hidden Faces of the American Earth Spirit*. Published by the author, 1983.

Budge, E. A. Wallis. *The Cave of Treasures*. The Religious Tract Society, 1927.

Burkert, Walter. *Greek Religion*. Harvard University Press, 1985.

Campbell, Joseph. *Occidental Mythology*. Penguin, 1987.

Campbell, Joseph. *Primitive Mythology*. Penguin, 1987.

Carmichael, Alexander. *Carmina Gadelica, Volume One*. T. A. Constable, 1900.

Castaneda, Carlos. *The Teachings of Don Juan: A Yaqui Way of Knowledge*. Simon and Schuster, 1973.

Cato. *The Complete Works of Cato the Elder*. Translated by Fairfax Harrison. Delphi Classics, 2019.

Cato and Varro. *Roman Farm Management*. Translated by Fairfax Harrison. Macmillan, 1913.

Collins, David J. "Scholasticism and High Medieval Opposition to Magic." In *The Routledge History of Medieval Magic*, edited by Sophie Page. Routledge, 2019.

Columella. *On Agriculture*. Translated by E. S. Forster and Edward H. Heffner. Harvard University Press, 1954.

Columella. *On Agriculture*. Translated by Harrison Boyd Ash. Harvard University Press, 1960.

Connor, Clifford. *A People's History of Science*. Nation Books, 2005.

Copenhaver, Brian. *Hermetica*. Cambridge University Press, 1992.

Copenhaver, Brian. *Magic in Western Culture*. Cambridge University Press, 2015.

Corbin, Henry. "Mundus Imaginalis, or the Imaginary and the Imaginal." In *Colloquium on Symbolism*. Paris, June, 1964.

Cowan, David R., and Chris Arnold. *Ley Lines and Earth Mysteries: An Extraordinary Journey into the Earth's Natural Energy System*. Adventures Unlimited, 2003.

Crowley, Aleister. *Magick in Theory and Practice*. Lecram Press, 1929.

Crowley, Aleister. *777 and Other Kabbalistic Writings of Aleister Crowley*. Weiser, 1973.

Cumont, Franz. *After Life in Roman Paganism*. Yale University Press, 1922.

Cumont, Franz. *Astrology and Religion Among the Greeks and Romans*. Dover, 1960.

Cusack, Carole M. *The Sacred Tree: Ancient and Medieval Manifestations*. Cambridge Scholars, 2011.

Davidson, Hilda Ellis. *Gods and Myths of Northern Europe*. Penguin Books, 1964.

Davidson, Hilda Ellis. *Roles of the Northern Goddess*. Routledge, 1998.

Day, John. *Psalms*. Sheffield Academic Press, 1948.

Day, John. *Yahweh and the Gods and Goddesses of Canaan*. Sheffield Academic Press, 2002.

della Porta, Giambattista. *Magia Naturalis (Natural Magick)*. Young and Speed, 1658.

De Mattos Frisvold, Nicholaj. *Trollrún: A Discourse on Trolldom and Runes in the Northern Tradition*. Hadean Press, 2021.

De Rola, Stanislas Klossowski. *The Golden Game: Alchemical Engravings of the Seventeenth Century*. Thames and Hudson, 1997.

Diagoras of Melos. "Derveni Papyrus: A New Translation." Translated by Richard Janko. Fragment VI. *Journal of Classical Philology* 96, no. 1 (2001): 20–21.

Dioscorides. *The Herbal of Dioscorides the Greek*. Ibidis Press, 2000.

Disel et al. "Poisoned After Dinner: Dolma With *Datura Stramonium*." *Turkish Journal of Emergency Medicine* 15, no. 1 (2015): 51–55.

Dorcey, Peter F. *The Cult of Silvanus: A Study in Roman Folk Religion*. E. J. Brill, 1992.

Duffy, Martin. *Anathema Maranatha: Christianity and the Imprecatory Arts*. Three Hands Press, 2022.

Dumézil, Georges. *Archaic Roman Religion, Volume One*. Johns Hopkins University Press, 1996.

DuQuette, Lon Milo. *Enochian Vision Magick*. Red Wheel/Weiser, 2008.

Eliade, Mircea. *Occultism, Witchcraft, and Cultural Fashions*. University of Chicago Press, 1976.

Eliade, Mircea. *Shamanism: Archaic Techniques of Ecstasy*. Bollingen Foundation, 1964.

Eliade, Mircea. *The Forge and the Crucible*. University of Chicago Press, 1962.

Elliott, J. K. *The Apocryphal New Testament*. Oxford University Press, 1993.

Ellis, Normandi. *The Ancient Tradition of Angels*. Inner Traditions, 2023.

Evans-Wentz, W. Y. *The Fairy-Faith in the Celtic Countries*. Carol Publishing, 1994.

Ficino, Marsilio. *Three Books on Life*. Translated by Carol V. Kaske and John R. Clark. Medieval and Renaissance Texts and Studies, 2019.

Flattery, David, and Martin Schwartz. *Haoma and Harmaline: The Botanical Identity of the Indo-Iranian Sacred Hallucinogen "Soma" and its Legacy in Religion, Language, and Miiddle-Eastern Folklore*. University of California Press, 1989.

Fortune, Dion. *Psychic Self Defense*. Red Wheel/Weiser, 1997.

Frazer, James George. *Folklore in the Old Testament*. Macmillan, 1918.

Frazer, James George. *The Golden Bough*. Macmillan, 1950.

Friends of the Wissahickon. "Devil's Pool: Take a Hike Through History." Friends of the Wissahickon website. Accessed February 22, 2024.

Gary, Gemma. *Traditional Witchcraft: A Cornish Book of Ways*. Troy Books, 2008.

Gauntlet, Arthur. *The Grimoire of Arthur Gauntlet*. Edited by David Rankine. Avalonia Press, 2011.

Geller, Markham J. *Melothesia in Babylonia*. Walter De Gruyter Inc., 2014.

Ginzburg, Carlo. *Ecstasies: Deciphering The Witches' Sabbath*. University of Chicago Press, 2004.

Giovino, Mariana. *The Assyrian Sacred Tree*. Vanderhoeck & Ruprecht, 2007.

Gordon, Richard, Dominique Joly, and William Van Andringa. "A Prayer for Blessing on Three Ritual Objects Discovered at Chartres-Autricum." In *Magical Practice in the Latin West*, edited by Francisco Marco Simón and Richard Gordon. Brill, 2010.

Graves, Robert. *The Greek Myths*. Penguin Random House, 2017.

Graves, Robert. *The Greek Myths, Volume One*. Pelican, 1955.

Graves, Robert. "Mushrooms, Food of the Gods." *The Atlantic*, August 1957.

Greenbaum, Dorian Gieseler. *The Daimon in Hellenistic Astrology: Origins and Influence*. Brill, 2016.

Greer, John Michael. *Circles of Power*. Aeon Books, 2017.

Greer, John Michael. *The Dolmen Arch*. Volumes 1–2. Azoth, 2023.

Greer, John Michael. *The Druid Magic Handbook*. Red Wheel/Weiser, 2007.

Greer, John Michael, and Christopher Warnock. *The Astral High Magic: De Imaginibus of Thabit Ibn Qurra*. Renaissance Astrology, 2011.

Grendon, Felix. *Anglo-Saxon Charms. Journal of American Folklore*, 1909.

Grimm, Jacob. *Teutonic Mythology*. Translated by James Steven Stalleybrass. George Bell and Sons, 1882.

Hall, Alaric. *Elves in Anglo-Saxon England: Matters of Belief, Health, Gender, and Identity*. Boydell, 2007.

Harms, Daniel, Joseph Peterson, and James Clark. *The Book of Oberon*. Llewellyn, 2021.

Harner, Michael. *Hallucinogens and Shamanism*. Oxford University Press, 1973.

Harrison, Jane. *Epilogomena to the Study of Greek Religion and Themis*. University Books, 1962.

Harrison, Jane. *Prolegomena to the Study of Greek Religion*. Meridian Books, 1959.

Hartley, Dorothy. *Lost Country Life*. Pantheon Books, 1979.

Heath, Catherine. *Elves, Witches & Gods: Spinning Heathen Magic in the Modern Day*. Llewellyn, 2021.

Heim, Gerhard. *The Celts*. St. Martin's Press, 1976.

Hekhalot Rabbati. Translated by Morton Smith. Corrected by Gershom Scholem. Transcribed and edited by Don Karr. 2009.

Hesiod. *Theogony and Works and Days*. Translated by M. L. West. Oxford University Press, 1998.

Hesiod. *Works and Days*. Edited and translated by Glenn W. Most. Harvard University Press, 2006.

Hohman, John George. *The Long Lost Friend*. Edited and translated by Daniel Harms. Llewellyn, 2020.

Homer. *The Homeric Hymns*. Translated by Jules Cashford. Penguin, 2003.

Hudson, Charles. *The Southeastern Indians*. University of Tennessee Press, 1978.

Hutton, Ronald. *Queens of the Wild*. Yale University Press, 2022.

Hutton, Ronald. *Stations of the Sun*. Oxford University Press, 2001.

Iamblichus. *On the Mysteries*. Translated by Emma C. Clark, John M. Dillon, and Jackson P. Hershbell. Society of Biblical Literature, 2003.

Janko, Richard. "The Derveni Papyrus: A New Translation." *Classical Philology* 96 (January 2001): 1–32.

Johnson, Jerry Alan. *Chinese Medical Qigong Therapy Vol. 3*. The International Institute of Medical Qigong, 2005.

Johnson, Jerry Alan. *Daoist Weather Magic and Feng Shui*. The International Institute of Daoist Magic, 2007.

Johnston, Sarah Iles. *Restless Dead: Encounters Between the Living and the Dead in Ancient Greece*. University of California Press, 1999.

Jung, Carl Gustav. *Aion: Researches into the Phenomenology of the Self*. Bollingen, 1978.

Jung, Carl Gustav. *Alchemical Studies*. Bollingen, 1970.

Jung, Carl Gustav. *Mysterium Coniunctionis*. Translated by R. F. C Hull. Bollingen, 1970.

Jung, Carl Gustav. *Synchronicity: An Acausal Connecting Principle*. Princeton University Press, 1973.

Jung, Carl Gustav. *The Undiscovered Self*. Mentor Books, 1958.

Kaplan, Aryeh. *Sefer Yetzirah: The Book of Creation*. Samuel Weiser, 1990.

Kerenyi, Carl. *Eleusis. Archetypal Image of Mother and Daughter*. Bollingen, 1967.

Kieckhefer, Richard. *Forbidden Rites*. Penn State Press, 1997.

Kieckhefer, Richard. *Magic in the Middle Ages*. Cambridge University Press, 2014.

Kiesel, William. *Magic Circles in the Grimoire Tradition*. Three Hands Press, 2012.

King, Charles W. *The Ancient Roman Afterlife: Di Manes, Belief and the Cult of the Dead*. University of Texas Press, 2020.

Kingsley, Peter. *Ancient Philosophy, Mystery and Magic: Empedocles and the Pythagorean Tradition*. Oxford University Press, 1995.

Kinsella, Thomas. *The Táin*. Oxford University Press, 2002.

Kirk, Robert. *The Secret Commonwealth*. Dover Publications, 2008.

Kluckhohn, Clyde. *Navaho Witchcraft*. Beacon Press, 1944.

Kraig, Donald Michael. *Modern Magick: Eleven Lessons in the High Magickal Arts*. Llewellyn, 1988.

Lao Tzu. *Tao Te Ching*. Translated by Gia Fu Feng and Jane English. Vintage Books, 1989.

Larson, Jennifer. *Greek Nymphs: Myth, Cult, Lore*. Oxford University Press, 2001.

Lecouteux, Claude. *Demons and Spirits of the Land*. Translated by Jon E. Graham. Inner Traditions, 2015.

Leitão, José. *Opuscula Cypriani*. Hadean Press, 2019.

Leitch, Aaron. *The Essential Enochian Grimoire*. Llewellyn, 2021.

Lewis, C. S. *The Discarded Image*. Cambridge University Press, 1964.

Lilly, John. *The Center of the Cyclone*. Bantam, 1971.

Lucan. *Pharsalia*. Translated by A. E. Housman and J. W. Duff. Harvard University Press, 1962.

Luck, Georg. *Arcana Mundi: Magic and the Occult in the Greek and Roman Worlds*. Johns Hopkins University Press, 2006.

Magnus, Albertus (pseudonym). *The Book of Secrets of Albertus Magnus*. Edited by Michael R. Best and Frank H. Brightman. Weiser Books, 2004.

Majercik, Ruth. *The Chaldean Oracles*. The Prometheus Trust, 2013.

Mathers, S. Liddell MacGregor. *The Greater Key of Solomon*. De Lawrence, Scott and Co., 1914.

McEvilley, Thomas. *The Shape of Ancient Thought*. Allworth Press, 2002. Kindle.

McKenna, Terence. *Food of the Gods*. Bantam Books, 1992.

McKenna, Terence. *The Archaic Revival*. Harper San Francisco, 1991.

McNeill, F. Marian. *The Silver Bough*. Canongate, 1956.

Michael, Coby. *The Poison Path Herbal*. Park Street Press, 2021.

Michell, John. *The Earth Spirit*. New York: Crossroad, 1975.

Morgan, Lee. *Sounds of Infinity*. The Witches' Almanac, 2019.

Munn, Henry. "The Mushrooms of Language." In *Hallucinogens and Shamanism*, edited by Michael Harner, Oxford University Press, 1973.

Muraresku, Brian C. *The Immortality Key: The Secret History of the Religion With No Name*. St. Martin's Press, 2020.

National Aeronautics and Space Administration. "Moon Facts." NASA website. Accessed January 27, 2024.

National Radio Astronomy Observatory. "Variability of the Moon's Apparent Motion Through the Sky." NRAO website. Accessed January 27, 2024.

Needham, Joseph, and Ling Wang. *Science and Civilization in China, Vol. 3*. Cambridge University Press, 2005.

ní Mheallaigh, Karen. *The Moon in the Ancient Greek and Roman Imagination*. Cambridge University Press, 2020.

Ogden, Daniel. *Magic, Witchcraft, and Ghosts in the Greek and Roman Worlds*. Oxford University Press, 2009.

Ogden, Daniel. *The Dragon in the West*. Oxford University Press, 2021.

Opsopaus, John. *Summary of Pythagorean Theology*. Published by the author, 2002–4.

The Other Jamestown. "Algonquian Life and Customs." virtual-jamestown website. Accessed February 26, 2024.

Ott, Jonathan. *Ayahuasca Analogues*. Natural Products Co., 1994.

Ott, Jonathan. *Pharmacotheon*. Natural Products Co., 1993.

Otto, Walter F. "The Meaning of the Eleusinian Mysteries." In *The Mysteries: Papers From the Eranos Yearbooks, Volume 2*, edited by Joseph Campbell. Bollingen, 1955.

Ovid. *Fasti*. Translated by James George Frazer. Harvard University Press, 1959.

Parke-Taylor, G. H. *Yahweh: The Divine Name in the Bible*. Wilfrid Laurier University Press, 1975.

Parpalo, Simo. "The Assyrian Tree of Life: Tracing the Origins of Jewish Mysticism and Greek Philosophy." *Journal of Near Eastern Studies* 52, no. 3 (July 1993): 161–208.

Pendell, Dale. *Pharmako/Poeia*. Mercury House, 1995.

Pendell, Dale. *Pharmako/Gnosis*. Mercury House, 2005.

Peterson, Joseph H. *Secrets of Solomon*. Twilit Grotto Press, 2018.

Peterson, Joseph H. *The Sixth and Seventh Book of Moses*. Ibis Press, 2008.

Peterson, Joseph H. *The Sworn Book of Honorius.* Ibis Press, 2016.

Philo of Alexandria. *The Works of Philo.* Translated by C. D. Yonge. Hendrickson Publishing, 1993.

Philosophy and Art Collaboratory. "Qi: Breath and Vital Energy." Website. Accessed February 9, 2024.

Pindar. *The Odes of Pindar.* Translated by Sir John Sandys. The Macmillan Co., 1915.

Pingree, David. "The Sabians of Harran and the Classical Tradition." *International Journal of the Classical Tradition* 9, no. 1 (summer 2002): 8–35.

Pirtea, Adrian. "From Lunar Nodes to Eclipse Dragons: The Fundamentals of the Chaldean Art (CCAG V/2, 131-40) and the reception of Arabo-Persian Astrology in Byzantium." In *Savoirs prédictifs et techniques divinatoires de l'Antiquité tardive à Byzance*, edited by Paul Magdalino and Andrei Timotin. Pomme d'Or, 2019.

Plato. *The Collected Dialogues of Plato.* Edited by Edith Hamilton and Huntington Cairns. Princeton University Press, 1961.

Plotinus. *The Enneads.* Translated by Lloyd Gerson. Cambridge University Press, 2019.

Pócs, Éva. *Fairies and Witches at the Boundary of South-Eastern and Central Europe.* Academia Scientiarum Fennica, 1988.

Poetic Edda, The. Translated by Jackson Crawford. Indianapolis: Hackett Publishing, 2015.

Ponzi, Marco. "Trinity College MS 0.2.48 — The Hermetic Lunatica." *ViridisGreen* (blog). *Medium*, Published May 27, 2017.

Pope, Alexander. "Epistles to Several Persons: Epistle IV." The Poetry Foundation. Accessed January 24, 2024,

Posthumus, David C. *All My Relatives: Exploring Lakota Ontology, Belief and Ritual.* University of Nebraska Press, 2018.

Price, Neil. *The Viking Way: Magic and Mind in Late Iron Age Scandinavia.* Oxbow Books, 2019.

Pseudo-Aristotle. *De Mundo.* Translated by E. S. Forster. Oxford University Press, 1914.

Pseudo-Dionysius. *The Complete Works.* Translated by Colm Liubheid. Paulist Press, 1987.

Ptolemy, Claudius. *Tetrabiblos, Book 1.* Translated by Robert Schmidt and edited by Robert Hand. The Golden Hind Press, 1994.

Puckett, Newbell Niles. *Folk Beliefs of the Southern [Omitted].* University of North Carolina Press, 1926.

Rand, William Lee. *Reiki: The Healing Touch.* International Center for Reiki Training, 1991.

Randles, W. G. L. *The Unmaking of the Medieval Christian Cosmos: From Solid Heavens to Boundless Æther.* Routledge, 1999.

Rankine, David. *The Grimoire Encyclopedia, Volume 2.* Hadean Press, 2023.

Riviere, J. Marques. *Tantrik Yoga.* Weiser, 1970.

Robinson, James, ed. *The Nag Hammadi Library.* Harper Collins, 1990.

Roebuck, Valerie J. *The Upanishads.* Penguin Books, 2003.

Rohde, Eleanour Sinclair. *The Old English Herbals.* Longman's Green and Co., 1922.

Romain, William F. *Archaeology of the Sacred.* The Ancient Earthworks Project, 2015.

Romain, William F. *Mysteries of the Hopewell: Astronomers, Geometers, and Magicians of the Eastern Woodlands.* University of Akron Press, 2000. Kindle.

Romain, William F. *Shamans of the Lost World: A Cognitive Approach to the Prehistoric Religion of the Ohio Hopewell.* Altamira Press, 2009.

Saif, Liana. "From Ġāyat al-ḥakīm to Šams al-maʾārif: Ways of Knowing and Paths of Power in Medieval Islam." *Arabica* no. 64 (2017): 297–345.

Sallust. *On the Gods and The World.* Translated by Thomas Taylor. Strigoi Publishing, 2017. Kindle.

Salmon, Emily, and John Salmon. "Tobacco in Colonial Virginia." Encyclopedia Virginia. Last updated August 26, 2024.

Sayce, Archibald. *Lectures on the Origin and Growth of Religion as Illustrated by the Religion of the Ancient Babylonians.* Williams and Norgate, 1888.

Scaraoschi, Alex. "Scaraoschi's Books of Sorcery." Website. Accessed February 6, 2024.

Schaaf, Phil. *The Complete Works of the Church Fathers.* 2016. Kindle.

Schama, Simon. *The History of the Jews, Volume One.* Harper Collins, 2013.

Schnieders, Paul. *The Books of Enoch: Complete Edition.* Translated by R. H. Charles. International Alliance Pro-Publishing, LLC., 2019.

Scholem, Gershom. *Kabbalah.* Keter Publishing House, 1974.

Scholem, Gershom. *Major Trends in Jewish Mysticism.* Schocken Books, 1974.

Schulke, Daniel. *Veneficium.* Three Hands Press, 2017.

Schulke, Daniel. *Viridarium Umbris.* Xoanon, 2005.

Schultes, Richard Evans, and Albert Hofmann. *Plants of the Gods.* Healing Arts Press, 1992.

Scot, Reginald. *The Discoverie of Witchcraft.* Dover, 1970.

Scot, Reginald. *The Discoverie of Witchcraft.* Elliot Stock, 1886.

Scott, Walter. *Hermetica, Volume 1.* Oxford University Press, 1924.

Sheldrake, Rupert. *The Sense of Being Stared At.* Park Street Press, 2013. Kindle.

Sheldrake, Rupert. "Is the Sun Conscious?" *Journal Of Consciousness Studies* 28, no. 3–4 (2021): 8–28.

Signe, Isager, and Jens Erik Skydsgaard. *Ancient Greek Agriculture.* Routledge, 1995.

Silvestris, Bernardus. *The Cosmographia of Bernardus Silvestris*. Translated by Winthrop Weatherbee. Columbia University Press, 1990.

Singer, Charles, "Early English Magic and Medicine." *Proceedings of the British Academy, Vol. IX*. January 28, 1920.

Sir Albert Howard. *The Soil and Health: A Study of Organic Agriculture*. University of Kentucky Press, 2006.

Spence, Lewis. *British Fairy Origins*. Watts and Co., 1946.

Steiner, Rudolf. *Agriculture Course: The Birth of the Biodynamic Method*. Translated by George Adams. Rudolf Steiner Press, 1958.

Steiner, Rudolf. "Part Three: The Plant World and the Elemental Nature-Spirits." Lecture in *Man as Symphony of the Creative World*. November 2, 1923, Dornach.

Storms, Godfrid. *Anglo-Saxon Magic*. Martinus Nijhoff, 1948.

Stratton-Kent, Jake. *Geosophia: The Argo of Magic, Vol. I*. Scarlet Imprint, 2010.

Stratton-Kent, Jake. *The Sworn and Secret Grimoire*. Hadean Press, 2021.

Stratton-Kent, Jake. *The Testament of Cyprian the Mage*. Scarlet Imprint, 2014.

Sturlusson, Snorri. *Edda*. Translated by Anthony Faulkes. Everyman, 1995.

Swart, Jacobus. *The Book of Sacred Names*. Sangreal Sodality Press, 2011.

Tacitus. *The Agricola and the Germania*. Translated by H. Mattingly and revised by S. A. Handford. Penguin Books, 1970.

Tarnas, Richard. *Cosmos and Psyche*. Plume, Penguin Group, 2006.

Taylor, Thomas. *The Mystical Hymns of Orpheus*. Bertram Dobell, 1896.

Theophrastus. *Enquiry Into Plants*. Putnam, 1916.

Thompson, Reginald Campbell. *Devils and Evil Spirits of Babylonia*. Luzac and Co., 1903.

Thompson, Reginald Campbell. *Reports of the Magicians and Astrologers of Nineveh and Babylon, Vol. 2*. Luzac and Co., 1900.

Thompson, Reginald Campbell. *Semitic Magic: Its Origins and Development*. Luzac and Co., 1908.

Thorndike, Lynn. *A History of Magic and Experimental Science, Volume Two*. Columbia University Press, 1923.

Tompkins, Peter, and Christopher Bird. *The Secret Life of Plants*. Harper Perennial, 1973.

Turner, Nancy. *Ancient Pathways, Ancestral Knowledge*. McGill-Queen's University Press, 2014.

Villavicencio, H. Andres, and I. Alejandro Virgile. *The Power of Asteria*. Asteria Books, 2021.

Virgil. *The Georgics of Virgil*. Translated by David Ferry. Farrar, Straus, and Giroux, 2005.

Virgil. *The Poems of Virgil: The Georgics*. Translated by James Rhoades. Encyclopaedia Britannica, 1952.

Wasson, R. Gordon. "The Hallucinogenic Fungi of Mexico." In *The Psychedelic Reader*, edited by Gunther Weil, Ralph Metzner, and Timothy Leary. The Citadel Press, 1973.

Wasson, R. Gordon, Albert Hofmann, and Carl Ruck. *The Road to Eleusis: Unveiling the Secret of the Mysteries*. North Atlantic Books, 2008.

Watts, D. C. *Dictionary of Plant Lore*. Academic Press, 2007.

Webb, William S., and Raymond S. Baby. *The Adena People, No.2*. The Ohio Historical Society, 1975.

Wigginton, Eliot, ed. *The Foxfire Book*. Random House, 1972.

Wilbert, Johannes. *Tobacco and Shamanism in South America*. Yale University Press, 1987.

Wilby, Emma. *Cunning Folk and Familiar Spirits*. Sussex Academic Press, 2005.

Wili, Walter. "The Orphic Mysteries and the Greek Spirit." In *The Mysteries: Papers From the Eranos Yearbooks, Volume 2*, edited by Joseph Campbell. Bollingen, 1955.

Williamson, Craig. *The Complete Anglo-Saxon Poems*. University of Pennsylvania Press, 2017.

Winroth, Anders. *The Conversion of Scandinavia*. Yale University Press, 2012.

World Heritage Ohio. *Guide to the Hopewell Ceremonial Earthworks* (pamphlet). 2019.

Yeats, William Butler. *Ideas of Good and Evil*. A. H. Bullen, 1903. Kindle.

Young, Francis. *Twilight of the Godlings*. Cambridge University, 2023.

Yronwode, Catherine. *Hoodoo Herb and Root Magic: A Materia Magica of African American Conjure*. Lucky Mojo Curio Company, 2002.

INDEX

Note: Page numbers in *italics* refer to illustrations.